AN INTRODUCTION TO
Ornithology

AN INTRODUCTION TO

Ornithology

THIRD EDITION

GEORGE J. WALLACE

Ph.D., Sc.D.

Professor Emeritus of Zoology
Michigan State University
East Lansing

HAROLD D. MAHAN

Ph.D.

Director
Cleveland Museum of Natural History
Cleveland

MACMILLAN PUBLISHING CO., INC.

New York

COLLIER MACMILLAN PUBLISHERS

London

MACMILLAN PUBLISHING CO., INC.
866 Third Avenue, New York, New York 10022
COLLIER-MACMILLAN CANADA, LTD.

Library of Congress Cataloging in Publication Data

Wallace, George John, (date)
  An introduction to ornithology.

  Bibliography: p.
  1. Ornithology. I. Mahan, Harold D., joint author.
II. Title.
QL673.W3 1975   598.2   74-480
ISBN 0-02-423980-1

Printing: 1 2 3 4 5 6 7 8   Year: 5 6 7 8 9 0

TO

Martha Cooper Wallace

AND

Mary Jane Mahan

FOR PATIENCE, ENCOURAGEMENT,
AND MATERIAL HELP

PREFACE

THE WIDESPREAD college use of the first (1955) and second (1963) editions of *An Introduction to Ornithology* points to the critical need to bring it up to date again. The past ten years have witnessed an explosion of ornithological literature, making the 1963 edition almost hopelessly out of date. This new version attempts to maintain much the same level and scope of the earlier editions — introductory ornithology at the college or university undergraduate level — but trying to incorporate the wealth of new literature has inevitably resulted in a more comprehensive and somewhat more technical text, and a more condensed style of presentation.

In recognition of recent trends in research, the chapters on external and internal features have been considerably expanded and partly rearranged, with new sections on molt in relation to breeding and migration, adaptations for food gathering, torpidity, salt glands, and water balance. The chapters on the annual cycle remain much the same except for updating to include some of the more pertinent studies of the past ten years. The treatment of migration has been altered to incorporate new findings, and the section on distribution has been divided into its purely geographical and its more ecologically oriented aspects. Necessarily much of the chapter on conservation had to be revised or rewritten to take into account rapidly changing environmental conditions.

Part II of the earlier editions was largely dispensed with, but certain features, such as the Orders and Families of Living Birds, were transferred to other chapters. Three appendices present abbreviated lists of ornithological organizations and their journals, a description of ornithological collections (including preparation of bird skins), and lists of endangered and declining species.

Citing literature poses a problem. As in previous editions short lists of Selected References are included at the ends of the chapters, mainly of books and monographs pertaining to the subject matter of those chapters. Then there is a longer Literature Cited section in the back of the book. More than 150 references cited in the second edition were deleted to make way for newer citations, in spite of which the bibliographies, even with continued pruning, threatened to become too burdensome for a beginning text. Hundreds of useful new references consulted were dropped to save on space and publication costs. In the selection of references two factors were considered: (1) choosing the most recent publication with a good bibliography that would guide the student in his quest for further literature on the subject, and (2) citing papers that would explore a particular topic more comprehensively than could be done here. Hence, if we state that a robin eats worms, we hope we will not be sued for plagiarism for failing to cite the several hundred papers verifying that robins eat worms. The North American literature and many Old World publications were surveyed through 1972; revision of most of the manuscript in 1973 precluded using many papers for that year. The student, in looking for the references cited in a particular chapter, should first consult the list at the end of that chapter; if not found there the reference is in Literature Cited.

Common names of North American birds follow the latest (1957), American Ornithologists' Union (A.O.U.) checklist and supplements. Scientific names of North American birds are not used in the book (except for lists in the Appendices) because they are available in the identification guides almost universally used in college courses. However, standardized common names for foreign birds often are not available; hence their scientific names are included. Ordinal and family names and their derivatives are used freely, for early familiarity with these terms is useful to the student. In conformance with the policy followed by all ornithological journals, bird names are capitalized when the full species name is used but not when the name is nonspecific (e.g., Common Crow and Eastern Meadowlark are capitalized but crow and meadowlark are not).

In the preparation of this new edition the senior author was largely responsible for revision of the text, except for the new material dealing with behavior and ecology (Chapters 8 and 14), for which the junior author was responsible. The latter was also responsible for securing most of the new illustrations.

For help with this revision we are indebted to many unnamed persons, particularly students, who have praised, criticized, or condemned portions of the text, including Macmillan's reviewers, who suggested many of the revisions and desirable inclusions. An unexpected critic was a former student (1948), William F. Davis, M.D., of Ashtabula, Ohio, who entirely unsolicited went through the 1963 edition with great care and pointed out errors and inconsistencies, even to a few misspellings, that had stood unchanged for many years. We are also grateful to Diane Pierce, the artist providing the new drawings; to Bruce Frumker for aid with photographs; to Merry Andersen for typing, clerical help, and assistance in various ways; and to Elisabeth H. Belfer and others at Macmillan for careful editing of manuscript and proofs. And as before the senior author is indebted to his wife for typing most of the new material, for proofreading, evaluations, and useful suggestions, and for help with the bibliographies — the unrewarding labors associated with the development of a new book. Dr. Sylvia Wallace McGrath was also very helpful for her painstaking and critical review of some of the chapters.

G. J. W.
H. D. M.

CONTENTS

1. Introduction: Learning About Birds   1
   The Living Bird   1
   Historical Background   2
   Ornithology Today   11
   SELECTED REFERENCES   23

2. Origin, Speciation, and Classification   24
   Fossil History   24
   Speciation   30
   The Taxonomic Categories   34
   Nomenclature   36
   Orders and Families of Living Birds   40
   SELECTED REFERENCES   54

3. Feathers; Molts and Plumages   56
   Feathers   56
   Molts and Plumages   6
   Other Integumentary Structures   82
   Feather Maintenance by the Oil Gland   83
   SELECTED REFERENCES   85

4. Bones, Muscles, and Locomotion   86
   The Skeleton   86
   The Muscular System   92
   Locomotion on Land and Water   97
   Bird Flight   106
   SELECTED REFERENCES   117

5. Digestion, Food, and Feeding Habits   118
   The Digestive System   118
   Feeding Techniques   126
   Food Habits   131
   SELECTED REFERENCES   146

6. Other Metabolic and Physiological Functions   147
   The Respiratory System   147
   The Circulatory System   152
   Excretion   160
   Reproductive Structures   164
   SELECTED REFERENCES   166

xi

7. Coordination: The Brain, Sense Organs, and
   Hormones   168
   The Brain   168
   The Sense Organs   169
   Endocrine Glands   183
   SELECTED REFERENCES   186

8. Bird Behavior   187
   Definitions and Concepts   187
   Communication and Displays   192
   Breeding Behavior   198
   Behavior and Feeding Habits   207
   Maintenance Activities   213
   Social Organization   215
   SELECTED REFERENCES   219

9. The Annual Cycle: Arrival, Territory, Courtship,
   and Mating   220
   Spring Arrival   220
   Territories   223
   Courtship and Mating   233
   Types of Pairing Bonds   240
   Replacement of Lost Mates   247
   Helpers at the Nest   248

10. The Annual Cycle: Nests, Eggs, Incubation, and
    Hatching   249
    The Nest   249
    The Eggs   266
    Incubation   272
    Hatching   282

11. The Annual Cycle: Postnatal Life   287
    Early Growth and Development   287
    Parental Care   296
    Nest Leaving   306
    Postnest Life   307
    Mortality and Nesting Success   311
    Longevity   316
    Some Late Summer Activities   317
    The Fall Journey   317
    Wintering Habits   318
    SELECTED REFERENCES PERTAINING TO THE
    ANNUAL CYCLE   320

12. Migration 324

Definitions and Migration Patterns 324
Seasonal or Annual Migrations 328
Origin and Causes of Migration 334
Some Mechanics of Migration 339
Orientation 343
Routes of Migration 349
SELECTED REFERENCES 356

13. Distribution 358

Origin and Dispersal 358
Zoological (Faunal) Regions and Subregions 364
Life Zones and Biotic Communities 373
SELECTED REFERENCES 385

14. Ecology 386

Habitats and Niches 386
Succession 390
Bird Populations 392
Ecological Adaptations 395
SELECTED REFERENCES 398

15. Birds in Our Lives 399

Birds as Food 399
Plumages 402
Other Economic Products 404
Food Habits in Relation to Man 406
Miscellaneous Damage by Birds 410
The Hand of Man 415
Diseases 419
SELECTED REFERENCES 420

16. Conservation: To Save or Not to Save 421

Extinct Birds 421
Endangered Species 429
Legislation 441
Parks, Sanctuaries, and Refuges 443
Diseases and Parasites 447
Man-made Mortality: The Modern Dilemna 451
SELECTED REFERENCES 460

Appendices

I Ornithological Organizations and Their Journals 465

National Organizations   465
State Organizations   469
Foreign Journals   472

II Ornithological Collections   474
Directions for Preparing Bird Skins   475

III Endangered Species Lists   483
U.S. Endangered Species   483
Arbib's Blue List for 1973   484
World Endangered Species   485
Literature Cited for Endangered Species   489

Literature Cited   490

Index   529

1

Introduction:
Learning about Birds

S TUDYING birds has always been a popular pursuit, from prehistoric to modern times. Birds appeal to people for many reasons.
Among these are their attractive, often breath-taking colors, their
joyous spring songs, their spectacular migrations and fascinating habits,
and, perhaps most of all, the challenge that the identification of so
many kinds affords the imaginative mind, the never-ending quest of
discovering something new. Who does not thrill at the sight of the first
Scarlet Tanager in spring, or hearing a Whip-poor-will at dusk, or
watching an albatross at sea? Small wonder that birds have become so
popular, both for recreational enjoyment and for scientific studies.

Offsetting these attractive advantages, however, is the fact that increasing demands for living space for an expanding human population, industrial spread, and continuing use of pesticides tend to restrict
bird life more and more and to make good birding areas less available.
At most colleges and universities students have to go farther afield each
year for birds. Even public lands ostensibly dedicated to the preservation of natural resources are under constant pressure for greater economic development. Our advanced and rapidly expanding technology
now threatens the survival of many animal species, perhaps including
man.

The Living Bird

Birds comprise one of the seven *classes* of vertebrates. Each class is distinguished, in part, by its integumentary covering: *dermal scales* in bony
fishes, a *scaleless glandular skin* in amphibians, scales of a new type
(*epidermal*) in reptiles, *hair* in mammals, and *feathers* in birds (Class
Aves). Feathers are unique to birds, thus making them among the most
easily defined and readily recognized categories of animals. No other
animal has feathers, and no bird is without feathers, except temporarily
in the few that are hatched with no natal down.

Later chapters deal in more detail with the anatomy and biology of birds, but some of the chief avian characteristics are summarized here. Well-known avian features, in addition to feathers, are the development of the forelimbs as wings, usually for flight; a feathered tail that serves for balancing, steering, and lift; and a toothless horny beak. The skeleton exhibits many unique adaptations, mainly for flight and bipedal locomotion. A comparatively large four-chambered heart permits complete separation of arterial and venous blood, and an exceedingly rapid heart rate and high body temperature provide for the rapid metabolism characteristic of birds. Compact lungs closely appressed to the ribs are assisted by a remarkable system of air sacs; a special voice box, the *syrinx*, is developed at the base of the trachea. The digestive system lacks teeth for mastication of food, but often has a crop for storing food items and a muscular gizzard for grinding hard materials. The brain of a bird is comparatively large, with well-developed optic lobes (associated with vision) and cerebellum for locomotor control and coordination, but the large cerebrum is relatively smooth and unfissured. Sight and hearing are exceedingly well developed, while taste and smell are comparatively degenerate. The nests, eggs, specialized care of the young, and often long migrations are well-known features associated with birds.

Birds vary in size, form, and wingspread (Table 1.1). They range in size from the diminutive hummingbirds to the massive Ostrich. A Cuban Bee Hummingbird (*Mellisuga helenae*) is about 2.5 inches in length, including bill and tail, and weighs less than 2 grams, while a male Ostrich may exceed 300 pounds. Some fossil birds were larger than any birds living today. Among the small songbirds (Oscines) the smallest may be a "bush-tit" (*Psaltria exilis*) in Java, but a babbler (*Micromacronus leytensis*) and a stubby-tailed flowerpecker (*Dicaeum pygmaeum*), both from the Philippines, are close competitors (Amadon, 1962), depending on what criteria are used for judging size.

Some "unbirdlike" forms include the kiwis (see Fig. 7.2), the penguins (see Fig. 4.13), and some of the early fossils (see Figs. 2.1–2.3). In wingspread (one of the criteria for hunters' tall tales) birds vary from the many essentially wingless forms to the albatrosses (10 to 12 feet), condors (9 to 12 feet), and several of the larger storks (10 feet or more). At least one fossil bird (page 30) is believed to have had a wingspread of 15 or more feet.

Historical Background

That observations of birds began long ago is shown by the crude drawings associated with ancient cultures. The oldest known records are from the Paleolithic (Old Stone) Age. Sketches of a crane or heron,

TABLE 1.1. SOME COMPARATIVE SIZES OF BIRDS
(DATA FROM STANDARD REFERENCES)

Species	Weight (lb)	Total Length (ft)	Wingspread (ft)
Emperor Penguin	57–94 (av. 70)	3.5–3.8	
Wandering Albatross	15–20 (av. 17)	4.5	max. 11.5, av. 10.1*
White Pelican	17	4–6	8–nearly 10
Great Blue Heron	6–8	3.5–4.3	5.4–6.2
Trumpeter Swan†	♂ max. 38, av. 27.9		
	♀ max. 24.5, av. 22.6	5–6	8–10 (or more?)
Whistling Swan†	♂ max. 18.6, av. 15.8		
	♀ max. 18.3, av. 13.6	4–4.6	6–7.3
Canada Goose†	♂ max. 13.8, av. 8.4		
	♀ max. 13.0, av. 7.3	3–3.5	5–5.5
Andean Condor			12.0
California Condor	20–23 (av. 21.5)	3.7–4.6	8.9–9.7
Golden Eagle	7–14.7	♂ 2.5–3	♂ 6.2–7
		♀ 3–3.5	♀ 6.8–7.8
Bald Eagle	7.2–11.5	♂ 2.5–2.8	♂ 6–7.1
		♀ 2.9–3.1	♀ 6.6–7.5
Wild Turkey†	♂ max. 23.8,‡ av. 16.3		
	♀ max. 12.3, av. 9.3	4(♂)	5(♂)
Whooping Crane	8.7–17.3	4.2–4.5	7.7
Sandhill Crane (Greater)	8.0–11.0 (av. 9.5)	3.3–4.0	6.6–7.5
Great Bustard	♂ max. 37, av. 24 (♀ smaller)	3.8	
Ruby-throated Hummingbird	3–4 (grams)	3.1–4.0 (in.)	4.0–4.75 (in.)
Bee Hummingbird	2 (grams)	2.5 (in.)	
Micromacronus	(see text)	3 (in.)	

* Twelve-foot wingspread possible, 13-foot or over a myth (Murphy, 1936).

† Weights from Nelson and Martin (1953).

‡ MacDonald (1961) reported a 28-lb ♂ from New Mexico.

a stork, and an owl have been found on the walls of caves inhabited by the Aurignacians in southern France and Spain some 17,000 to 18,000 years ago (Fisher, 1954). Later (9000 to 6000 B.C.) the Magdalenians, the last of the Old Stone Age, left remarkable examples of art work, including a swan carved on a pebble, two probable Great Auks sketched on a rock, and a reindeer antler (Fig. 1.1) carved in the form of a Capercaillie (a European grouse). Then came Neolithic man, also in Spain, who illustrated a greater variety of birds. A dozen recogniz-

FIG. 1.1. *An example of Magdalenian (Old Stone Age) art: a grouse carved on a reindeer antler.* [Redrawn by Diane Pierce; original sketch from Museé de St. Germain.]

able kinds have been identified among their drawings, including a flock of alert geese ready for the take-off (Allen, 1951).

The ancient Sumerians in Mesopotamia left a new type of art work. One building dating about 3100 B.C. depicted a row of birds, probably doves, carved in limestone, and a copper relief of an eagle clutching two stags. Almost contemporaneous (about 3000 B.C.), but more advanced in the use of bird symbols, were the ancient Egyptians, whose hieroglyphic writings employed many bird characters, and whose carvings, paintings, and mummified specimens were left on monuments and in temples (Fig. 1.2); those of the highly venerated Sacred Ibis (*Threskiornis aethiopica*) remain as testimony to a bird now nearly exterminated from the land where it once was worshiped. Among the early Chinese, also, birds figured in legends and appeared on stamps, tapestries, and wall paintings, although art work in tapestries reached a higher development in France at a much later date (Fig. 1.3L).

The Mayas of Central America, who had developed a high degree of skill in wood and stone art long before white men arrived, left impressive carvings of the King Vulture, Ocellated Turkey, and Great Horned Owl (Fig. 1.3R). Both the Mayas and ancient Aztecs worshiped the Quetzal—formerly the death penalty was inflicted for killing one,

FIG. 1.2 *Egyptian Vulture, an example of ancient Egyptian art.* [Drawing by Diane Pierce from several sources.]

but now it faces extermination by poachers snaring birds at their nests for zoos (Kern, 1968). Sculptures in stone in South America, are also remarkable. In an archaeological park in St. Augustin, Colombia, are many fine examples, probably carved by ancient Indians. Alaska is also noted for its wood carvings by early Indians (Fig. 1.4).

Biblical references to birds are numerous. Some 273 birds or bird passages are discussed in detail by Holmgren (1972). Doves are referred to most frequently—the Dove of Peace with an olive branch is still symbolic—but eagles and other birds of prey, as well as ravens, apparently appealed to the people of Israel. Moses compiled a list of "clean" (fit to eat) and "unclean" fowl. One passage (Numbers 11:31), whose interpretation has caused some controversy, relates: "And a wind went forth from the Lord and it brought quails from the sea, . . . so that they covered the ground to the height of two cubits." If the people in the wilderness numbered "600,000 footmen" (v. 21) and the "least gathered ten homers" (v. 32), the harvest would be 66 million bushels of quail (Allen, 1948).

Except for the allusion to Migratory Quail (*Coturnix*) quoted above, perhaps the oldest published reference to migration is Job 39:26, "Doth the hawk fly by thy wisdom and stretch her wings toward the south?" Equally famous is the passage (Jeremiah 8:7), "The stork in the heaven knoweth her appointed times and the turtle [dove] and the crane and the swallow observe the time of their coming." Sometimes, in spite of 2,000 years of study, it seems that we have not progressed much beyond this astute observation.

More classical study on birds began with Aristotle and other Grecian scholars, more than three hundred years before the Christian Era. Aristotle devised a simple classification for the known birds, conducted anatomical studies by actual dissection, and even studied bird embryos, long before the invention of the microscope. In his classification he recognized or accounted for about 170 kinds, probably more than were known in any of the European countries during the following century. Roman scholars, particularly Pliny the Elder, compiled additional information on birds, but none had the discriminating acumen of Aristotle; many of the writings of Pliny, in fact, are of unreliable data furnished by travelers about strange birds in strange lands.

Later, during the early Christian era, ornithology, like all sciences except perhaps medicine, suffered a relapse, when scientific pursuits were considered antireligious and sometimes resulted in persecution. Further development of bird lore awaited the coming of the Renaissance in the twelfth century, when an awakening appreciation of living things and consequent studies of bird life blossomed into fuller flower in England, France, Germany, and other European countries. Several notable works on birds, including the first really scientific classification

(Willughby, 1676) appeared long before Linnaeus' binomial system of nomenclature (Beddall, 1957).

Birds played an important role in the early history of America. Though we now know from archaeological investigations that prehistoric Indians made frequent use of birds for food, wearing apparel, and ornaments (Miller, 1957; Mayfield, 1972b), the first published reference pertains to Columbus. With his rebellious crew on the verge of mutiny, he took new hope at the sight of land birds out at sea and actually changed his course to follow the birds to land (Tooke, 1961). Conceivably, birds may have altered early American history, as otherwise Columbus might have landed elsewhere or not at all. Other navigators often depended on birds for subsistence and, unfortunately, often visited and ruthlessly plundered relatively helpless island colonies; the extinct Dodo and Great Auk stand as mute testimony of this. The Cahow or Bermuda Petrel, once fabulously abundant, is known to

FIG. 1.3. OPPOSITE *A heron as depicted in a tapestry of The Hunt of the Unicorn series from southern France in the fifteenth century.* [Photo courtesy of Metropolitan Museum of Art, The Cloisters Collection, Gift of John D. Rockefeller, Jr., 1937.] ABOVE *Owl motif on a yoke found on the Gulf Coast near Veracruz, Mexico; from the "classic" period of Olmec civilization.* [Drawing by Diane Pierce.]

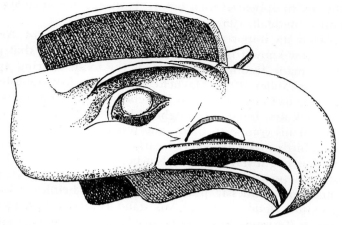

FIG. 1.4. *Eagle mask from northwest coast of Alaska, carved by early Indians (ca. 1700's).* [Drawing by Diane Pierce from specimen in Cleveland Museum of Art.]

have saved some of the early colonists from starvation during the famine of 1614–1615, but such exploitation and later settlement by man and his introduced pests nearly exterminated the petrels (Murphy and Mowbray, 1951). Hopes of restoration are now frustrated by DDT in their oceanic prey; Wurster and Wingate (1968) predict complete reproductive failure in the late 1970's if the observed decline continues.

The first settlers in America found a rewarding abundance of game birds. Though the Turkey is the most symbolic of these, there were many native grouse, flocks of wild pigeons said to number billions of birds, and a seemingly inexhaustible supply of waterfowl. In the gold-rush era, the sale of murre eggs to feed the West Coast's swollen population precipitated gun battles over lucrative egg-collecting rights in the murre colonies (Ferry, 1952).

Not all the relations of early settlers in this country with birds were for food. Both explorers and settlers often took a keen interest in observing new birds. Lists were compiled, records were kept, and drawings and engravings of birds as well as several natural-history books were on the market long before Audubon. Many specimens, eviscerated and heavily salted, were packed with botanical collections, which were then in great demand, and shipped to Europe, though few of the bird specimens long survived such treatment. Then came the more critical and elaborate works of such men as Mark Catesby (1682–1749?), who produced a two-volume treatise on *The Natural History of Carolina, Florida and the Bahama Islands;* Alexander Wilson (1766–1813), who literally drove himself to an early death in the desperate but not quite successful attempt to complete his nine-volume work of *American Ornithology* (his biographer, George Ord, completed the eighth and ninth volumes); and John James Audubon (1785–1851), whose greater artistic skill and other achievements overshadowed his accomplished predecessors (Fig. 1.5).

History records many practical uses for birds. Domestic fowl, now the most abundant birds in the world, were introduced into Egypt from India about 2000 B.C. (Coltherd, 1966). Cormorants were trained for fishing by the Japanese in the sixth century, a practice apparently adopted by the Chinese some 500 years later. Rings or straps were placed about the cormorants' necks to prevent them from swallowing their prey, and they were trained (or forced by use of leashes) to return their catch to their owner's boat. Less generally known is the ancient partnership between certain natives in Africa and their avian honey-guides (Fig. 1.6); the birds lead the hunters to bee trees and both birds and man share the spoils. Guano deposits of certain sea birds are a valuable asset in some regions; those of the Guanay Cormorant on islands off the coast of Peru (see Fig. 15.2) bring the Peruvian government an annual revenue of millions of dollars.

FIG. 1.5. LEFT *John James Audubon, pioneer ornithologist and artist (1785–1851), whose incomparable* Birds of America, *an elephant folio of 435 hand-colored lithographs, sometimes sells for $100,000 or more.* [From Nat. Audubon Soc.] RIGHT *Gyrfalcon, an example of the artistic skill and meticulous detail that went into Audubon's work.* [Photo from an Audubon painting.]

FIG. 1.6. *The honeyguides (Indicatoridae) of Africa are used by some natives to lead the way to bee trees. When the honey is taken, some comb is left for the bird as a reward. This ancient partnership between birds and man is being lost in the more Westernized parts of Africa.* [Drawing by Diane Pierce.]

Bird plumages have long been used both for ornament and for comfort; in fact, the use of feathers for millinery purposes once threatened the extermination of some of our most colorful birds (see page 441).

The beneficial role of birds in relation to agriculture has received wide recognition; one concrete evidence of this is the monument to the gulls in Salt Lake City, erected by the appreciative Mormons after California Gulls had wiped out a plague of Mormon crickets (*Anabrus simplex*) that threatened to destroy their crops. A similar memorial in stone to a colony of swallows in Japan recognizes their economic value (Inoue, 1954), but the monument to the Passenger Pigeon in Wisconsin (see Fig. 16.4) is more a symbol of regret for the passing of a species.

Thus Americans have a rich background for the study of birds. The high standards set by Wilson and Audubon have been continued by a distinguished line of workers until museum cabinets and library shelves are well stocked with reference material. Birds are the best known of any of the animal groups. In sharp contrast to the insects, for instance, there are probably few undiscovered species of birds, even in the remote corners of the earth. A good bird library numbers hundreds, or even thousands, of volumes. New bird books probably average one a week, both in England and America, and the number of journal articles and bulletins on birds runs to more than 4,000 per year (Baldwin and Oehlerts, 1964). Thus, although there are open fields and many unanswered questions in ornithology, extensive library resources, great museum collections, and works of art provide a rich and attractive background for further investigation.

Ornithology Today

Ornithology is a curious mixture of a popular pastime and a precise science. Millions of Americans (11,200,000 according to a U. S. Census Bureau poll) watch birds more or less seriously: casual watchers at window feeders, genuine nature lovers, meticulous list-keepers, and dedicated scientists. Ornithologists are sometimes classified as "amateurs" (those for whom bird study is a hobby) and "professionals" (those who make their living working with birds), but there is no distinct line between them. Often serious amateurs have a broader and more extensive knowledge of birds than the professional specialist.

The contributions of amateurs to ornithology are truly great; some are incidental bits of useful knowledge, others long-term scholarly projects. To appreciate the latter one need only examine A. C. Bent's *Life Histories of North American Birds*, a 23-volume shelf of books compiled and largely written by a former businessman in Massachusetts; or Mrs. M. M. Nice's *Studies in the Life History of the Song Sparrow*, a two-volume classic based on 12 years of study by a housewife in Ohio and Illinois; or the more technical treatises of C. H. Greenewalt (a chemical engineer with, and later president of, Du Pont) on hummingbird flight and the acoustics of bird song.

Some of the more notable ornithological activities, participated in by both amateurs and professionals, are characterized below.

IDENTIFICATION AND BIRD LISTS

Careful identification is an indispensable prerequisite for serious bird study. Beginning classes spend nearly all of their field periods learning to recognize birds. A persistent instructor may stress other important fundamentals: that a bird being observed may have been around all year (*permanent resident*), may have just arrived from more southern wintering grounds (a *transient* or *summer resident*), or, early in the season, may be a *winter visitor* that will soon depart, like the transients, for more northern breeding grounds. (See page 330 for further definition of these status groups.) In conjunction with these field trips the student should keep careful notes and consult a field guide frequently.

Most ornithologists started identification and listing in their early years; exceptions — encouraging to an instructor — are those who "catch fire" in an ornithology class. In beginning classes at Michigan State

University we have usually shown pictures of the 50 most common or most easily recognized birds to see how many are known by students at the beginning of the course. Invariably one or more students correctly identify all 50 species and, from the standpoint of identification, are not neophytes.

Beginning bird students are usually required to keep lists; if the lists are to be useful for future references, they *must* include the date and place of observations and *should* include additional notes. Usually the student makes a list of the birds seen on each class trip and a cumulative list for the term, semester, or year. Many older ornithologists, even if busy in other occupations, have kept annual lists for many years. A good example is Aretas Saunders, who kept records, at least intermittently, for about 70 years. His "Forty Years of Spring Migration in Southern Connecticut" (1959), giving arrival dates (early, average, and late) of summer residents, departure dates for winter visitors, arrival and departure dates for transients, as well as a discerning analysis of the data, is a model example of the value of keeping annual lists.

The ultimate in record keeping for many people is the life list—the number of birds a person has identified in the field, locally, state-wide, in North America, or in the world. This has long been a somewhat competitive business, but was further stimulated by the formation of the American Birding Association in 1969, which exists mainly for the purpose of promoting birding as a hobby and a sport, and to assist its members in the pursuit of this popular activity. The Association provides information on where to find the "most wanted" birds and publishes an up-to-date roster of the leading listers for each state, for North America, and for the world. Several life lists now exceed 3,000 species, a pinnacle set by Ludlow Griscom, former dean of bird listers, in 1958. Those who have identified 600 or more species in North America north of Mexico (129 persons in 1973) belong to the elite "600 Club." Well-organized ornithological safaris to all parts of the world now assist ambitious birders in their quest for new birds.

THE CHRISTMAS COUNTS

Audubon Christmas counts, initiated on a modest scale in 1900 (27 persons at 25 stations) by Frank M. Chapman of the American Museum of Natural History, have become the nation's most popular social birding event, participated in by many thousands of observers (18,798 in 1971) in all states and most provinces of Canada (Fig. 1.7). Originally visualized as a potential census of the country's birdlife, it soon developed into a competitive listing game, until Aldo Leopold, among others, pointed out that it was not a true census, but merely a count of the birds found in the most productive habitats within the prescribed 15-mile-diameter circle by groups of observers numbering from one to

FIG. 1.7. *Bird watchers in the Sacramento Valley, California, study a flock of geese, identify the species (mainly White-fronted and Snow Geese), and try to estimate numbers.* [Photo by Allan D. Cruickshank.]

a hundred or more. Counts are made on a selected date, usually sometime during the last two weeks in December. The National Audubon Society dictates the rules and regulations to be followed and issues instructions for the submission of the counts to its national office in New York.

The extent of the scientific usefulness of the counts has never been fully realized. Although they are primarily for "fun," the past 73 years of records provide an extraordinary backlog of potentially useful data. Some analysts (see Arbib, 1967, for example) have tried to point out how to use the counts most effectively, but at best there are great difficulties. For more information on the nature and magnitude of the counts, readers should consult Allan Cruickshank's discerning, and often amusing, introductions to the Christmas Counts for the 17 years (1955–1972) he served as editor.

AUDUBON FIELD NOTES AND AMERICAN BIRDS

Perhaps more useful statistically are the other records published by the National Audubon Society in collaboration with the U. S. Fish and

Wildlife Service in *Audubon Field Notes* (1947–1970) and its successor *American Birds* (1971–   ). These other records comprise the breeding bird surveys and additional reports for each of the four seasons. In the breeding bird surveys observers select an area of relatively uniform habitat (such as a beech-maple woodlot, a marsh, a grassland, or a desert) and count singing males by taking several censuses in June each year. Preferably the census area should be covered by the same observer in successive years to determine population changes, but many such projects were terminated when the study area became a suburb or factory site. Some of the difficulties in the interpretation of these data have been evaluated by Brewer (1972).

Modifications of these census techniques have been used in many ways: winter inventories and breeding surveys of waterfowl populations, actual counts or estimates of endangered species, and projection of estimated populations for a whole state or country. Great Britain, for example, estimates its breeding land bird population at 67 million pairs. Detailed censuses in Illinois taken 50 years apart (1906–1909 and 1956–1958) showed fairly comparable summer populations for the two periods, but striking differences in species composition with changing habitats. For instance, Graber and Graber (1963) found sharp decreases in some insectivorous and predatory birds and an influx of millions of Starlings not present during the earlier period. Stewart and Kantrud (1972) calculated the total breeding population for each of North Dakota's 50 leading species (which comprised 91 percent of the total population) and a combined total of all species of about 26 million pairs—a little lower density per acre than found in Illinois (both states have many acres of low density habitat) and considerably less than found in Great Britain (before it demolished its hedgerows). (See Chapter 14 for more specific data on bird populations.)

NORTH AMERICAN NEST-RECORD CARD PROGRAM

This continent-wide survey of nesting birds is carried out with the cooperation of many hundreds of observers in the United States and Canada. Cooperators fill out specially provided nest-record cards with as complete nesting data as are feasible. Cornell University maintains the computer analysis of the records; in time these should yield voluminous data on individual species. Some states have their own units cooperating with the project; others (Maryland, Massachusetts, and Michigan) are utilizing the Cornell data in mapping the breeding distribution of the birds within their own boundaries. Both the nest-record card idea and the mapping (ornithological atlas) projects stem from earlier ventures in Great Britain. Gochfeld (1972) proposes a similar tropical register for the heretofore neglected neotropical birds and solicits cooperators for the project.

BIRD BANDING

Several individual banding or marking projects were under way in this country in the 1800's. J. J. Audubon, for instance, marked a brood of Eastern Phoebes with silver threads in 1803 and found that two of the birds returned to the same locality the following year. By 1920 so many uncoordinated banding projects were under way that the United States Biological Survey (now Fish and Wildlife Service), realizing the potential value of such activities, took over supervision of all banding programs. Qualified persons (there are necessary rules and regulations) can secure permits to band birds. A federal permit, renewable periodically (or revocable) and usually a state permit in the state or states in which the bander operates are required. The Bird Banding Laboratory at Patuxent Wildlife Research Center in Laurel, Maryland, provides bands of assorted sizes (Fig. 1.8), a Bird Bander's Manual of instructions, and occasional Newsletters. Operators provide their own traps, nets, and tools for banding and are required to submit detailed reports on their banding activities to the Bird Banding Laboratory. Special permits are required to use mist nets, color bands, or other marking devices.

Since 1920 about 20 million birds have been banded in this country and about a million are added each year. Nearly 2 million *returns* (a bird returning to the station where banded) and *recoveries* (a bird re-trapped or found dead in another area) have been obtained. Many remarkable discoveries, too numerous to summarize here, have been revealed by bird banding. Bird-Banding magazine (see Appendix I) publishes many of these findings. Some of these important data will be called to the attention of the student at appropriate places in the text.

ATTRACTING BIRDS

Among the many people who enjoy birds are those who try to attract them to their homes by providing food in winter and/or nest boxes (Fig. 1.9) and by planting fruit-bearing shrubs and sheltering evergreens. Plantings are probably the most effective means of permanent attraction, in effect creating a sanctuary; nesting boxes will attract only a few species, and winter feeding has its limitations. Some have even questioned the ethics of winter feeding, since it tends to concentrate birds in small areas with the possibility of spreading diseases. Corn-fed wild ducks are more susceptible to lead poisoning than those on a more natural diet of aquatic plants (page 448), and sparrow die-offs have been noted, apparently from a virus disease, in concentrated feeding areas. Winter feeding, particularly in suburbs, also tends to build up populations of "nuisance" birds—pigeons, Starlings, House Sparrows,

FIG. 1.8 ABOVE *U. S. Fish and Wildlife Service bands come in assorted sizes from 0 to 8. Shown are No. 2's, serially arranged on wire as shipped from the government office; No. 5's, placed on convenient wire holder for carrying in field box; No. 8, opened, ready for placement on bird.* [Photo by Philip G. Coleman.] LEFT *A banded bird with a numbered aluminum band on its right leg and a color band (not recommended except for detailed studies) on its left leg.* [Photo courtesy of *National Geographic Magazine.*]

FIG. 1.9 [OPPOSITE]. *One of the chief means of attracting birds to the home grounds is to put up boxes for hole-nesting species. The House Wren, shown here, is quick to adopt man-made substitutes for the natural cavities in which it otherwise nests.* [Photo by Allan D. Cruickshank.]

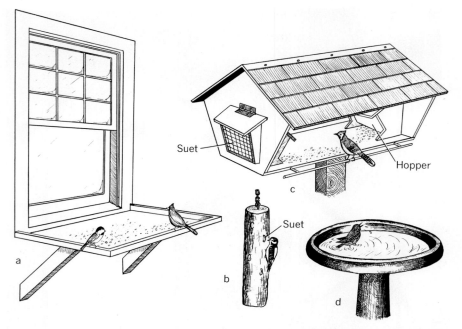

FIG. 1.10. *Devices for attracting birds.* a. *A simple window shelf, with raised border to keep seeds from blowing off.* b. *Suet log with suet for insectivorous birds.* c. *Feeder deluxe, with suet tray, feed hopper, and perches.* d. *Bird bath.* [Drawings by Homer D. Roberts.]

and blackbirds. Enabling them to survive a hard winter thwarts one of nature's means of population control.

All in all, however, attracting birds gives many people much enjoyment and provides food and shelter for many desirable birds that otherwise might not survive. Figure 1.10 illustrates some of the devices used to attract birds. Many books (see Terres, *Songbirds in Your Garden,* for instance) and bulletins are available for further information on the much-practiced art of attracting birds.

BIRDS AS SPORT

Prominent among the many people concerned with birds from the recreational standpoint are 15 million hunters. Although bag limits and trophies are the professed goal of many hunters, others (perhaps the majority according to a recent National Wildlife Federation questionnaire) hunt for the enjoyment of being out of doors. Trained biologists in game departments are usually enthusiastic sportsmen; often that is the main reason for choosing a career in wildlife. Waterfowl hunters in particular have a good opportunity of getting more than bag limits out of their sport, since ducks on the wing, or on the water, are a pleasure to watch and pose challenging identification difficulties.

Other popular recreational uses of birds are for falconry, pigeon racing, and cockfighting; this last, thought by some to be proof of our bloodthirsty nature, is now illegal in this country, although carried on surreptitiously in some states. In Mexico and parts of the Orient it is still a highly competitive "sport"; fighting cock owners often use lethal metal spurs on their birds and stake large bets on the outcome of the fight. In India less bloodthirsty natives use francolins and bulbuls in contests that are harmless to the birds.

Falconry (Fig. 1.11) is one of the most ancient of sports, originating in China some 4,000 years ago and subsequently spreading to other parts of Asia and Europe where it was employed chiefly by the nobility. Now scarcity of falcons—most of them are considered endangered species (page 436)—and legal restrictions hamper the activities of would-be falconers. Often bitter controversies have developed between the

Fig. 1.11. *Once a sport restricted largely to the nobility, falconry is now practiced by many people, in spite of legal restrictions and scarcity of suitable falcons. Here a student works with a trained Prairie Falcon.* [From L. C. Pettit, *Introductory Zoology,* copyright 1962 by The C. V. Mosby Company, St. Louis; courtesy of Mich. State Univ. Mus.]

dedicated falconers who maintain that continuance of their sport, if properly managed, assures greater protection as well as restocking possibilities for falcons, and the "protectionists" who would banish falconry entirely because plundering nests for "falcons" pose a threat to all predatory birds.

Training and racing homing pigeons is now primarily a sport, but from Roman times through World War II, pigeons as well as other birds, such as frigatebirds in Polynesia, were trained to carry messages during wartime, sometimes quite effectively, but sometimes the enemy intercepted the bird and the message. Homing pigeons are now widely used in orientation studies (page 347).

EDUCATION AND RESEARCH

University and municipal museums serve a dual role. The larger municipal museums (see Appendix II) have elaborate exhibits, viewed by millions of visitors, of birds from around the world in simulated natural habitats. These and many smaller public museums also have well-organized educational and conservation-oriented programs. Some museums have large research collections. The largest, the American Museum of Natural History in New York, has nearly a million specimens of birds. Harvard University and the University of Michigan vie for the largest, and most utilized, university collections. Obviously, museum collections are indispensable for many types of studies by staff and graduate students.

Biological stations, usually university operated, serve the purpose of getting students out into the field to study birds (Fig. 1.12). One of the oldest (1909) and best known, which also has a well-organized teaching and research program, is the University of Michigan Biological Station at Douglas Lake in northern Michigan. In 1954 Michigan State University established its W. K. Kellogg Station at Gull Lake in the southern part of the state. Both stations, the latter including the Kellogg Bird Sanctuary and the Kellogg Forest, have a wide variety of habitats and good facilities for studying birds. Other summer stations, offering both beginning and advanced work with birds, include the University of Virginia Biological Station in the Appalachians, the Lake Itasca Forestry and Biological Station in northern Minnesota, the University of Oklahoma Biological Station at Lake Texoma, the Rocky Mountain Biological Station in Colorado, and the Montana State Biological Station at Flathead Lake. Many important publications by graduate students and members of the instructional staffs have emanated from these field stations.

Several other field stations, used primarily for research, are located in strategic areas favorable for research on birds. Among these are the Bowdoin (Maine) Scientific Station on Kent Island, New Brunswick, for

FIG. 1.12. *A class in advanced ornithology looking for birds at the W. K. Kellogg Gull Lake Biological Station of Michigan State University.* [Photo by Robert L. Fleming, Jr.]

studies on sea birds; the Delta Wildlife Research Station in the prairie province of Manitoba, for waterfowl research; the Southwestern Research Station of the American Museum of Natural History in the Chiricahua Mountains of Arizona (six different life zones available); the Hastings Natural History Reservation in Carmel Valley, California; the Smithsonian Institution's Canal Zone Biological Laboratory on Barro Colorado Island in Panama; the Asa Wright Nature Center in Trinidad; and the Arctic Research Station at College, Alaska. These places offer unique opportunities for studies of unusual birds in unusual environments.

In a different category are Patuxent Wildlife Research Center, operated by the U. S. Fish and Wildlife Service at Laurel, Maryland; the Kalbfleisch Field Research Station on Long Island, operated by the American Museum of Natural History; the Manomet Bird Observatory in Massachusetts; Carnegie Museum's Powdermill Nature Reserve in the mountains of southwestern Pennsylvania; and, perhaps the ultimate in facilities, the Laboratory of Ornithology at Sapsucker Woods in Ithaca, New York, operated by Cornell University.

Other opportunities for field studies, primarily in Central America, are afforded by the Organization for Tropical Studies, a cooperative enterprise with 25 sponsoring institutions. Students are encouraged, and sometimes subsidized, to spend a term or semester in the tropics.

In addition to the more technical stations enumerated above, the National Audubon Society operates summer camps and nature centers for both laymen and students. Summer camps with instructional programs (classes) are located in Maine, Connecticut, Wisconsin, and Wyoming. Audubon Centers, extensively used by visiting groups of teachers and children, are located in Connecticut, Ohio, Wisconsin, and California.

Students interested in expanding their ornithological background should spend a session at one of these biological stations. Those interested in employment in state (see Fig. 1.13) or federal service should consult Day's *Making a Living in Conservation: A Guide to Outdoor Careers.*

In summary, it should be obvious that birds afford attractive and often ideal subjects for study. Their widespread availability, a rich

Fɪɢ. 1.13. *State Game Division personnel trap a Sharp-tailed Grouse for later examination and release. Box operates by a trap door arrangement on top through which birds fall after they climb sloping sides for corn bait.* [Photo by Mich. Dept. Nat. Res.]

background of literature, extensive museum resources, and many active ornithological organizations at local, state, and national levels (see Appendix I) have much to offer an aspiring student. And in spite of this formidable background, there are still many unanswered and challenging questions about birds.

SELECTED REFERENCES

ALLEN, E. G. 1951. The history of American ornithology before Audubon. Trans. Amer. Phil. Soc., 41(3):387–591.

CRUICKSHANK. A. D. 1955–1972. Audubon Christmas bird counts. Audubon Field Notes, 9–26 (1972 count in American Birds).

DAY, A. M. 1971. *Making a Living in Conservation; A Guide to Outdoor Careers.* Harrisburg, Pa.: Stackpole Books.

FISHER, J. 1954. *A History of Birds.* Boston: Houghton.

GRABER, R. R., and J. W. GRABER. 1963. A comparative study of bird populations in Illinois, 1906–1909 and 1956–1958. Bull. Ill. Nat. Hist. Sur., 28:383–528.

HICKEY, J. J. 1953. *A Guide to Bird Watching.* Reprint ed. Garden City, N.Y.: Garden City.

HOLMGREN, V. C. 1972. *Bird Walk Through the Bible.* New York: Seabury Press.

NYE, A. G., Jr. 1966. Falconry, pp. 164–173 in *Birds in Our Lives* (A. Stefferud, ed.). Fish and Wildlife Service, U.S.D.I.

PETERSON, R. T. 1957. *The Bird Watcher's Anthology.* New York: Bonanza.

TERRES, J. K. 1953. *Songbirds in Your Garden.* New York: Thomas Y. Crowell Co.

2

Origin, Speciation, and Classification

Fossil History

SCIENTIFIC evidence indicates that birds are derived from reptiles. Not only do the anatomy and embryology of reptiles and birds denote close kinship, but some noteworthy fossils actually link the two groups. It may seem paradoxical that a warmly feathered, highly organized, flying bird could stem from a cold-blooded, earth-bound reptile, but of course both the birds and reptiles of early times were much different from those of today; in fact, birdlike reptiles and reptilelike birds existed together during the Mesozoic geological era, more than 100 million years ago.

The exact point of departure of birds from early reptiles is not clear from the fossil record. Apparently a group of stem reptiles (*cotylosaurs*) gave rise to the turtles, mammals, and several extinct reptilian lines at a very early date—the turtles in the Carboniferous Period and the mammals in the Triassic (Table 2.1). But they also gave rise to an exceedingly plastic and generalized group known as the *Pseudosuchia* (*thecodonts*), which are ancestral to all the later Jurassic and Cretaceous dinosaurs, the modern reptiles, and the birds. Heilmann (1927) evokes a hypothetical group, the *Proaves*, as a connecting link between the rather generalized pseudosuchians and the first birds. Apparently the nearest living relatives of the birds are the crocodiles.

Among other Mesozoic reptiles that have been suggested as possible ancestors to the birds are

1. The *pterosaurs* or *pterodactyls*, buoyant, flying forms that, of necessity for flight, had many avian features.
2. The *Saurischia* (reptile-pelvis), which included some light-bodied, fleet-footed forms like *Struthiomimus* ("ostrich-mimic")
3. The *Ornithischia* (bird-pelvis), which included bipedal runners like the "duck-billed" dinosaurs.

TABLE 2.1. GEOLOGIC TIME TABLE

Era	Periods and Epochs	Estimated Years Since Each Began	Birds and Other Life Characteristic of Periods
CENOZOIC (*Age of Mammals*)	Quarternary Recent	20,000	Modern birds (8,600 species); modern man
	Pleistocene	1,000,000	About 1,000 fossil species, half modern forms; primitive man; widespread extermination by climatic factors
	Tertiary Pliocene	12,000,000	Birds may have reached maximum abundance; some modern species
	Miocene	28,000,000	Some passerines; most nonpasserine types represented
	Oligocene	40,000,000	Many gruiform types; buteos and a few modern genera
	Eocene	60,000,000	Diatrymids; ancestral(?) ostrich; primitive ardeid, anatid, tetraonid, scolopacid types; primitive hawks, owls, and vultures
MESOZOIC (*Age of Reptiles*)	Cretaceous	120,000,000	Toothed divers (*Hesperornis, Ichthyornis*); primitive flamingos, loonlike, grebelike, rail-like, and charadriiform birds; last of dinosaurs
	Jurassic	155,000,000	First birds (*Archaeopteryx*); dinosaurs dominant
	Triassic	190,000,000	Rise of ruling reptiles; first mammals; bony fishes
PALEO- ZOIC (*Age of Invertebrates*)	Permian	215,000,000	Life transitional between Paleozoic and Mesozoic
	Carboniferous	300,000,000	Age of amphibians; first reptiles
	Devonian	350,000,000	Age of fishes; first amphibians
	Silurian	390,000,000	First sharks (Chrondichthyes); first land animals
	Ordovician	480,000,000	First vertebrates (fishlike ostracoderms)
	Cambrian	550,000,000	Most invertebrate phyla present
PROTERO- ZOIC		925,000,000	Simple marine invertebrates; sponge spicules
ARCHEO- ZOIC		1,500,000,000	Presumptive origin of life, no fossils

Apparently all these groups, though possessing many birdlike features, had already become highly specialized and died out without descendants during the Cretaceous, by which time birds were already well established. The loss of the clavicle (wishbone) in all these groups (except the ancestral *Pseudosuchia*) disqualifies them as ancestors of the birds; that is, since all flying birds possess a clavicle, it is very unlikely that their immediate ancestors would be without one.

FOSSIL BIRDS

Unfortunately, bird remains are not as well represented among known fossils as the reptiles and mammals. Bird skeletons, because of their delicate structure, do not fossilize readily, and the living habits of birds are not conducive to preservation of specimens. So it is fortunate that several birds that lived during the Jurassic, perhaps 130 million years ago, have been recovered as fossils. One of these, a headless specimen now in the British Museum, was found in 1861 in the lithographic slate quarries in Bavaria then being mined extensively for writing stones. The bird was named *Archaeopteryx* ("ancient wing"). In 1877 a more complete specimen was found in the same region. It was named *Archaeornis* ("ancient bird") and is housed in the Berlin Museum. In 1956 a third specimen and still later a fourth specimen were found near the site of the 1861 fossil. Unfortunately, the latter were in a poor state of preservation and add little to our knowledge but do help to clarify several formerly obscure points. All four specimens (plus an isolated feather found in 1861) are now usually referred to as *Archaeopteryx* and are considered conspecific (the same species) by some (de Beer, 1954; Swinton, 1960), but not by others (Wetmore, 1960; Savile, 1957a).

Archaeopteryx (Fig. 2.1) was a medium-sized bird about the size of a pigeon but with a long lizardlike tail with *lateral* tail feathers, an arrangement unlike that of any modern bird. Short, rounded wings with feathers probably enabled it to make short gliding flights, perhaps even limited flapping flight, and claws on the wings suggest arboreal habits. The structure of the feet likewise indicates perching adaptations. The three fingers were separate, not fused as in modern birds. Other reptilian features include unfused pelvic bones and a lizardlike beak with teeth set in the jaws. Restorations (Fig. 2.1 lower) depict the bird almost completely clothed with feathers but with scales over the head and neck.

Except for two unidentified feathers from Australia the next oldest known bird fossils are from the Cretaceous geological period, some 35 million years later than the time of *Archaeopteryx*. These include rich finds in the shale beds of western Kansas and other prairie states, where inland seas inundated the land and left deposits when the water receded. Among these deposits was *Hesperornis* ("western bird"), a

FIG. 2.1. TOP *Fossil remains of* Archaeopteryx *(ancient wing), showing claws on digits of forelimbs and long tail.* BOTTOM *Restoration of* Archaeopteryx, *about 20 inches long.* [Courtesy of Amer. Mus. Nat. Hist.]

F I G. 2.2. *A restoration of* Hesperornis, *showing teeth, laterally directed legs with lobed or webbed toes for swimming, the peculiar (but largely imaginary) plumage, and absence of wings.* [Drawing by Diane Pierce.]

large, cormorantlike bird (Fig. 2.2) perhaps 5 feet in length. *Hesperornis* was apparently an aquatic bird, specialized for swimming and diving and presumably subsisting on fish. Like its Jurassic predecessors, it had teeth, which were unevenly set in the jaws. The neck was long and flexible, the tail shorter than in *Archaeopteryx* and more like that of modern birds, and the wings were reduced to a single pair of slender bones.

Another "find" from the same Cretaceous shales of Kansas was *Ichthyornis* ("fish-bird"), so called from its amphicoelous vertebrae. Originally *Ichthyornis* was believed to have had teeth, but Gregory (1952) concluded that the skulls assigned to *Ichthyornis* actually were those of Mososaurs (reptiles) associated in the same beds. Now Gingerich (1972) believes that a toothed jaw in the Peabody Museum collection at Yale really belongs to *Ichthyornis*. However, some doubt still persists about either *Hesperornis* or *Ichthyornis* having teeth. Apparently *Ichthyornis* resembled a gull somewhat in size and habits, was a strong flier with a keeled sternum and well-developed wings, and presumably lived on fish. Both *Ichthyornis* and *Hesperornis* disappeared without leaving descendants when the inland seas receded.

According to Brodkorb's *Catalogue of Fossil Birds* (1963–1971), the Cretaceous period has so far yielded about 33 species belonging to 16 genera: 4 hesperornithiform birds, 9 ichthyornithiform types, 3 "flamingos," 1 "pelican," 4 loonlike, 2 grebelike, 3 rail-like, and 7 charadriiform birds. However, considerable juggling around of specimens has been done. Baird (1967) believes that several other birds should be transferred from the Eocene period to the Cretaceous, and Cracraft

(1972) would reassign the rail-like *Telmatornis* to the charadriiform birds. And of course the fossils discovered to date are only a fraction of those that must have occurred during the period; Brodkorb (1960) calculates that as many as 110,000 species may have evolved and died out during the 60 million years of the Cretaceous.

During the Eocene epoch, dating back some 60 million years, bird remains, though still relatively scarce, show much better representation, including both strange types long extinct as well as the beginnings of definite lines that have persisted to the present day. An oft-cited example of an extinct line is *Diatryma*, a 7-foot flightless monster (Fig. 2.3L), first recovered from the Eocene of Wyoming. Later discoveries revealed additional specimens from Wyoming, New Mexico, New Jersey, and Europe. Though suggestive of the Ostrich in size and perhaps habits, these giant fossils are believed to be more closely related to the rails (Brodkorb, 1967).

More modern types also prevailed in the Eocene. Discoveries include fossils assigned to herons, ducks, vultures, hawks, grouse, cranes, rails, sandpipers, and owls. In later deposits, particularly in the Miocene and Pliocene, examples of nearly all nonpasserine families in North American have been found. No modern species, or even genera, date back as far as the Eocene, but a larklike bird (*Palaeospiza bella*) from the

FIG. 2.3. LEFT Diatryma, *a heavy-bodied giant from the Eocene of Wyoming, probably remotely related to the rails.* RIGHT Phororhacos, *giant bird of prey, suggestive of and probably remotely related to the Secretarybird.* [Redrawn by Diane Pierce from original by Amer. Mus. Nat. Hist.]

Upper Miocene of Colorado and a "finch" (*Palaeostruthus hatcheri*) from the Lower Pliocene of Kansas show that some advanced types were in existence more than 10 million years ago.

Bizarre forms also continued to appear and disappear throughout the Tertiary. Examples of these are *Phororhacos* (Fig. 2.3R) and other cranelike giants from the Miocene and Oligocene of Patagonia (Argentina). These were large-headed, heavy-bodied, cursorial birds of prey, equaling or exceeding the Ostrich in size. Probably bird life reached its maximum development in the late Miocene and Pliocene periods before unfavorable climatic conditions caused widespread extinction over much of the world.

Best known of the Pleistocene fossils are those from the asphalt pits at Rancho la Brea, California, where many birds as well as mammals became entombed. Fossil Lake in eastern Oregon and several beds in Florida have also yielded valuable Pleistocene remains, including several giant hawks or vultures, one of which — *Teratornis incredibilis* — is believed to have been the largest flying bird (Howard, 1952).

Brodkorb (1971) figures the total fossil species (exclusive of the passeriform birds, which he has not yet catalogued) discovered to date to be 1,522: 898 *paleospecies* (those now extinct) and 624 *neospecies* (species still living). However, known fossils are but a small fraction of the avifauna that must have existed in the past. Brodkorb (1960) calculates that more than a million species have evolved since the origin of the class Aves, of which less than one percent exist today. Moreau (1966) challenges these figures.

Speciation

In preceding pages we have spoken freely of *species* — the real structural units in any classification — and implied that species have evolved and died out by the hundreds of thousands in the past. But we have not defined a species or explained how species originate, how long they live, or how they are separated from each other. The following paragraphs attempt the impossible by trying to do so.

What is a species? The layman's definition, which is almost correct, is: a species is any plant or animal distinguishable from other plants or animals — a robin, a crow, or a chicken. But this definition poses problems. Anyone can tell a Plymouth Rock from a Rhode Island Red; yet they are not separate species. All domestic breeds of chickens are regarded as a single species derived from *Gallus gallus,* the ancestral Red Jungle Fowl; presumably the separation of the different varieties from the ancestral stock began millions of years ago and then was hastened by cross-breeding and hybridization in captivity thus producing the many varieties we now know. Even the identification of crows poses problems. The Northwestern Crow (*Corvus caurinus*), formerly regarded as a full species and currently on many "life lists," is probably merely a variety of the Common Crow (*Corvus brachyrhynchos*) and not deserving of specific status. Moreover, the Carrion Crow (*Corvus corone*) of Europe may be a geographic form of the Common Crow, eliminating another "life bird" for many people. The Northwestern Crow interbreeds freely with the Common Crow where they overlap and probably the Carrion Crow and the Common Crow would interbreed, if they could get together.

Hence, whether or not two closely related forms interbreed successfully is a fundamental criterion in defining a species. The two American crows mentioned above can and do interbreed. Whether or not the European and American forms could might be determined by breeding them in captivity, but would not prove that they would do so in the wild because other barriers (ecological, nesting season, etc.) might prevent it.

The criterion of interbreeding gives a biological or genetic concept to the earlier morphological definitions of a species. This biological concept was advanced by Mayr's (1942) comprehensive (but not quite all-inclusive) definition of a species as "groups of actually or potentially interbreeding natural populations, which are reproductively isolated from other such groups." Hence, forms that interbreed freely, even though distinguishable in the field (such as the "Yellow-shafted" and "Red-shafted" Flickers) are now regarded as *subspecies* or *geographic races*. And, more rarely, forms that are indistinguishable in the field are regarded as full species if they are reproductively isolated and do not interbreed; the Alder Flycatcher (*Empidonax alnorum*) and Willow Flycatcher (*E. traillii*), distinguishable by song but not by plumage, are an example (Stein, 1963).

Since subspecies are usually geographically separated, it follows that new subspecies can originate by a nonmigratory species forming pockets of isolated populations that readily interbreed where they overlap but are reproductively and geographically isolated in other parts of their range. Eventually, perhaps in thousands of years, these "pockets" of subspecies might fail to interbreed because of some ecological, physical, or physiological barrier and then would be regarded as full species. Conversely, if the geographic barriers were ineffective in separating two or more populations, they would remain as full species. Sedentary game birds are good examples of subspeciation: 19 species of grouse divisible into 102 subspecies, 183 phasianid species into 546 subspecies. Ducks are a good example of the other extreme. They are very mobile; a drake reared in coastal marshes might pair with a hen from the interior on their common wintering grounds and go to the prairie provinces for nesting, thus hindering or preventing subspecific differentiation.

Species not divisible into subspecies are called *monotypic species*. Many North American ducks are monotypic; some are not even separable from Old World populations. For example, the Common Teal (*Anas crecca crecca*) of Europe has been combined with our Green-winged Teal (*A. crecca carolinensis*) to form a single species, *A. crecca*. (See Nomenclature, page 36). Conversely, species divisible into two or more subspecies are called *polytypic*. Good examples, in addition to the galliform birds cited above, are the Screech Owls (18 subspecies north of Mexico, Fig. 2.4), Horned Larks (22 subspecies, including the "Shore Lark" of Europe), and Song Sparrows (36 subspecies at the last, but probably not final, count). Mayr (1946) says that about 75 percent of the known species of birds are polytypic.

FIG. 2.4. *Screech Owls, like Song Sparrows, are divisible into numerous subspecies; 18 are recognized in North America north of Mexico. This one is presumed to be an Eastern Screech Owl* (Otus asia asio) *because it was found in Michigan, but its subspecific identification could be determined only by careful study, which usually would include comparing it with other specimens in a good reference collection.* [Photo by Dale A. Zimmerman.]

Species or populations with broadly overlapping or identical breeding ranges are called *sympatric,* those with noncontiguous or completely separated ranges are called *allopatric.* But sympatry and allopatry are not necessarily permanent. Barriers separating allopatric forms might break down, allowing them to intermingle and, perhaps rarely, merge into a single species. This may be taking place in the Ross' Goose–Lesser Snow Goose complex. The rarer Ross' Goose, with the smaller gene pool, may be absorbed ("genetic swamping") by the more abundant Lesser Snow Goose which now overlaps with it and results in increased interspecific contacts (Trauger et al., 1971). Similarly the Mallard–Black Duck–Mexican Duck trio may eventually become one species (Aldrich, 1970; Johnsgard, 1960); some regard this consolidation as already attained. More often, however, two allopatric forms remain effectively isolated long enough to be able to come together without interbreeding, Fisher (1972) thinks that Laysan and Black-footed Albatrosses are similar genetically, but have been isolated so long that now they occur together on some islands without interbreeding. They are ecologically isolated by nesting and breeding behavior: timing, voice, posture, courtship dances, and coloration. Hybridization between the two species is rare, and known hybrids are infertile. To complicate matters, some forms, such as redpolls (Wynne-Edwards, 1952), can act as distinct species in some parts of their range, if effectively isolated, and as subspecies in other parts of their range where they still interbreed. Such forms are sometimes called "incipient species"; in time they may become full species.

Just how species originate in nature is difficult to answer. Apparently complete transformation of one species into another by subtle evolutionary changes is possible, but, as implied above, recognizable speciation is geographical in origin: two conspecific populations becoming geographically separated long enough to become full species, incapable of interbreeding if they come together again. How long this takes is a moot question; perhaps several thousand years for subspecies formation, perhaps a million for a full species. However, speciation is more rapid in the smaller, more advanced, and more plastic forms. The House Sparrow, for example, is believed to be undergoing rapid evolutionary changes in this country (Packard, 1967), but penguins in the Antarctic are not. The longevity of a species—its lifespan from origin to extinction—is also a moot question. It is longer in the less active, older forms than in active passerines with a high rate of metabolism. Some think that the longevity of a species might be in the neighborhood of a million years, perhaps an ominous forewarning that man's time has come.

As already implied, some species and groups are more plastic than others. Presumably the penguins, the Ostrich, and oceanic birds are quite stable; possibly they will not undergo much further evolutionary change or will do so more slowly. But some birds, such as the ducks, woodpeckers, and hummingbirds, are more plastic and still undergoing change. Interspecific mingling and hybridization are common in these groups. Hybridization is also frequent in our wood warblers. A well-known example is in the regular interbreeding of Golden-winged and Blue-winged Warblers, where isolating mechanisms break down and two well-known hybrids, the Brewster's and Lawrence's Warblers, result. Most hybrids, like the proverbial mule, are sterile, but these warbler hybrids continue breeding to some extent by back-crossing with either parent.

The Taxonomic Categories

Biologists have a propensity for classifying things; hence, they have attempted to arrange all plants and animals in an orderly sequence based on apparent kinship by grouping them into categories such as species, genera, families, orders, classes, and phyla. Sub and super categories are also established for most of these groups. Phylogenetic trees (Fig. 2.5) are one method of trying to show these relationships, but such trees are virtually impossible to construct accurately with our present state of knowledge. The larger categories of birds—orders and families—have not been rearranged much in recent years, but genera and species are continually being shuffled around. The first half of this century witnessed a prodigious buildup of subspecies—the bread of life for many collectors and museum workers. Now the tendency is to minimize or even drop *trinomials* (subspecies names), but to continue studying geographic variations without applying names to them. And a new vogue, "numerical taxonomy"—that is, applying numbers rather than names to animals—threatens to undo all the meticulous work of past generations of taxonomists. (See Sokal and Sneath, 1963, for a defense of numerical taxonomy, and Amadon, 1966, Blackwelder, 1967, and Mayr, 1965, for a defense of conventional taxonomy.)

In defining taxonomic units there are increasing attempts to utilize all available data, including morphological, ecological, and genetic. Details in life histories, such as nests, eggs, development of young, and breeding behavior, sometimes are useful in determining relationships. In recent years electrophoretic patterns of egg-white proteins, eye-lens proteins, and blood plasma have been widely used as taxonomic characters. Sibley and co-workers in particular (see Sibley, 1960, 1970, for bibliographies) have worked painstakingly in these difficult fields. Sometimes their results suggest changes in classification; sometimes (reassuringly) they support the present arrangements.

The smallest unit recognized by names in our classification is the

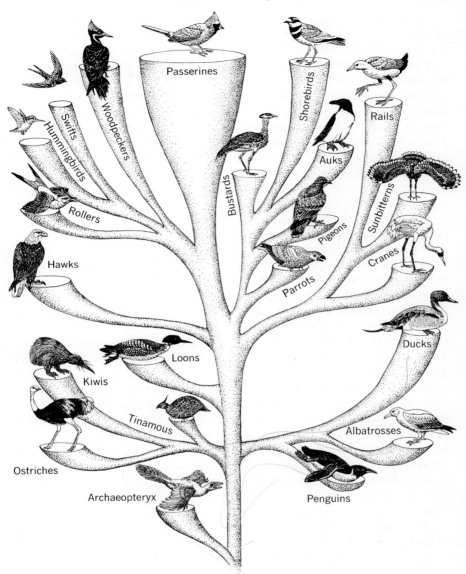

Fɪɢ. 2.5. *Hypothetical tree, showing probable relationships among some avian orders and families. Many of the relationships are uncertain.* [Drawing by Diane Pierce, after L. C. Pettit.]

subspecies or *geographic race.* These, as the second name implies, are geographic, or sometimes ecological, variations in populations — eastern, western, northern, southern, coastal, desert, montane — and differ only in inconstant and often overlapping measurements and intensity of color (darker, paler, brighter, etc.). Ornithologists describing new "millimeter races" are often called "splitters," as opposed to "lumpers" who tend to combine subspecies. The present trend is toward more lumping, not only of subspecies but also of genera. Formerly many genera were monotypic, containing only one species; now they are polytypic, often containing many species. A good example is in the duck genus *Anas,* which combines six or more formerly recognized genera into one. Robins and Schnell (1971) suggest combining six genera of "grassland sparrows" into two.

The number of genera currently recognized is probably of little concern to beginning ornithologists, but the number of species is important. Current estimates of the number of species in the world are about 8,600, with enough named subspecies to bring the total of species and subspecies to about 30,000. The former number will probably remain relatively constant except for re-evaluation of the specific status of previously described species, as there are probably few undiscovered species. But the number of recognized subspecies fluctuates greatly: a rapid increase when taxonomists were describing new subspecies, now some decreases with current lumping. It is improbable that a new species will be discovered in North America because the last new species — the Colima Warbler, discovered belatedly in the Chisos Mountains in Texas — was described in 1889. (The Cape Sable Sparrow, described in 1919, is now relegated to subspecific status, and "Sutton's" Warbler, discovered in 1939 in West Virginia, is considered a hybrid.) In other parts of the world, however, new species are found quite frequently, two of the latest almost in our backyard: a new hummingbird in Mexico and, surprisingly, a new warbler, the Elfin Woods Warbler (*Dendroica angelae*), in Puerto Rico in one of the most-studied tropical regions in the world (Kepler and Parkes, 1972). Other discoveries in the past decade include a new barbet in Rhodesia, a new bird of paradise in New Guinea, a new babbler in the Philippines, a new bulbul in Madagascar, and a new tanager and two new icterids in South America.

Nomenclature

Giving names to things is called *nomenclature,* as opposed to *taxonomy* or *systematics* (terms used more or less interchangeably), which is the study or science of classification and our attempt to arrange groups in a true phylogenetic sequence. Our system of binomial (two-name) nomenclature dates back to 1758, when the Swedish botanist Linnaeus devised the plan of applying a scientific name (genus and species) to every plant and animal. In describing subspecies a new term, the *trinomial* (subspecies designation) has to be added. Formerly, when fewer species were known, this system worked fairly well, but with continued naming of new forms, particularly insects and other invertebrates, the system is becoming quite burdensome and may yet "degenerate" into numerical taxonomy, perhaps for bacteria and insects but, hopefully, not for birds.

Necessarily, there are rules and regulations for naming birds. A Law of Priority states that the first *valid* name applied to a bird in a published description must stand, even though seemingly inappropriate. Often, in exploring the past literature, we find old, long-forgotten names and have to replace the newer names. An International Commission on Zoological Nomenclature meets occasionally to act on nomenclatural problems. In North America a committee of the American Ornithologists' Union (A.O.U.) decides what names, both common and scientific, should be used for North American birds. The nomenclature currently in use in this country is prescribed in the A.O.U. *Check-list of North American Birds,* Fifth edition, 1957, and the Thirty-second Supplement, 1973.

The scientific name of a bird consists of the generic name, capitalized and put in italics (or underlined in script), followed by the species name and then the subspecies, if there are subspecies. Thus the full name for the Atlantic Song Sparrow is *Melospiza* (which designates the generic group to which the Song Sparrows and several other sparrows belong) *melodia* (for the species group or whole Song Sparrow complex) *atlantica* (for the coastal race), followed by the name of the describer, in this case Todd (*Melospiza melodia atlantica* Todd). If one wishes to speak of Song Sparrows without designating a particular race, only the binomial *Melospiza melodia* is used. It is now approved practice for field observers to use only the binomial, as the particular subspecies involved is usually difficult to determine, except by assumption on purely geographical grounds (see Fig. 2.4).

FAMILIES

The next higher unit of classification beyond the genus is the family, perhaps the most used and useful category except for the species. About 170 families of living birds are listed in most classifications. An additional 41 fossil families have been listed by Wetmore (1960), a number that will be considerably augmented when Brodkorb's *Catalogue of Fossil Birds* is complete. The large number of avian families,

a. Turkey Vulture

b. Red-shouldered Hawk

c. Osprey

d. Prairie Falcon

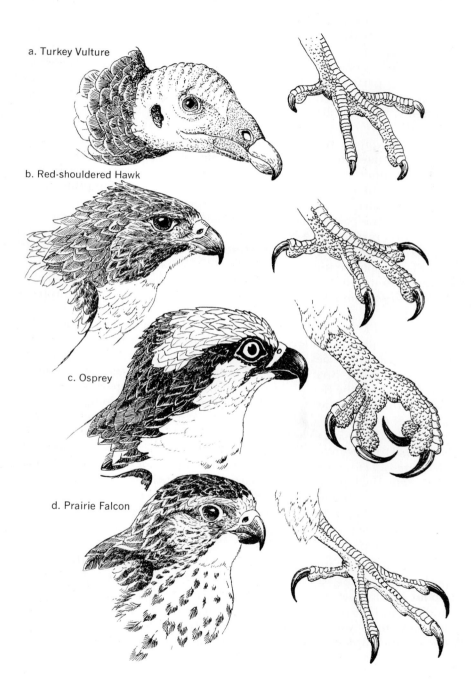

especially in the Passeriformes, reflects the desire for dealing with reasonably small groups; otherwise many of the passerine families would be combined.

Family distinguishing characters, in addition to those that are internal and thus not obvious, include variations in the bill (the general structure is usually an ordinal character), the scutellation (scales) of the tarsus, and number and comparative length of the primaries. In the tube-nosed swimmers (Procellariiformes), for example, the four families are separated largely by the position and structure of the nostrils. In the Falconiformes, variations in bills and feet are helpful in setting up the five recognized families; Figure 2.6 illustrates some of these variations in the four North American families. In the less variable passerine groups, however, the bills and feet are not always sufficiently distinctive, but the length (or absence) of the outer (10th) primary is sometimes useful in separating songbird families.

Misleading criteria in setting up families, as well as other taxonomic categories, are *convergent* and *divergent evolution*. An example of the former is found in the similar bills of the loons, herons, and king-fishers, which do not express genetic relationship but merely similar (convergent) food habits. Conversely, the dissimilar bills of the closely related mergansers and other diving ducks reflect different food habits (the mergansers catch fish and the other divers are vegetarian or sift out invertebrates from bottom ooze).

Family names of birds, as in all animal families, end in *-idae*. (Similarly, subfamily names end in *-inae* and superfamilies in *-oidea*.) Family names can be and often are converted for common usage into such terms as *fringillids* for the Fringillidae family, or *tyrannids* for the tyrant flycatchers (Tyrannidae), or *parulids* for the New World warblers (Parulidae). Such derivatives are more accurate designations than merely "warblers" or "flycatchers," of which there are unrelated Old World groups, while "sparrow" or "finch" family is very misleading because it includes buntings, grosbeaks, cardinals, and many other fringillids.

ORDERS

American classifications, following Wetmore (1960) and Peters' *Check-list of Birds of the World* (1931–1972), divide the class Aves into

FIG. 2.6. *Bill and foot characteristics of four falconiform families.* a. *Turkey Vulture (Cathartidae), showing perforate nostril and comparatively weak foot with reduced hind toe.* b. *Red-shouldered Hawk (Accipitridae), showing imperforate nostril and more strongly developed grasping foot.* c. *Osprey (Pandionidae), showing strongly hooked beak, long claws, scaly sole, and reversible outer toe.* d. *Prairie Falcon (Falconidae), showing circular nostril with bony central tubercle and beak with notch and tooth.* [Drawings by Diane Pierce from specimens in Cleveland Mus. Nat. Hist.]

about 27 orders of living birds, with about 6 others known only from extinct or fossil forms. Some Europeans, however, notably Stresemann (1959), use smaller ordinal units and recognize 51. Of the 27 orders used here, 20 are represented in North America. Three of these barely reach our borders (oceanic or tropical), so that most states have representatives of 17.

Avian orders often lack the sharp and sometimes obvious distinguishing characters associated with other vertebrates. Among the reptiles, for instance, the turtles belong to a well-defined reptilian order; and the carnivores, rodents, bats, and marsupials comprise well-known mammalian orders with easily observable distinguishing features, such as teeth and toes. Ordinal characters for birds, however, are often based on less obvious internal features, such as the structure of the palate and the arrangement of the bones of the skull. Internal features, especially of the skeleton, are presumed to be more stable and less subject to adaptive change than external features. The general structure of the bills and feet (not minor variations) is very useful, but has to be used with caution, and in combination with other characters, because of adaptive convergence due to feeding habits.

Ordinal names of birds (in this country) end in *-formes*. As in the case of families, the names are often converted for convenient usage into anglicized terms such as *passeriform* (*passerine,* also used, stems from the older name Passeres) for perching birds, *columbiform* for pigeonlike, and *falconiform* for hawklike birds.

The following classification lists the 27 orders and 170 families of living birds of the world, along with the number of species in each family, distribution of each, and a few distinguishing characters for the nonpasserine families and the North American passerine families. The beginning student, rather than trying to memorize these, should gradually familiarize himself with those found in his locality. Regional check-lists are usually available for this purpose.

Orders and Families of Living Birds

(North American families are in **boldface.**)

1. SPHENISCIFORMES, Penguins.
    1. Spheniscidae, 17 species, 22 forms;[1] Antarctic, north to Galápagos; heavy-bodied (up to 94 lb); flightless; wings flipperlike for swimming; plumage dense of scalelike feathers completely investing the body (no apteria except in embryos); body well insulated with subcutaneous fat.

---

[1] *Forms,* as used here, denotes total species and subspecies.

2. STRUTHIONIFORMES, Ostriches.[2]
    1. Struthionidae, 1 species, 6 forms; Africa, southeast Asia (Arabia); large (up to 300 lb in ♂), flightless, cursorial; wings degenerate, with numerous fluffy plumes; plumage loose (no interlocking), sparse on neck and absent on thighs; 2 toes.
3. RHEIFORMES, Rheas.
    1. Rheidae, 2 species, 6 forms; South American; large (up to 50 lb), flightless, cursorial; wings degenerate, with soft loose plumes; no rectrices; head and neck feathered; 3 toes.
4. CASUARIIFORMES, Cassowaries, Emus, 2 living families (another fossil); wings much reduced; plumage coarse, hairlike, drooping; long aftershafts; no real rectrices.
    1. Casuariidae, Cassowaries, 3 species, 30 forms; North Queensland and adjacent islands (mostly New Guinea); beak compressed; bony casque on head, often with gaudy wattles on featherless neck; remiges reduced to 4–6 stiff black quills; legs stout.
    2. Dromiceidae, Emus, 1 species (another extinct), 3 forms (2 others extinct); Australia; beak depressed; neck feathered, no casque or wattles; no quills on wings.
5. APTERYGIFORMES, Kiwis (*Apteryx*).
    1. Apterygidae, 3 species, 5 forms; New Zealand; beak long, decurved, with nostril near tip; no apparent wings; plumage hairlike; legs short and stout; strong toes terminating in sharp claws.
6. TINAMIFORMES, Tinamous.
    1. Tinamidae, 45 species, 118 forms; neotropical (Mexico through South America); bobwhite to fowl size; bill usually somewhat depressed and arched; wings developed for flight, short and concave; mainly cursorial.
7. GAVIIFORMES, Loons.
    1. **Gaviidae,** 4 species, 8 forms; circumpolar; bill strong, straight, compressed; wings pointed; legs far back on body; tarsus flattened, bladelike; front toes fully webbed.
8. PODICIPEDIFORMES, Grebes.[3]
    1. **Podicipedidae,** 20 species, 38 forms; world-wide, all continents and large islands; wings short (2 species flightless); plumage soft and dense, toes lobed with flattened nails; tail reduced to tuft of feathers.

---

[2] Bock (1963) and Sibley and Frelin (1972) recommend combining the three orders of ratites (Ostriches, Rheas, and Cassowaries and Emus).

[3] Probably the loons and grebes are not as closely related as this arrangement indicates (see Storer, 1956).

9. PROCELLARIIFORMES, Tube-nosed Swimmers, 4 families; nostrils tubular; hooked bill composed of horny plates; nasal glands large; wings long and narrow (large number of secondaries); plumage compact, "oily"; feet palmate.

　　1. **Diomedeidae,** Albatrosses, 14 species, 19 forms; wide-ranging oceanic birds, breeding mainly on islands in southern hemisphere; nostrils lateral, widely separated by culmen.

　　2. **Procellariidae,** Shearwaters, Fulmars, 56 species, 117 living forms; oceans, islands, and coasts throughout the world; nostrils dorsal, fused but imperforate.

　　3. **Hydrobatidae,** Storm Petrels, 18 species, 40 forms; oceans, islands, and coasts throughout the world; nostrils dorsal, fused and perforate (one tube).

　　4. Pelecanoididae, Diving Petrels, 5 species, 9 forms; southern oceans; nostrils opening vertically; wing bones shortened, flattened; alcidlike.

10. PELECANIFORMES, Totipalmate Swimmers, 6 living families (6 other fossil families); totipalmate foot unique (4 toes joined in common web); nostrils rudimentary or absent; gular sac.

　　1. **Phaëthontidae,** Tropicbirds, 3 species, 14 forms; tropical oceans; nostrils small, perforate; gular sac rudimentary; middle rectrices long and filiform.

　　2. **Pelecanidae,** Pelicans, 6 species, 12 forms; world-wide, except polar extremes; nostrils obsolete; gular pouch very large; bill large, long and hooked; huge wingspread; legs short.

　　3. **Sulidae,** Boobies, Gannets, 9 species, 21 forms; nearly world-wide; nostrils obsolete; gular pouch small; bill strong, conical, not hooked; lores, chin and upper throat bare.

　　4. **Phalacrocoracidae,** Cormorants, 29 living species (1 extinct), 58 forms; world-wide; nostrils obsolete; gular sac small; bill hooked; long neck and body.

　　5. **Anhingidae,** Snake-birds, 2 species, 6 forms; all continents, chiefly tropical or subtropical; long, narrow, finely serrate bill; long neck with kink (eighth cervical articulates at angle); plumage glossy; tail long, fluted in ♂.

　　6. **Fregatidae,** Frigatebirds (Man-o'-war Birds), 5 species, 14 forms; temperate and tropical oceans; bill long and strongly hooked; throat bare with inflatable crimson pouch in ♂; wings extremely long and pointed (7-foot wingspread); tail long and deeply forked.

11. CICONIIFORMES (Ardeiformes of some authors), Long-legged Waders, 7 living families (2 fossil); neck long, often folded in flight; many species with decorative plumes or bare areas about

head and neck; long legs; toes unwebbed except in flamingos (which some authors put in the next order).

1. **Ardeidae,** Herons, Egrets, Bitterns, 58 species, 159 forms; world-wide; bill long, straight, acute; neck extremely long with lengthened sixth cervical causing kink; nuptial plumes and "aigrettes" often highly developed; powder-down tracts.

2. Cochleariidae, Boat-billed Heron, 1 species, 3 forms; neo-tropical; bill broad, flat with prominent keel; head and eyes large.

3. Balaenicipitidae, Whale-headed Stork, 1 monotypic species; Africa; large bill swollen at base, strongly hooked at tip.

4. Scopidae, Hammerheads, 1 species, 2 forms; Africa, Arabia, Madagascar; bill moderately long, compressed; head large and crested; neck short.

5. **Ciconiidae,** Wood Ibises, Storks, Jabirus, 17 species, 24 forms; nearly world-wide; bill variable, stout at base; bare areas about head and neck; long list of minor characters, including rudimentary penis, 2 coats of down, and neck pouches.

6. **Threskiornithidae,** Ibises, Spoonbills, 28 species, 49 forms; nearly world-wide; bill long, slender, decurved (ibises) or straight, broad and spatulate (spoonbills).

7. **Phoenicopteridae,** Flamingos, 6 monotypic species; mainly pantropical, also high Andes; bill large, lamellate, bent downward at middle; front toes short and fully webbed; hallux reduced or wanting.

12. ANSERIFORMES, Screamers, Waterfowl, 2 living families (1 fossil); ordinal characters chiefly internal, not very diagnostic.

1. Anhimidae, Screamers, 3 monotypic species; South America; bill not ducklike; 2 long sharp spurs on wing; tibia partly bare; front toes with small basal webs; no uncinate processes, their absence an archaic feature found elsewhere only in *Archaeopteryx*.

2. **Anatidae,** Ducks, Geese, Swans, 145 species, 228 forms; world-wide; bill flat and lamellate (or long and "toothed" in mergansers); plumage dense with undercoat of down (apteria reduced); front toes fully webbed.

13. FALCONIFORMES, Diurnal birds of prey, 5 living families (2 fossil); bill strong, hooked, with basal cere; nostrils usually imperforate; wings with large sail area; feet raptorial.

1. **Cathartidae** (Vulturidae of some authors), New World Vultures, 6 species, 11 forms; New World; nostrils large, oval and perforate; head largely bare, often colored; feet relatively weak with reduced hallux; no syringeal muscles.

    2. Sagittariidae, Secretarybird, 1 monotypic species; Africa; bill without tooth, legs extremely long (hawk head with heron legs); peculiar "quills" on back of head and neck.

    3. **Accipitridae,** Kites, Hawks, Eagles, Old World Vultures (Aegypiidae of some authors), Harriers, 205 species, 506 forms; world-wide; typically raptorial (strong beaks, feet, and claws); broad expanse of wings and tail.

    4. **Pandionidae,** Osprey, 1 species, 5 forms; nearly world-wide (except New Zealand); set off from Accipitridae largely by tarsal and toe structure; tarsus short, stout, reticulate; toes powerful, with prickly scales, outer toe reversible.

    5. **Falconidae,** Caracaras, Falcons, 58 species, 177 forms; world-wide; bill strongly hooked, with prominent tooth in true falcons; nostrils usually circular with bony tubercle in center; wings usually long and pointed (caracaras somewhat aberrant).

14. GALLIFORMES,[4] Megapodes, Gallinaceous Birds, Hoatzin, 7 living families (1 fossil); bill short, obtuse, culmen decurved and bent over lower; wings short, concave, with stiff remiges; aftershaft prominent; feet and claws strong; well-developed crop, gizzard, and caeca.

    1. Megapodiidae, Megapodes, 10 species, 40 forms; Australasia; feet exceptionally large and strong; aftershaft reduced; very peculiar nesting habits (page 281).

    2. **Cracidae,** Curassows, Guans, Chachalacas, 38 species, 90 forms; Texas (Rio Grande) through South America; bare areas about head and neck, often with crests, casques or wattles; trachea elongated and looped in some species; arboreal.

    3. **Tetraonidae,** Grouse, Ptarmigans, 18 species, 102 forms; Holarctic; nostrils feathered; tarsus feathered wholly or in part, including the toes in *Lagopus;* toes more or less pectinate; sides of neck with modified feathers and sometimes inflatable pouches.

    4. **Phasianidae,** New World Quails, Old World Partridges, Pheasants, Jungle Fowl, Peafowl, 165 species, about 545 forms, many of questionable status; mostly Old World (except New World quail); characters very variable, but often with spectacular plumages in males; wattles, spurs, and bare areas in some species.

    5. Numididae, Guineafowl, 7 species, 35 forms; Africa and Mad-

---

[4] The 7 galliform families have been juggled around by various authors, some being reduced to subfamily status. Some authors place the Hoatzin in a separate order (Opisthocomiformes), to which Cracraft (1971) would add a new fossil family. (Some authors consider the Hoatzin a cuculiform bird.)

agascar; head and neck largely bare and colored, often with wattles, casques and hackles; plumage dense, dark, and spotted.

6. **Meleagrididae,** Turkeys, 2 species, 6 forms; central and southern United States, Mexico, northern Central America; wattles prominent on bare head; feathers lustrous, squarish; tail broad, rounded.

7. Opisthocomidae, Hoatzins, 1 monotypic species; South America; head crested, with scanty feathers; plumage rather loose; arboreal, with claws on wings in young; crop muscular.

15. GRUIFORMES, "Marsh" Birds, 12 living families (10 fossil); ordinal characters very variable and not diagnostic (group of "misfits").

1. Mesitornithidae, Roatelos, Monias, 3 monotypic species; Madagascar; weak wings, flightless or nearly so; long body, tail, and toes.

2. Turnicidae, Bustardquails, 15 species, 51 forms; Africa, Australasia; quail-like wings and body, but long tarsus; no hallux.

3. Pedionomidae, Plainwanderers (Collared Hemipodes), 1 monotypic species; Australia; short hallux, tail very short.

4. **Gruidae,** Cranes, 14 species, 23 forms; all continents except South America; large birds, with long sharp bill and long (heronlike) legs, but short toes; some with ornamental plumes or crowns on head; trachea convoluted, folded in sternum.

5. **Aramidae,** Limpkins, 1 species, 5 forms; southern United States through South America; long, slightly decurved, heavy bill; short wings and long legs (cranelike and rail-like).

6. Psophiidae, Trumpeters, 3 species, 6 forms; South America; gallinaceous bill, rail-like legs; plumage soft and fluffy.

7. **Rallidae,** Rails, Coots, Gallinules, 133 species, 324 forms; world-wide; body compressed; wings short (some species flightless); toes long with incumbent hallux.

8. Heliornithidae, Sungrebes, 3 species, 6 forms; pantropical; long bodies (neck, wings, and tail); legs short, toes lobed.

9. Rhynochetidae, Kagus, 1 monotypic species; New Caledonia; head large and crested; eyes large; powder down all over.

10. Eurypygidae, Sunbitterns, 1 species, 2 forms; neotropical; "sandpiper" with "bittern" head and neck; powder-down feathers.

11. Cariamidae, Cariamas, Seriemas, 2 monotypic species; South America; long tufts of feathers around nostrils; long legs and tail.

12. Otididae,[5] Bustards, 23 species, 48 forms; all Old World con-

[5] The most divergent of the gruiform families; Hendrickson (1969) would remove them from the Gruiformes.

tinents; large, heavy-bodied (up to 30 lb), long-necked, long-legged terrestrial birds; inflatable neck pouches in some; no hallux.

16. CHARADRIIFORMES, Shorebirds, Gulls and Terns, Alcids, 16 living families (2 fossil); ordinal characters very variable; pterylosis similar throughout.

1. **Jacanidae,** Jaçanas, 7 species, 17 forms; pantropical, north to Texas; frontal shield prominent; sharp spur or knob on bend of wing; toes extremely elongate (characters somewhat intermediate between Gruiformes and Charadriiformes; classed by some with Gruiformes).

2. Rostratulidae, Painted Snipe, 2 species, 3 forms; southern South America and Old World tropics; bill with nerve endings; trachea looped over clavicle and breast muscles.

3. **Haematopodidae,** Oystercatchers, 6 species, 21 forms; all continents; bill large, compressed, blunt at tip, bright red; toes partly webbed; no hallux.

4. **Charadriidae,** Plovers, Turnstones,[6] Surfbirds,[6] 63 species, 102 forms; world-wide; bill short, tapering or swollen at tip; plumage often with contrasting patterns of bands or collars and rump and wing patches.

5. **Scolopacidae,** Sandpipers and allies, 82 species, 118 forms; world-wide; bill usually long and slender, flexible at tip, sometimes decurved; tarsus usually scutellate (plovers usually reticulate); hallux usually present.

6. **Recurvirostridae,** Avocets, Stilts, 7 species, 12 forms; all continents; bill long, slender, straight (stilts) or recurved (avocets); legs very long.

7. **Phalaropodidae,** Phalaropes, 3 monotypic species; Holarctic in breeding distribution, wintering in southern hemisphere; bill moderate, needlelike or fairly stout; toes lobate or margined.

8. Dromadidae, Crabplovers, 1 monotypic species, shores of Indian Ocean; bill deeply cleft, compressed; claw of middle toe pectinate; hallux present.

9. Burhinidae, Thick-knees, 9 species, 26 forms; all continents except North America; bill stout; eyes large (crepuscular habits); legs stout, especially at tibiotarsal joint ("thick-knees"); hallux absent.

10. Glareolidae, Pratincoles, Coursers, 17 species, 44 forms; all Old World continents, but most forms in Africa; bill short (pratincoles) or long and tapering (coursers), with oblong, impervious

---

[6] Austin (1961) transfers the turnstones to the Scolopacidae, and Jehl (1968) does likewise for the Surfbird.

nostrils; wings long and pointed (pratincoles) or short and broad (coursers).

11. Thinocoridae, Seedsnipe, 4 species, 12 forms; South America; bill finchlike; legs short; tail graduated.

12. Chionididae, Sheathbills, 2 species, 5 forms; islands off South African and South American coasts; somewhat galliform; bill with horny sheath; spur on wing; legs and feet strong.

13. **Stercorariidae,** Skuas, Jaegars, 4 species, 10 forms (3 monotypic jaegars, 1 polytypic Skua); cold oceans, both hemispheres; bill complex, of 4 parts, with cere; wings and tail long and pointed; species separable by tail characters; front toes fully webbed with strong claws.

14. **Laridae,** Gulls, Terns, 82 species, 185 forms; world-wide; bill hooked (gulls) or straight and pointed (terns); tarsus moderate (gulls) or short (terns); feet small, webbed.

15. **Rynchopidae,** Skimmers, 3 species, 6 forms; pantropical; bill strongly compressed, bladelike, with upper mandible shorter; tarsus short; feet small.

16. **Alcidae,** Auks, Murres, Puffins, 22 species, 36 forms; northern hemisphere oceans; bill variable, extreme in puffins; legs far back (sit on tarsi), feet fully webbed; compact plumage.

17. COLUMBIFORMES, Pigeonlike Birds, 2 living families (1 recently extinct); ordinal characters mainly plumage peculiarities; feathers dense, loosely set in skin; well-developed crop.

1. Pteroclidae, Sandgrouse, 16 species, 45 forms; Eurasia, Africa; bill grouselike, without operculum; tarsus short, fully feathered; toes short with hexagonal scutes, sometimes feathered; plumage noniridescent.

2. **Columbidae,** Pigeons, Doves, 289 species, 835 forms (many questionable); world-wide, but mainly Old World; "pigeonbilled," with bare basal cere or operculum overhanging nostril; tarsus bare; plumage often iridescent. [The Raphidae, Dodos and Solitaire, consisting of 3 species, 1 each on the islands of Mauritius, Réunion, and Rodriguez, are now extinct. The extinct Raphidae were pigeonlike (or rail-like), heavy-bodied flightless forms with tufted tail. Storer (1970) would place the Solitaire from Rodriguez in a separate monotypic family, Pezophapidae.]

18. PSITTACIFORMES, Parrots, Lories, Macaws.

1. Psittacidae, 315 species, 768 forms (many of questionable status); pantropical; many peculiarities: "parrot" bill (maxilla movable, hinged to skull); tongue thick and fleshy; foot zygodactylous, with reversible fourth toe; plumage harsh, usually brightly colored.

19. CUCULIFORMES, Plantain-eaters, Cuckoos, 2 families; zygodactylous foot, fourth toe reversible; skin thin and tender.
   1. Musophagidae (elevated to ordinal rank by some), Plantain-eaters, 20 species, 43 forms; Africa; crested, long-tailed arboreal birds, with stout, serrate bill; unique color pigments (turacin and turacoverdin).
   2. **Cuculidae,** Cuckoos, Anis, Roadrunners, 127 species, 359 forms; all continents; bill variable, but more or less decurved and compressed; tail long and graduated.
20. STRIGIFORMES, Owls, 2 living families (1 fossil); well-defined order; large head with fixed eyes and facial disc, often asymmetrical ears; short neck; feathered tarsus, often including toes; raptorial foot; soft, lax plumage.
   1. **Tytonidae,** Barn and Grass Owls, 11 species, 60 forms; nearly world-wide; facial disc triangular, bordered by deep cleft; plumage comparatively sparse; toes bare, middle toe pectinate.
   2. **Strigidae,** Typical Owls, 123 species, 531 forms; world-wide; facial disc circular; plumage dense; notched primaries.
21. CAPRIMULGIFORMES, Goatsuckers and allies, 5 families; owl-like head and plumage, but typically weak bill and feet; hind toe short; gape enormous.
   1. Steatornithidae, Oilbirds, 1 monotypic species; northern South America; bill hooked and notched, not especially flattened; 12 long rictal bristles on each side; large aftershaft.
   2. Podargidae, Frogmouths, 12 species, 29 forms; Australia and vicinity; enormous head and gape and wide bill; rudimentary aftershaft; powder-down patches on rump.
   3. Nyctibiidae, Potoos, 5 species, 14 forms; neotropical; bill small, narrow, decurved at tip; no rictal bristles.
   4. Aegothelidae, Owlet-frogmouths, 8 species, 17 forms; Australian region; bill small, flat; long bristles on lores.
   5. **Caprimulgidae,** Goatsuckers, 67 species, 203 forms; all continents; bill small and weak but with wide gape; usually long rictal bristles; eyes large; claw of middle toe pectinate.
22. APODIFORMES, Swifts, Hummingbirds, 3 living families (1 fossil); long wings due to elongaged manus (short humerus); elevator muscles (pectoralis minor) of wing large; weak feet.
   1. **Apodidae,** Swifts, 76 species, 219 forms; world-wide; bill short, small, wide gape; wing flat, curved, with short secondaries and long primaries; tail short, often spine-tipped; plumage dull.
   2. Hemiprocnidae, Crested Swifts, 3 species, 15 forms; Indian region; head crested; eyes large; tail long, forked, the outer feathers attenuated.

3. **Trochilidae,** Hummingbirds, 320 species, 688 forms; neo-tropical; bill usually long and slender, very variable; plumage usually metallic, iridescent; sexes extremely dimorphic in many cases.

23. COLIIFORMES, Colies (Mouse-birds).

1. Coliidae, 6 species, 29 forms; Africa; long-tailed, crested, arboreal birds, with weak, rounded wings and dense, compact plumage; toes directed forward and hallux reversible.

24. TROGONIFORMES, Trogons.

1. **Trogonidae,** 34 species, 103 forms; pantropical; bill short, flat, decurved, and serrate in some species; tail long and graduated with squarish feathers; plumage soft and lax, often brilliantly colored; feet small and weak.

25. CORACIIFORMES, Kingfishers, Todies, Motmots, Bee-eaters, Rollers, Hoopoes, Hornbills, 10 families; typically large-headed, large-billed birds with metallic bright plumages; toes syndactylous, the anterior toes joined in various combinations.

1. **Alcedinidae,** Kingfishers, 87 species, 335 forms; world-wide; large heads, sometimes crested, and strong beaks; feet small and weak, with third and fourth toes joined; tarsus short, tibia partly bare.

2. Todidae, Todies, 5 monotypic species; West Indies; head smaller; bill flattened and weakly serrate; rictal bristles prominent; short wings and tail; metallic plumage.

3. Momotidae, Motmots, 8 species, 45 forms; neotropical; bill large, decurved, serrate; plumage soft, loose-webbed; tail usually long, often racquet-tipped.

4. Meropidae, Bee-eaters, 24 species, 50 forms; Old World tropics into temperate regions; bill long, decurved, pointed; tail long, central rectrices elongate.

5. Coraciidae, Rollers, 11 species, 32 forms; Old World tropics; bill short, stout, hooked; 3 anterior toes joined at base.

6. Brachypteraciidae, Groundrollers, 5 monotypic species; Madagascar; short wings, long tarsus and tail (otherwise as above).

7. Leptosomatidae, Cuckoo-rollers, 1 species, 3 forms; Madagascar and adjacent islands; wings long and pointed; tail long, square-tipped.

8. Upupidae, Hoopoes, 1 species, 9 forms; Eurasia, Africa; bill long, slender, decurved; nostrils round; head crested with long, black-tipped feathers; plumage not metallic.

9. Phoeniculidae, Woodhoopoes, 6 species, 27 forms; Africa; nostrils elongate; head not crested; tail long and graduated.

10. Bucerotidae, Hornbills, 45 species, 104 forms; Africa, southern

Asia, and adjacent islands; bill enormous, with casque or helmet on culmen; tarsus short, rough and scaly; anterior toes joined for various lengths, forming broad sole.

26. PICIFORMES, Jacamars, Puffbirds, Barbets, Honeyguides, Toucans, Woodpeckers, 6 families; foot zygodactylous, with distinctive arrangement of flexor tendons; bill very variable, but usually long, strong or well developed.

   1. Galbulidae, Jacamars, 15 species, 38 forms; neotropical; bill long and straight, with angular gonys; plumage soft and lax, usually metallic.

   2. Bucconidae, Puffbirds, 30 species, 76 forms; neotropical; bill shorter, stout and rounded; head large (head feathers erectile).

   3. Capitonidae, Barbets, 73 species, 255 forms; pantropical; bill large, stout, swollen, and often bearded at base; toes with long claws; stocky, inactive birds with gaudy colors.

   4. Indicatoridae, Honeyguides, 11 species, 36 forms; Ethiopia, Asia; bill short, finchlike with ridged and laterally swollen maxillae; nostrils with narrow membrane, not bristly; sober plumages.

   5. Ramphastidae, Toucans, 37 species, 87 forms; neotropical; bill enormous, light and spongy, with serrate edges; face partly bare; wings weak; plumage lax.

   6. **Picidae,** Woodpeckers, Piculets, Wrynecks, 210 species, 853 forms; world-wide; bill strong and chisel-like (except in wrynecks); nostrils concealed by bristly feathers; tongue extremely long and extensible; rectrices stiff-pointed (except in piculets and wrynecks); legs and feet strong; fourth toe permanently reversed.

27. PASSERIFORMES, Perching Birds, 69 living families (2 fossil); a large assemblage of more than 5,000 species with very variable external features (bill, wings, tail, and plumage), but "perching" foot characteristic (hallux well developed, incumbent, with long claw; 3 anterior toes directed forward). A few characters for North American families only are listed here.

   1. Eurylaimidae,[7] Broadbills, 14 species; India, Malaysia, Philippines, Africa.

   2. Dendrocolaptidae, Woodhewers, 48 species; neotropical.

   3. Furnariidae, Ovenbirds, 215 species; neotropical.

   4. Formicariidae, Ant-thrushes, 222 species; neotropical.

   5. Conopophagidae, Antpipits, 11 species; neotropical.

   6. Rhinocryptidae, Tapaculos, 26 species; Costa Rica to Argentina.

[7] See Olson (1971) for a suggested rearrangement of this and the following 12 families.

7. **Cotingidae,** Cotingas, 90 species; southern Arizona to Argentina.
8. Pipridae, Manakins, 59 species; southern Mexico to Argentina.
9. **Tyrannidae,** Tyrant Flycatchers, 365 species; chiefly neotropical but north to Canada and Alaska; bill usually flattened (wider than high at base) and slightly hooked, with prominent rictal bristles at base; tarsus short, rounded behind; feet small and weak.
10. Oxyruncidae, Sharpbill, 1 species; Costa Rica to Brazil.
11. Phytotomidae, Plantcutters, 3 species; South America.
12. Pittidae, Pittas, 23 species; southeastern Asia, Australia, Africa.
13. Acanthisittidae, New Zealand Wrens, 4 species; confined to New Zealand.
14. Philepittidae, Asities, False Sunbirds, 4 species; Madagascar.
15. Menuridae, Lyrebirds, 2 species; Australia.
16. Atrichornithidae, Scrubbirds, 2 species; Australia.
17. **Alaudidae,** Larks, 75 species; mainly Old World, 1 North American species; bill moderately long and pointed, not flat; legs, feet, and toes well developed, with long, sharp, straight hind claw.
18. **Hirundinidae,** Swallows, 75 species; world-wide; bill small, flattened, with wide gape; wings long and pointed; legs short; feet and toes small and weak.
19. Dicruridae, Drongos, 20 species; Africa, southeastern Asia to Australia.
20. Oriolidae, Old World Orioles, 28 species (including 2 *Irena*); Old World.
21. **Corvidae,** Crows, Magpies, Jays, 100 species; nearly world-wide; bill rather large and strong; nostrils covered by stiff feathers; feet fairly large and strong.
22. Cracticidae, Bell Magpies, Australian Butcherbirds, 10 species; Australia.
23. Grallinidae, Magpie-larks, 4 species; Australia.
24. Ptilonorhynchidae, Bowerbirds, 18 species; Australia, New Guinea.
25. Paradisaeidae, Birds of Paradise, 44 species; New Guinea. Some authors combine the bowerbirds and birds of paradise into one family.
26. **Paridae,** Titmice, 65 species; mainly Holarctic; bill small, but stout and sharp in most species; wings rather short and rounded; nostrils concealed by feathers.
27. **Sittidae,** Nuthatches, 22 species (includes 5 Australian Neosittidae); mostly Old World; bill slender, straight or slightly recurved; tail short, not used for support; toes and claws long.

28. Hyposittidae, Coral-billed Nuthatch, 1 species, Madagascar.
29. **Certhiidae,** Creepers, 17 species; northern hemisphere, mostly Old World; bill long, slender and decurved in North American species; tail long and stiff-pointed for support; toes and claws long.
30. Paradoxornithidae, Parrotbills, Suthoras, 18 species; Eurasia.
31. **Chamaeidae,** Wrentits, 1 species; Oregon to Lower California; bill short and stout; wings short and tail long; plumage soft and lax. (Sometimes included in the Timaliidae.)
32. Timaliidae, Babblers, 264 species; Old World tropics.
33. Campephagidae, Cuckoo-shrikes, 71 species; Old World (Africa, Asia, Australia).
34. Pycnonotidae, Bulbuls, 110 species; Old World tropics.
35. Chloropseidae, Leafbirds, 12 species; Oriental Region.
36. **Cinclidae,** Dippers, 5 species; nearly world-wide; bill straight and slender, compressed, with notch near tip; tail short; tarsus and toes long and strong; plumage dense, compact.
37. **Troglodytidae,** Wrens, 63 species; mainly New World tropics; bill slender, usually somewhat decurved; wings short and rounded, usually barred; tail barred.
38. **Mimidae,** Thrashers, Mockingbirds, 30 species; New World; bill variable, but strong, decurved in some species; tail long; rictal bristles present.
39. **Turdidae,** Thrushes, 305 species; world-wide; bill variable, but usually straight and strong, notched near tip; tail mostly fairly short and square; tarsus "booted" (not scutellate); juvenal plumage spotted.
40. Zeledoniidae, Wrenthrushes, 1 species; Costa Rica, Panama. A puzzling species that has been juggled around; egg-white proteins and field studies (Hunt, 1971) suggest affinities with the Parulidae.
41. **Sylviidae,** Old World Warblers, Gnatcatchers, Kinglets, 398 species; nearly world-wide; large family with very variable characters; North American gnatcatchers with sharp slender bills and rounded tail longer than wings; tarsus scutellate, except in the 5 kinglets.
42. Muscicapidae, Old World Flycatchers, 328 species; Old World.
43. **Prunellidae,** Accentors, Hedge-sparrows, 11 species; Europe, northern Asia to Alaska.
44. **Motacillidae,** Wagtails, Pipits, 48 species; nearly world-wide; bill sharp; slender; tail long; toes long, with long claw on hallux.
45. **Bombycillidae,** Waxwings, 3 species; Holarctic; prominently crested birds with soft, brownish plumage; secondaries (irre-

spective of sex or age) often tipped with red waxlike spots; tail with terminal yellow (or red) band.

46. **Ptilogonatidae.** Silky Flycatchers, 4 species; southern United States through Central America; nearly unicolored (black, brown, or gray), crested birds with short wings and long tail, and soft, silky plumage.

47. Dulidae, Palmchats, 1 species; Santa Domingo.

48. Artamidae, Woodswallows, 10 species; Africa, Asia, Australia.

49. Vangidae, Vanga Shrikes, 12 species; Madagascar.

50. **Laniidae,** Shrikes, 72 species; Old and New World; bill strong and hooked, notched near tip; feet fairly strong with sharp claws.

51. Prionopidae, Woodshrikes, 13 species; Old World.

52. Cyclarhidae, Peppershrikes, 2 species; neotropical.

53. Vireolaniidae, Shrike-vireos, Greenlets, 3 species; neotropical.

54. Callaeidae, Wattled Crows, Huias, Saddlebacks, 2 living species (Huia extinct); New Zealand.

55. **Sturnidae,** Starlings, 104 species; Old World, introduced in New World; bill fairly long, straight, and strong in most species; wings long and pointed (in *Sturnus*); tail usually short and square; legs and feet strong; plumage often glossy and metallic.

56. Meliphagidae, Honey-eaters, 160 species; Australia, Pacific Islands.

57. Nectariniidae, Sunbirds, 104 species; Old World tropics.

58. Dicaeidae, Flowerpeckers, 54 species; Asia, Australia.

59. Zosteropidae, White-eyes, 80 species; Africa, Asia to New Zealand.

60. **Vireonidae,** Vireos, 37 species; New World; bill stout (compared to parulids), slightly notched and hooked; 10th (outer) primary short.

61. **Coerebidae,** Honeycreepers, 36 species (Mayr); neotropical, to Florida. Some authors delete this family and redistribute the 36 species.

62. Drepanididae, Hawaiian Honeycreepers, 22 species; confined to Hawaii.

63. **Parulidae,** Wood Warblers, 110 species (Mayr); New World; bill usually slender and pointed, not notched or hooked; 9 primaries; sexual dimorphism in plumage often pronounced.

64. **Ploceidae,** Weaverbirds, 263 species (Mayr); Old World, introduced in New World; bill short, stout, conical; 10th primary rudimentary.

65. **Icteridae,** Blackbirds, Troupials, 88 species (Mayr); New World; bill very variable, but typically rather long, pointed and

conical, often prolonged backward into a casque; 9 primaries; plumage with black often predominating.

66. Tersinidae, Swallowtanagers, 1 species; northern South America.

67. **Thraupidae,** Tanagers, 197 species (Mayr); New World, mainly neotropical; hard to characterize, but bill typically rather short and stout, swollen, with prominent tooth in North American species; plumage often strongly dimorphic.

68. Catamblyrhynchidae, Plush-capped Finches, 1 species; northern Andes.

69. **Fringillidae,** Grosbeaks, Finches, Buntings, Sparrows, 425 species (Mayr); world-wide except Australia; characters variable, but bill typically short and thick, often massive (grosbeaks); feet and toes strong; 9 primaries. Several authors recognize the Emberizine "buntings" as a separate family (Emberizidae). Most of the puzzling 9-primaried Oscines (families 63–69) have been juggled around considerably.

SELECTED REFERENCES

A.O.U. *Check-list of North American Birds.* 5th ed., 1957. Lancaster, Pa.: Amer. Ornith. Union.

AUSTIN, O. L., JR. 1961. *Birds of the World.* New York: Golden Press.

BEECHER, W. J. 1953. A phylogeny of the Oscines, Auk, 70:270–333.

BRODKORB, P. 1963–1971. Catalogue of fossil birds. Bull. Fla. State Mus., Vol. 7, No. 4; Vol. 8, No. 3; Vol. 11, No. 3; Vol. 15, No. 4 (Parts I–IV).

DE BEER, SIR G. 1954. *Archaeopteryx lithographica.* London: British Mus. Nat. Hist.

HEILMANN, G. 1927. *The Origin of Birds.* New York: Appleton.

MAYR, E. 1942. *Systematics and the Origin of Species.* New York: Columbia Univ. Press.

———— and D. AMADON. 1951. A classification of recent birds. Amer. Mus. Novitates, No. 1496:1–42.

———— and L. L. SHORT. 1970. Species taxa of North American birds. Publ. Nuttall Ornith. Club, No. 9.

PETERS, J. L. 1931–1972. *Check-list of Birds of the World.* Cambridge, Mass.: Harvard Univ. Press, Vols. 1–14.

ROMER, A. S. 1945. *Vertebrate Paleontology.* Univ. Chicago Press. Chapter 13.

SIBLEY, C. G. 1960. The electrophoretic patterns of avian egg-white proteins as taxonomic characters. Ibis, 102:215–284.

————. 1970. A comparative study of the egg-white proteins of passerine birds. Bull. Peabody Mus. Nat. Hist., 39.

SOKAL, R. R., and P. H. A. SNEATH. 1963. *Principles of Numerical Taxonomy.* San Francisco and London; W. H. Freeman.

STORER, R. W. 1960. The classification of birds. Chapter III in A. J. Marshall's *Biology and Comparative Physiology of Birds.* New York: Academic.

SWINTON, W. E. 1960. The origin of birds. Chapter I in A. J. Marshall's *Biology and Comparative Physiology of Birds*. New York: Academic.

TORDOFF, H. B. 1954. A systematic study of the avian family Fringillidae based on the structure of the skull. Univ. Mich. Mus. Zool. Misc. Publ., 81:1–42.

VAN TYNE, J., and J. BERGER. 1959. *Fundamentals of Ornithology*. New York: Wiley. Chapters 12 and 13.

WETMORE, A. 1960. A classification for the birds of the world. Smithsonian Misc. Coll., Vol. 139, No. 11:1–37.

3

<p style="text-align:center">~~~~~</p>

Feathers, Molts, and Plumages

A MAJOR concern in avian anatomy is the visual appearance of a bird: the external features, the coat of feathers, and the coloration. Thus this chapter, the first in a series devoted to anatomical and physiological considerations, deals with these external features and their adaptations.

In descriptions of birds, reference is commonly made to regions of the body and to features of the bill, wings, feet, and tail. These body regions (e.g., *nape, jugulum, crissum*), feather groups (e.g., *primaries, rectrices*), and associated structures (e.g., *mandibles, hallux*) constitute the *topography* of a bird (Fig. 3.1).

Feathers

ORIGIN AND DEVELOPMENT

Phylogenetically, feathers are believed to have been derived from reptilian scales in Mesozoic times, when the elongated loose scales that occurred on certain birdlike reptiles of that period may have developed fringed margins. Some such covering was indispensable for heat retention, and feathers probably developed more or less synchronously with warm-bloodedness, though some authors maintain that warm-bloodedness preceded feather development and occurred in the pterodactyls. Hence, development of feathers may have been an adaptational response accompanying accelerated metabolism. Actually, the transition from reptilian scales to feathers has not been adequately demonstrated, but Rawles (1960) states that such a transition has been observed in domestic fowl. Scales and feathers have a common developmental origin; both develop from similar germ buds, which, on the bare tarsus of the Bald Eagle, for instance, give rise to scales and on the tarsus of a Golden Eagle, grouse, or owl give rise to feathers. Thus, such germ buds are bipotential and may develop into a scale or a feather.

A feather begins its development as a *papilla* (Fig. 3.2), which thrusts

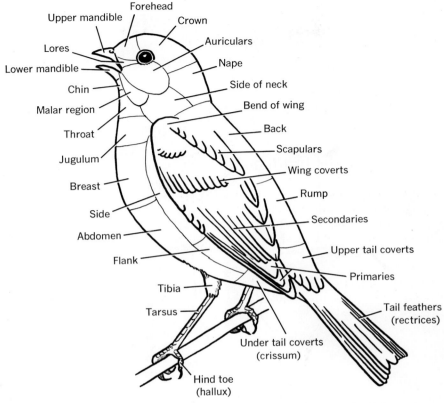

Forehead
Upper mandible
Crown
Lores
Auriculars
Lower mandible
Nape
Chin
Side of neck
Malar region
Bend of wing
Throat
Back
Jugulum
Scapulars
Breast
Wing coverts
Side
Rump
Abdomen
Secondaries
Flank
Upper tail coverts
Tibia
Primaries
Tarsus
Tail feathers
(rectrices)
Under tail coverts
(crissum)
Hind toe
(hallux)

F IG. 3.1 *Topography of a passerine bird.* [Drawing by Homer D. Roberts.]

up the overlying epidermis. The outer epidermal layer of stratum corneum forms the sheath of the feather; the inner basal or Malpighian
layer contains the future feather parts, which arise from a germinating
ring or collar at the base of the papilla. The central dermal pulp cavity
contains blood vessels that carry nutrients and color pigments to the
growing feather, but as the structure matures the surrounding sheath
is shed or preened off, the feather parts unfold and harden, and the
blood supply is shut off. The result is a structure composed entirely of
dead cornified epidermis known as *keratin*. The base of the feather is
imbedded in a pit or *follicle* from which another feather will arise when
the first is molted.

In early development two types of feathers are recognized: (1) *neossoptiles*, which constitute the downy plumage of newly hatched birds,
and (2) *teleoptiles*, the adult-type feathers (probably including adult
down in species that have adult down) that replace the neossoptiles.

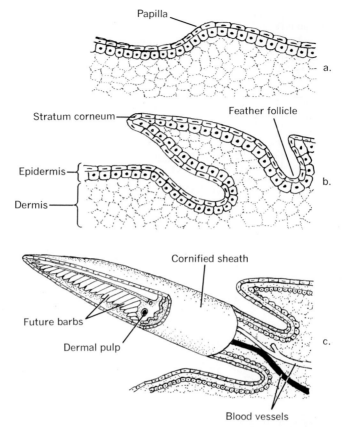

Papilla

Stratum corneum

Feather follicle

Epidermis

Dermis

a.

b.

Cornified sheath

Future barbs

Dermal pulp

c.

Blood vessels

Fig. 3.2 *Development of a feather.* a. *Origin of the feather papilla.* b. *Continued growth of papilla.* c. *Unfolding of maturing feather.* [Redrawn by R. M. Naik from Storer's *General Zoology,* copyright 1943 by McGraw-Hill Book Co., Inc., New York.]

Papers by Lillie and co-workers (1932–1944) should be consulted by those who wish further information on the physiology of feather development.

STRUCTURE

Figure 3.3a illustrates a flight feather and its parts. It consists of a flattened *vane* or *webs* (inner and outer) supported by a central *shaft* composed of a pith-filled *rhachis* (webbed portion) and a tubular *quill* or *calamus* (webless portion). The *inferior* (lower) *umbilicus* is the opening into the calamus through which nutrients and pigments were originally supplied to the developing feather. An inconspicuous *superior* (upper) *umbilicus* occurs at the junction of the calamus and rhachis. In some

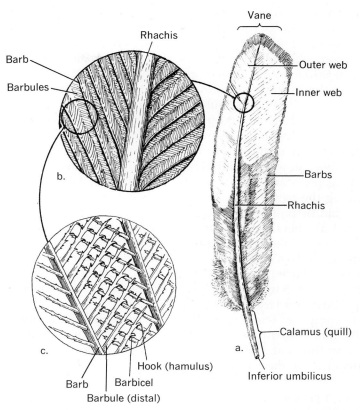

FIG. 3.3. a. *Flight feather* (remex) *of plantain-eater* (Musophagidae). *b. Enlargement of portion of vane, showing arrangement of barbs and barbules. c. Further enlargement of barbs and barbules, showing interlocking arrangement.* [Drawings a and b by R. M. Naik; drawing c by Diane Pierce.]

birds a prominent *aftershaft* (Fig. 3.4b) is present, notably in the Emu, where it is nearly as long as the main feather, and in such galliform birds as grouse, pheasants, and quail, where it undoubtedly has real insulation value. The aftershaft is believed to be of a primitive character, but it has been independently lost, reduced, or retained in so many unrelated groups of birds that its evolutionary significance is not evident.

The outer and inner webs of the primaries are unequal, with the outer web narrower than the inner. Both webs, but particularly the inner, are sometimes notched or emarginate toward the tip, a "slotting" device of importance for many birds in flight (page 109).

The webbed portion of the feather consists of a parallel series of closely spaced *barbs* (about 600 pairs in a pigeon primary) on either side of the rhachis, and each barb in turn gives rise to a series of *barbules* or *radii* (several hundred pairs per barb) that overlap and interlock with the barbules of adjacent barbs. Figure 3.3c shows this interlocking mechanism, which consists of a series of microscopic *barbicels* with little hooks or *hamuli* on the *distal* or *anterior* barbules (those directed toward the tip of the feather) which overlap and hook onto the rolled edge of the *proximal* or *posterior* barbules (those directed toward the base of the feather). Both the distal and proximal barbules have barbicels, but only the distal have hamuli. This somewhat limited sliding arrangement of barbules gives flexibility to the feather while at the same time making it sufficiently rigid to be relatively impervious to air and water. The interlocking parts are readily pulled apart, as can be demonstrated by separating the barbs on a chicken feather, but they can be slipped in place again by pulling them through closed fingers. Birds repair such damaged feathers in preening by drawing them through the bill.

Some loosely plumaged flightless birds, such as the ostrichlike birds and the kiwis, lack this interlocking mechanism; still other species have a scarcity of barbules in certain feathers. The dorsal plumes of the egrets (see Fig. 3.6), for instance, and the lax body feathers of many tropical birds are largely devoid of barbules.

The complete feather, then, is an exceedingly lightweight but durable and efficient structure, whose complexity is well illustrated by the many functional parts described above. A single pigeon primary consists of more than a million parts.

FUNCTIONS OF FEATHERS

Feathers serve a variety of important protective and ornamental functions. Feathers furnish the bird with an admirably lightweight yet durable covering; lightness is an obvious asset in flight, durability a protection against mechanical or physiological injury to the tender skin. Plumages are generally quite waterproof, for though a bird may deliberately soak itself in bathing, it can shed water effectively during rains. A duck's closely imbricated feathers, with the aid of secretions from the oil gland (page 83), are structurally waterproof. Some birds, however, do not have a waterproof plumage. Cormorants and snake-birds, when pursuing fish underwater, get thoroughly soaked and then perch with outstretched wings on rocks, piers, or trees to dry out. Balancing and thermoregulation (pages 156–59) may also be involved in the drying process. The related oceanic frigatebirds avoid contact with water (except for dipping feathers lightly to bathe) by gleaning food from the surface or by robbing other birds; if accidentally submerged, they may get waterlogged and drown. Tropical species with a lax plumage often get drenched, sometimes deliberately bathing, in the frequent showers characteristic of the rain forests, but dry out in the intermittent periods of sunshine. Leaf bathing serves the same function; the birds fly into and rub against wet leaves.

Perhaps the most important physiological function of feathers is heat retention. Feathers are the most efficient type of insulation known, for birds maintain extremely high body temperatures, commonly 104° to 112°F, even in subzero weather. The feathers of a bird are full of dead air spaces, especially when fluffed out. Fluffing out the feathers, an arrangement made feasible by special muscles in the skin, increases the depth of insulating material by adding to the air spaces within the feathery layers. Thus heat loss is reduced to a minimum. Conversely, in warm weather, the feathers are often depressed or held close to the body to permit loss of body heat. Feathers also function in courtship displays and for sex recognition (see Chapter 9).

KINDS OF FEATHERS

Feathers are commonly divided, on the basis of function and location, into the various kinds listed below.

1. *Contour* feathers—Typically, feathers with shaft and vane, covering body, wings, and tail.
   a. General body feathers (Fig. 3.4b).
   b. Wing feathers (*remiges*), divided into primaries and secondaries.
   c. Tail feathers (*rectrices*).
2. *Semiplumes*—Feathers with a prominent shaft but downy web, lacking the interlocking mechanism of contour feathers. Interspersed with and concealed by contour feathers.
3. *Down* feathers—Soft feathers with a minute shaft (except in adult duck down), the barbs usually arising as a fluffy tuft from the end of the quill; provide the natal covering of newly hatched birds and appear as an undercoat on some adult birds (Fig. 3.4a).
4. *Filoplumes*—Minute hairlike feathers with a slender barbless shaft and inconspicuous tuft of weak barbs and barbules at tip (Fig. 3.4c); occur among the contour feathers.
5. *Rictal* bristles—Hairlike feathers, with rudimentary barbs at base, around the *rictus* (base of bill) of such birds as flycatchers and goatsuckers; may be modified contour feathers.
6. *Powder downs*—Feathers producing a powdery substance best seen as paired pectoral and pelvic yellowish patches on herons and bitterns; probably derived from disintegrating down that persists throughout life.

Obviously the most conspicuous feathers of an adult bird are the contour feathers. Though the chief function of these is to provide a warm protective covering and to serve as implements of flight (wing

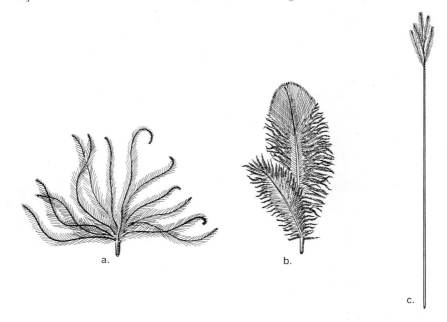

FIG. 3.4. a. *Down feather.* b. *Contour feather of Ruffed Grouse, showing aftershaft.*
c. *Filoplume.* [Drawings by Homer D. Roberts.]

and tail feathers), on some parts of the body they become modified for
a great variety of ornamental functions. On many birds the crown
feathers (Fig. 3.5) are modified into crests or topknots (Cardinal), ornamental plumes (Gambel's Quail) or colored "flash" patches (kinglets).
Other birds have special ruffs (Ruffed Grouse) or pinnae (Prairie
Chicken) about the neck (see Figs. 9.6 and 9.7) that are utilized in

FIG. 3.5 a. *Ornamental head plume of Gambel's Quail.* b. *Flash patch of Golden-
crowned Kinglet.* c. *Topknot of Cardinal.* [Drawings a and c by Dale A. Zimmerman, drawing b by Diane Pierce.]

displays. Special upper tail coverts in the peacocks, "aigrettes" or breeding plumes in certain herons and egrets (Fig. 3.6), highly specialized tail feathers in many birds (e.g., lyrebirds), and the fantastic decorations of the birds of paradise (see Fig. 9.9) serve a similar sexual purpose. Other modifications seem more practical, as in the spine-tipped rectrices of Chimney Swifts and woodpeckers that facilitate clinging to vertical surfaces; a long or broad tail fan may be both ornamental and useful in flight, as in pheasants and turkeys, and as in the lengthened outer rectrices of the Barn Swallow that presumably give added maneuverability.

Semiplumes are interspersed, rather sparingly in some birds, among the other body feathers, especially along the edges of the feather tracts. Sometimes they have been classed as contour feathers, sometimes as down, but various degrees of intergradation occur between the two types. When present in sufficient quantity, semiplumes add insulation and buoyancy.

Down feathers are of special importance as the natal covering in

FIG. 3.6. *Great Egret feeding young. The dorsal plumes (aigrettes) of egrets were once worth $32 an ounce in the millinery trade and nearly brought about extermination of the birds.* [Photo by Allan D. Cruickshank.]

birds that are born in an advanced state of development (pheasants, ducklings) and as a dense undercoat in many aquatic birds. In ducks a special down is grown by the females in spring and plucked for lining the nest.

Filoplumes are the fine hairlike feathers evident on the body of a plucked chicken. Filoplumes may be entirely hairlike or there may be a tuft of loose barbs and barbules at the tip (Fig. 3.4c). The conspicuous hairlike feathers on the necks of breeding cormorants and anhingas, often called filoplumes, are probably modified contour feathers (Van Tyne and Berger, 1959).

Rictal bristles (vibrissae), conspicuous at the base of the bill in many "flycatching" species, are also probably modified contour feathers. Lederer (1972), among others, doubts their commonly ascribed flycatching function, but they probably do have a tactile or sensory function.

Powder-down feathers, already adequately described, are found in quite a few diverse groups of birds besides the herons and bitterns. Young herons, which lack a functional oil gland at first, constantly rub the bill over the powder-down tracts, suggesting that they function as a substitute for the oil gland in such birds.

REMIGES AND RECTRICES

Of special importance to birds are the flight feathers of the wing (remiges) and the tail feathers (rectrices), both for flight and for display. The primaries in particular are also useful in classification for determining orders and families.

Flying birds have from 9 to 12 primaries, and the number is fairly constant within groups; in flightless cassowaries the number is reduced to 3 or 4 stiff quills and in the Ostrich increased to 16 fluffy, ornamental plumes. In general there has been a reduction in the number of primaries from the more primitive flying birds to the 9-primaried Oscines, which many authorities regard as the highest development among birds. Most other passerines and many nonpasserines have 10. This denotes an evolutionary trend from 12 to 9, but Stresemann (1963b) maintains that the ancestral number was 10 and that there was a subsequent increase in many aquatic nonpasserines and a reduction in the higher passerines. Primaries are usually numbered from the innermost (proximal) to the outermost (distal). Unfortunately, some Old World ornithologists reverse this numbering system, causing some confusion for workers studying primaries (Van Tyne and Berger, 1959).

The number of secondaries is more variable and less constant within taxonomic groups, hence less useful as a taxonomic tool. Numbers vary from 6 or 7 in hummingbirds, 9–11 in passerines, and 10 or more in most nonpasserines to a maximum of 25 in Andean Condors and 32 in the Wandering Albatross (Van Tyne and Berger, 1959).

The rectrices serve a variety of functions in balancing, steering, and lift (page 114) and for display purposes. Most birds have 12 rectrices (6 pairs), but there are some virtually tailless birds (e.g., grebes) with a sharply reduced size and number, and other birds with elaborately modified and increased rectrices.

Some tails are utilized in spectacular displays. Some shorebirds use their tail feathers to produce whistling and winnowing sounds during nuptial flights. Sharp-tailed Grouse vibrate and rattle their tail quills during their ceremonial dances. The fantastic displays of the birds of paradise utilize tail feathers or upper tail coverts. The long "train" of the peacock is familiar to everyone who has visited a zoo. Perhaps the peak of development is in the tail of the lyrebirds of Australia (Fig. 3.7). Young birds from one to four years of age have "plain" tails, but do display and sing a little. At about four years the plain tails become "filamentous" (by loss of barbules that connect barbs) and reach their full development at seven or eight years (Smith, 1965).

FIG. 3.7. *An Australian lyrebird, with his tail brought forward and lowered over his head, gives a spectacular performance on his dancing mound in the Tarra Valley, a tree-fern-covered sanctuary for lyrebirds in Victoria. See also Fig. 13.8.* [Courtesy of Australian Information Service.]

Number of Feathers

The number of feathers on a bird and the significance of numbers in relation to size, rate of metabolism, and season of year were largely neglected in biological studies of birds until 1933, when the Smithsonian Institution sponsored an ambitious feather-counting project. The initial counts, made on 153 birds of 79 species, mostly passerines, disclosed fewer feathers on the smaller birds, with the lowest number (940) on a Ruby-throated Hummingbird and the highest number (2,973) on an American Robin, which was one of the largest birds used in the count. A male Hairy Woodpecker had 375 more feathers than a male Downy Woodpecker. Two Eastern Phoebes, collected at about the same time, had a difference of only two feathers, though perhaps this was a mere coincidence. An interesting seasonal adaptation is that birds have their thinnest plumage in late summer (due to feather loss) and their densest covering in winter. An American Goldfinch in winter plumage has nearly a thousand more feathers than one in summer plumage.

Since the above studies were made chiefly on passerine birds, the few counts available for nonpasserines are of interest. One of the earliest of these was undertaken by a dairyman to settle an argument as to how many feathers there are on a chicken. He counted 8,325 on a Plymouth Rock. Another count on a Mallard hen disclosed 11,903 feathers. Apparently the highest count ever made was on a Whistling Swan, which had 25,216 contour feathers. However, 80 percent of these were the small feathers on the head and neck, so that the rest of the body plumage was not notably dense. In fact, smaller birds generally have more feathers per unit of body weight than the larger birds, an adaptation for heat retention in the smaller forms with a higher rate of metabolism. Table 3.1 gives some sample feather counts for both passerine and nonpasserine birds. High feather counts on such birds as the Pied-billed Grebe, Pintail, American Coot, and owls are of particular interest; the grebe had more than twice as many feathers as the larger, more coarsely feathered eagle, no doubt reflecting the need for better insulation in aquatic birds.

A few counts on Old World birds (Markus, 1965) appear quite comparable to those for New World forms: in 11 Laughing Doves (*Streptopelia senegalensis*) the counts ranged from 3,900 to 4,390 (av. 4,192); in 3 species of barbets (Capitonidae) the number of feathers ranged from 2,210 to 3,014; and a white-eye (Zosteropidae) had 3,307 feathers.

Pterylosis

The contour feathers of a bird, it may be noted by plucking a fowl or by examining a nestling robin, are not evenly distributed over the

TABLE 3.1. FEATHER COUNTS OF
REPRESENTATIVE BIRDS

Species	Sex	Date	Number of Feathers	Reference
Pied-billed Grebe	♀	Dec. 30	15,016	Brodkorb, 1949
Least Bittern	♂	May 20	3,867	Brodkorb, 1949
Whistling Swan	—	Nov. 5	25,216	Ammann, 1937
Mallard	♀	Mar. 19	11,903	Knappen, 1932
Pintail	♂	Jan. 28	14,914	Brodkorb, 1949
Bald Eagle	imm.	Jan. —	7,182	Brodkorb, 1955
Plymouth Rock	—	—	8,325	Wetmore, 1936
Coot	♂	Nov. 16	13,913	Brodkorb, 1949
Least Sandpiper	♀	Apr. 30	4,480	Brodkorb, 1949
Screech Owl	♂	Feb. 18	6,458	Brodkorb, 1949
Barred Owl	♂	June 24	9,206	Brodkorb, 1949
Common Nighthawk	♀	July 9	2,034*	Wetmore, 1936
Common Nighthawk	♀	Apr. 30	3,332	Brodkorb, 1949
Ruby-throated Hummingbird	♂	June 11	940*	Wetmore, 1936
Ruby-throated Hummingbird	♀	May 15	1,518	Brodkorb, 1949
Hairy Woodpecker	♂	Apr. 23	2,395	Wetmore, 1936
Downy Woodpecker	♂	Mar. 26	2,020	Wetmore, 1936
(Northern) Blue Jay	♂	Oct. 8	1,898*	Wetmore, 1936
(Southern) Blue Jay (juv.)	♀	July 25	3,773	Brodkorb, 1949
Brown Thrasher	♂	June 11	1,960*	Wetmore, 1936
Brown Thrasher (juv.)	♀	July 9	3,379	Brodkorb, 1949
House Sparrow	♂	July 2	1,359*	Wetmore, 1936
House Sparrow	♂	July 5	3,138	Staebler, 1941
Bobolink	—	—	3,235	Dwight, 1900
Brewer's Blackbird	—	—	4,915	Seton, 1940
Brown-headed Cowbird	♀	July 2	1,622*	Wetmore, 1936
Brown-headed Cowbird	♂	Feb. 21	4,297	Wing, 1952

* There appears to be no ready explanation of why Wetmore's counts run so much lower than subsequent counts on the same species.

whole body, but are arranged in special tracts. These feather tracts are called *pterylae* and the featherless areas between the tracts are called *apteria;* together they account for the distribution or arrangement of feathers (*pterylosis*) over the bird's body. Both the feathered and the featherless areas are named after the regions where they occur. There are eight (or more in many species) feather tracts and adjacent apteria (Fig. 3.8).

The pterylae can be subdivided further. The *capital tract,* for instance, can be divided into *frontal, coronal, occipital, loreal,* and *auricular* regions or elements to denote special parts of the head. In a few birds, such as the Ostrich and the penguins, apteria are lacking—that is, the feathers are uniformly distributed over the body—and in densely plu-

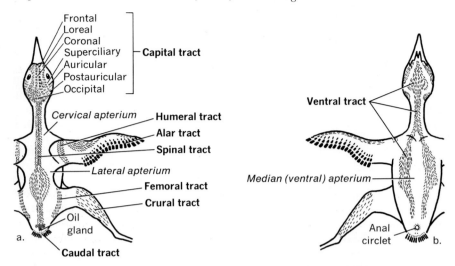

FIG. 3.8 *Feather tracts and apteria in the Common Yellowthroat. a. Dorsal view, showing subdivisions (regions) of the capital tract. b. Ventral view.* [Redrawn by Homer D. Roberts from R. E. Stewart, *Auk*, 69, 1952.]

maged birds the apteria are reduced and sometimes, as in the ducks, covered with down. Even in the ratites and penguins apteria are present in the embryo, indicating that that was the original arrangement (Clench, 1970).

Molts and Plumages

Birds periodically renew their plumage by shedding and replacing their old feathers in a process called *molting*. Nearly all birds have one complete annual molt, usually in late summer after nesting, and most birds have an additional complete, nearly complete, or partial molt in the spring before the breeding season. The various plumages worn by birds at different ages and seasons and the molts producing them are summarized as follows:

Plumage	Molt
Natal down	postnatal
Juvenal	postjuvenal
First winter (nonbreeding)	first prenuptial
First nuptial (breeding)	first postnuptial (annual)

These four molts and plumages carry the bird through its first year. Then its second winter plumage may, in many cases, differ in appearance from the first winter plumage, and its second nuptial plumage may likewise differ from the first nuptial. Many birds require more than two years to reach maturity and thus may have third, fourth, and even fifth winter and nuptial plumages before acquiring fully adult dress. The larger gulls commonly take three or four years to attain their full white and gray-mantled plumage, and in Bald Eagles it is three to five years or more before the symbolic white of the head and tail is complete.

In many colorful passerines the second-year coat of feathers (first breeding season) may be a mixture of new (colorful) and old (brownish) features. The Indigo Bunting in its second year may be mottled with brownish, rather than uniformly blue, and its remiges and rectrices are brown rather than black. The Rose-breasted Grosbeak, American Redstart, and Scarlet Tanager have mixed feathering in their second year. The spring plumage of all birds is usually referred to as nuptial regardless of whether or not they are breeding, or in full plumage.

Humphrey and Parkes (1959), citing inconsistencies in the terminology used for molts and plumages, have proposed a different classification. Inconsistencies include use of implied breeding condition (nuptial) and time or season (annual, winter) in describing molts and plumages, whereas these terms often do not apply on a world-wide basis. "Nuptial," for example, is used for the plumage worn in the spring regardless of whether or not there has been a feather change, and regardless of whether or not the species is breeding. Seasonal terms, devised largely for the northern hemisphere, do not always fit in other parts of the world. Pending further study, they propose to retain the terms "natal" and "juvenal," but to call the first winter plumage (and all subsequent plumages if there is no real feather change in the spring) the *basic plumage*, which is attained by a *prebasic* (rather than postjuvenal) molt. However, in birds that have a real feather change in the spring, an additional term, the *alternate plumage*, attained by a *prealternate molt*, is used. Hence, the sequence of plumages and molts proposed by Humphrey and Parkes is:

Natal down	postnatal molt
Juvenal plumage	first prebasic molt
First basic plumage	first prealternate molt
First alternate plumage	second prebasic molt

There has been much controversy over the merits of this new terminology. Many "older" authorities (Stresemann, 1963a; Miller, 1961; Amadon, 1966b) prefer to retain the long-standing terminology largely worked out by Dwight (1900) nearly a hundred years ago, but some "younger" ornithologists (e.g., Watson, 1963) defend the newer classification. Eventually some compromise may be reached; both systems have useful terms.

The Sequence of Molts and Plumages

This section explains the sequence and some of the variations in the molts and plumages of birds. The young are hatched with a covering of natal down (neossoptiles), completely investing the body in *precocial* birds like ducks and gallinaceous birds (Fig. 3.9), or more or less sparsely distributed in tracts, as in *altricial* birds born in a more immature stage like robins or sparrows. Some birds (e.g., woodpeckers, kingfishers, and some passerines) are born naked; usually these go directly into juvenal plumage without an apparent down stage. A few birds (the

F IG. 3.9. *Sharp-tailed Grouse five days old. The young of precocial birds are born with a complete covering of natal down and can usually run from the nest soon after hatching.* [Photo by Mich. Dept. Nat. Res.]

megapodes or mound-builders in Australia) pass through the down stage in the egg and have their juvenal plumage and are able to fly soon after hatching. For more detailed information on the natal plumages of North American passerines, see Wetherbee (1957).

The natal down is pushed outward (postnatal molt) on the tips of the incoming juvenal plumage, which first appears as stiff quills whose confining sheaths gradually rupture to unfold the juvenal feathers (Fig. 3.10). The familiar spotted plumage of a young American Robin, with wisps of natal down still clinging to the other feathers, is a good example of this stage. Then, usually in later summer (July to September), follows a postjuvenal (first prebasic) molt, in which the fluffy juvenal feathers are shed and replaced by the feathers of the winter or nonbreeding (first basic) plumage.

The postjuvenal molt usually involves only the general body feathers, not the remiges and rectrices which are retained in most species until the following year. This feature often makes it possible to distinguish

FIG. 3.10. *These Brown Thrashers are in juvenal plumage, but show wisps of natal down still adhering to the juvenal feathers.* [Photo by Edward M. Brigham, Jr., courtesy of *Jack-Pine Warbler*.]

the young from adults in the fall, as the somewhat worn wing or tail feathers of the young, acquired soon after hatching in spring or early summer, may be distinguishable from the relatively unfrayed wing or tail feathers of the adults, acquired during the late summer molt (Fig. 3.11). To use this slight distinction accurately, as in age determinations of game birds in the fall, one needs to be thoroughly familiar with the sequence of feather replacement in the species involved. According to Weeden (1961) age determinations based on the tips of the primaries are not reliable in the Rock Ptarmigan; there are too many inter-mediates.

There are some exceptions to the general rule of retention of the juvenal remiges and rectrices, however. Long-billed Marsh Wrens have a complete postjuvenal molt (and a complete prenuptial) including wing and tail feathers (Kale, 1966); such replacements may be neces-sary because of feather wear in the harsh marsh vegetation in which

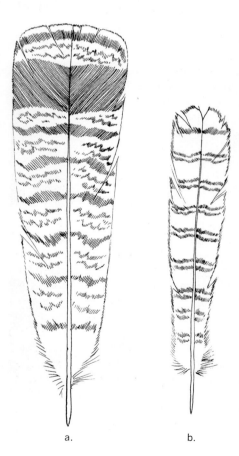

FIG. 3.11. *Tail feathers of adult* (a) *and immature* (b) *Ruffed Grouse, showing unfrayed feather of adult and worn or frayed feather of a young bird. Note notch in tip of juvenal feather.* [Drawings by Diane Pierce.]

a.   b.

they live. Early broods in the Cardinal (Scott, 1967) have a complete postjuvenal, but late broods do not molt all their rectrices. In the Red-winged Blackbird 28 percent of the females have a complete post-juvenal, but only 2 percent of the male Brown-headed Cowbirds have such a molt (Selander and Giller, 1960).

The next plumage change occurs in the late winter or early spring, when a so-called prenuptial molt produces the nuptial (breeding) plumage. Relatively few birds have a complete molt at this time, though many have a partial replacement of feathers. In other species, the spring dress is acquired by fading (no feather change) or by the wearing off of buffy feather tips worn in winter. The breeding plumage of the Blue Jay, for instance, is duller (less bright), because of fading, than that worn in fall and winter. The male Bobolink's handsome black and white nuptial plumage, though acquired by an early prenuptial (prealternate) molt in February or March, is largely concealed by buffy feather tips, which wear off before he reaches his summer home. Less perceptible is the wearing off of the gray-tipped feathers on the black throat of the male House Sparrow. The American Goldfinch is one of the less usual cases of actual plumage replacement in spring. New body feathers, but not new remiges and rectrices, account for the striking change in this species from its dull winter plumage to the black and gold of spring and summer.

After the breeding season, in July, August, or September for most northern hemisphere species, birds undergo a complete postnuptial (second prebasic) molt in which all of the feathers, including the re-miges and rectrices, are replaced. The molt is usually prolonged and gradual, lasting several weeks to several months. Many birds go into partial retirement at this time, cease singing, and become relatively inactive, even fasting for long periods in the case of penguins (Rich-dale, 1951). Growing a completely new plumage undoubtedly imposes a severe physiological drain on the bird, which generally restricts its physical activities. However, because of increased nutritional needs to grow new feathers and to replace heat loss from decreased insulation, many birds require more food while molting and may even time the molt, as in the case of the Bullfinch (*Pyrrhula pyrrhula*) in England (Newton, 1966), to coincide with a period of maximum food availability.

Typically the feathers are shed a few at a time, region by region, so that the molting bird may appear somewhat ragged and unkempt, a patchwork of old and new feathers. The flight feathers are usually shed one after another in a definite sequence, more or less synchronously from each wing, with new feathers replacing those lost rapidly enough so that a bird is usually not deprived of flight during the molt. Excep-tions occur among some aquatic birds not dependent on flight for escape; these shed all their flight feathers more or less simultaneously and are temporarily unable to fly.

The eclipse plumage in ducks, which is a dull, henlike plumage assumed by most drakes in the northern hemisphere during the summer, is quite exceptional among birds. It is usually considered an arrested winter plumage, brought on by an early postnuptial molt in May or June, which is merged or overlaps with a hastened prenuptial molt so that the drake attains its breeding plumage (and sometimes starts courtship and mating) in the fall or winter instead of waiting until spring. Some authors suggest, however, that the eclipse may be a special molt acquired by male ducks to conceal and protect them during their flightless period; it, or a similar type of molt, also occurs in several families of Old World passerines. There is considerable variation among different ducks, as well as among individuals of the same species, so that drakes may be observed in nearly all stages of plumage in the summer and fall. The hens follow a more normal sequence of molts; a complete postnuptial in the summer, after nesting, and a prenuptial (body feathers and tail) in late winter or early spring, including the shedding of old down and the acquisition of a special nest down.

To a certain extent, birds can shed and replace damaged feathers irrespective of the regular molts, but perhaps it is more usual to retain broken stubs until the time for the next molt. *Pulled* primaries in a captive bird, for instance, or lost tail feathers, are rather quickly replaced, but a *clipped* wing deprives the bird of flight feathers until the next wing molt, which is usually an annual affair. A *pinioned* bird has the manus (hand) amputated and thus can never grow primaries. *Brailing* merely binds the manus to the forearm and prevents unfolding the wing. It is customary to clip, pinion, or brail only one wing, so that the bird has difficulty getting off the ground, or flies in low circles if it can rise at all. Some strong-winged birds, however, can manage labored flight even when clipped, pinioned, or brailed.

MOLT IN RELATION TO BREEDING AND MIGRATION

Birds face the problem of timing their summer molts in a way that will interfere least with their breeding cycle and subsequent migration. In most birds both the postjuvenal (first basic) and postnuptial (second basic) molts take place in the period between the fledging of the young and fall migration, but there are many cases, especially in birds with a prolonged molting period, where the molt overlaps breeding and/or migration.

Most aerial feeders (caprimulgids, swifts, flycatchers, and swallows) postpone molting until they reach their wintering grounds, but both

Mengel (1952) and Traylor (1968) have shown that the Acadian Fly-catcher, unlike other Empidonaces, has an incomplete postjuvenal and a complete postnuptial molt before migrating. In Purple Martins, unlike other swallows, both adults and young initiate molting in July and August before migrating, then suspend molting during long overwater flights (not on short overland stages) and complete the molts on their wintering grounds (Niles, 1972). Among caprimulgids, adult Chuck-will's-widows in the southeastern states molt on the breeding grounds in lieu of raising a second brood which would otherwise be possible. Rohwer (1971) theorizes that if two broods were raised, the wintering grounds would not support them all and that a second brood would be largely wasted.

In some birds molting is a prolonged affair and overlaps with part or all of the breeding cycle. Clark's Nutcrackers in the western mountains begin the postnuptial molt while breeding, sometimes before the eggs are laid, and may be in various stages of molt for eight or nine months (Mewaldt, 1958). Glaucous Gulls in northern Alaska start their post-nuptial molt during the egg-laying period in late May. Replacing the primaries is a long process and both molting and breeding have to be accomplished during the short summer at that latitude (Johnston, 1961).

Arctic-nesting sandpipers make a variety of adjustments. Many are long-range, intercontinental migrants, but Red-backed Sandpipers wintering in California show considerable overlap in molting, breeding, and migration (Holmes, 1966). Molt is initiated during nesting, but the main molt of the flight feathers follows nesting and precedes migration. Some of the birds merely move to coastal areas for molting before migrating south. Farther north, at Barrows, Alaska, molting is tele-scoped into an even shorter summer; its timing and duration are regulated to some extent by the prevailing weather (Holmes, 1971). Other sandpipers (e.g., Baird's and Semipalmated) postpone molting until they reach their wintering grounds in South America. In Ptarmigans, which winter in Alaska, the males and unsuccessful females molt earlier than the successful females; the latter delay the molt until after the eggs hatch, even suppressing the molt if bad weather delays nesting (Weeden, 1966).

Birds made temporarily flightless by a simultaneous wing molt also have adjustments to make. Ducks, grebes, coots, and rails skulk and hide in the marshes during the flightless period, but most loons (not the Red-throated) retain their flight feathers until they reach their wintering grounds (Woolfenden, 1967). Loons are inefficient fliers with a heavy wing load; even the loss of a single primary might hamper their flight. European Dippers, and probably the American form, also have a flightless period, but are not seriously handicapped because swimming provides a means of escape and they do not have far to migrate (Sullivan, 1965).

There are other significant adjustments for timing the molt in relation to breeding. In the White Tern (*Gygis alba*), on Christmas Island in the Pacific (Ashmole, 1968), molting occurs between two successive breeding seasons regardless of the time of year because tropical oceanic birds are not dependent on a particular season for breeding. The Andean Sparrow (*Zonotrichia capensis*), in some areas of Colombia, manages two breeding periods and two molts annually, the short molts being telescoped in between the two breeding periods (Miller, 1959, 1961). In some other parts of its range the species only rarely exhibits the double cycle (Davis, 1971; Wolf, 1969), these differences probably dictated by the prevailing weather conditions (Wallace, unpublished).

These are but a few of the examples of birds that make major or minor adjustments in the timing of breeding, molt, and migration. The recent literature details many other examples.

PLUMAGE COLORATION

The extraordinarily varied plumage in birds, which runs a gamut of colors and color patterns surpassing in variety and vividness that found in any other vertebrates, is due to (1) pigments (*biochromes*), consisting chiefly of *carotenoids* and *melanins*, and (2) structural colors (*schemochromes*), which result from the breaking up and reflection of the different components of white light by the physical structure of the feather. "These two fundamentally different color sources are often so interrelated that, in most cases, the observed coloration is the result of a combination of one or more pigments with structural colors" (Rawles, 1960).

The carotenoids (also called *lipochromes*) are responsible for most of the bright red, orange, and yellow colors in birds (melanins in the red-yellow color range are duller). These occur in a great variety of combinations and intensities, from the brilliant red of the Scarlet Tanager to the "washed out" red of the Hepatic Tanager, or the intense yellow of the American Goldfinch compared to the pale yellow of some of the strains of domestic Canaries. Intensity of color may be determined by the concentration of pigments or by the interactions of different pigments; it may be monochromatic, as in the flamingos, or caused by a complex mixture of chemically closely related pigments (Brush and Seifried, 1968). The fat-soluble carotenoids are carried by body fluids into the feather at the time of its formation; after keratinization sets in, the fat solvent disappears, leaving the pigment concentrated in the barbs and barbules. The red tips on the secondaries of waxwings are carotenoid pigments in the expanded and flattened portions of the rhachis. Intensity of color can be controlled to some extent in domesticated birds by providing carotenoid-rich foods.

The melanins are the most abundant and widely distributed pigments in birds. They are responsible for the dark pigments (dark browns, grays, and blacks), as well as for the duller or lighter shades of red and brown (chestnut, tan). Melanins occur as granules, the darker shades usually in the form of rods or ovals, the lighter as smaller spheres. Harrison (1965) retains the use of the terms *eumelanin,* for gray and black pigments, and *phaemelanin,* for brown and buff colors, and suggests the term *erythromelanin* for the unnamed chestnut-red colors.

A third, less generally known group of pigments are called *porphyrins.* Two of these, found only in African plantain-eaters (Musophagidae), are of special interest. One is *turacin,* a deep red-copper pigment found primarily in the wing feathers; the other is *turacoverdin,* a green pigment found in the body plumage. Porphyrins also occur in the feathers of owls and bustards (Otididae).

Structural colors (schemochromes) are due to the physical properties of feathers, the various spectral or rainbow colors being either reflected or absorbed. The bright blue of the Blue Jay, various bluebirds (*Sialia*), and Indigo Bunting, for example, is due to blue light reflected from a layer of blue-producing cells that overlies the melanin-impregnated pigment cells found in the barbs of the feather (Fig. 3.12). Green

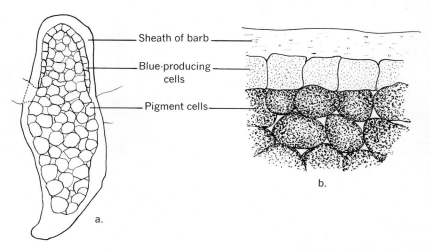

Sheath of barb

Blue-producing cells

Pigment cells

b.

a.

FIG. 3.12. a. *Cross section of barb of Blue Jay feather.* b. *Sagittal section of same, showing dorsal portion of sheath, layer of blue-producing cells, and underlying pigment cells containing melanin granules.* [From Gower, *Auk,* 53, 1936.]

feathers in parrots and other noniridescent greens are essentially the same except that a transparent *yellow* layer superimposed on the blue-producing cells modifies would-be blue to green. Looked at in *transmitted* light (instead of *reflected* light) of a microscope, blue feathers appear brown, and green feathers appear yellow, because only the true pigments are then evident. The so-called *cotingin,* formerly described for the violet colors in certain Cotingidae, is now believed to be caused by reflection of blue light from a red pigment (Brush, 1969).

White feathers are devoid of pigments, so that all the light falling on them is reflected as white light; if all were absorbed, the effect would be black. White plumage is of course normal for many birds—gulls, terns, swans, egrets, to mention only a few—but sometimes in other species the factors responsible for pigment production are lost and *albinism* results. This is particularly true of browns and blacks and is due to lack of melanin-producing cells (*melanophores*). Albinism may be *partial* (Fig. 3.13T), where the white is limited to certain feathers or regions; *total* (Fig. 3.13B), where the plumage is entirely white and the eyes red; or *imperfect,* where the pigment is very dilute, resulting in lighter than normal colors. Albinistic birds are usually short-lived because of physical and physiological defects, but Middleton (1960) reported a partial (nearly total) albino American Robin that returned to his banding station in Pennsylvania for eight successive years.

Examples of albinism in birds are so frequent that editors of ornithogical journals are reluctant to publish them. In a survey of the literature Gross (1965) found 1,847 recorded cases of albinism in North American birds in 276 species in 54 families. Most frequently reported cases were in robins, crows, sparrows, and red-wings, but albinism does occur, often undetected, in gulls, thus causing some problems in identification (Gross, 1964).

The production of an excess of dark pigments is known as *melanism.* It explains the dark phases found in birds like the Rough-legged Hawk, which has a light or normal phase and a dark phase. Apparently, increase in melanin production accounts for the darker geographic races in humid climates; Bowers (1959) has shown that the darker coastal race of the Wrentit in California has more and larger pigment granules than the lighter inland race. *Erythrism* (excess red pigment) is less common, but is well known because of its frequent occurrence in the Screech Owl, which has two color phases, red and gray, a condition known as *dichromatism.* Hrubant (1955) has shown that the gray phase is dominant, as matings of grays produced only grays (53

FIG. 3.13. TOP *Partial albino American Robin.* [Photo by Philip G. Coleman from specimen in Mich. State Univ. Mus.] BOTTOM *Total albino.* [Photo by L. H. Walkinshaw.]

cases) and no reds or intermediates, and matings between red parents were required to produce reds. These phases show a striking distributional pattern; the red phase is common in the northeast, rare in parts of the south, and unknown in the west (Owen, 1963).

Another color aberration is *xanthochroism* (abnormal yellow), rare in nature, but common in captive parrots. Harrison (1966) questions most cases of "alleged" xanthochroism, saying it is really loss of melanin, which leaves carotenoid pigments more evident.

Another familiar color phenomenon is *iridescence,* the interchanging colors so commonly observed on hummingbirds, grackles, and the necks of pigeons. This, in general, is due to interference or unequal dispersion and scattering of light from flattened and twisted barbules so that the angle of view is constantly changing, as in soap bubbles.

In addition to colors caused by pigments or feather structure, birds sometimes take on a superficial hue from chemicals, usually iron oxide, in the soil or water with which they come in contact (*adherent coloration*). Sandbill Cranes probing in bogs sometimes transfer a reddish brown iron oxide stain to their feathers in preening. Similarly, the Great White Pelican (*Pelecanus onocrotalus*) in Africa becomes tinged with an orange-brown stain when swimming in iron-rich waters (Baxter and Urban, 1970). However, the pinkish suffusion on the otherwise white feathers of tropicbirds and the yellow stain on certain hornbills is a secretion from the oil glands transferred in preening (*cosmetic coloration*).

USES OF COLORS

During the past century, from Darwin down to present times, there have been many attempts to explain the meaning of various colors in birds, but these always involve the risk of anthropomorphic interpretations. Many birds are protectively or concealingly colored (*cryptic coloration*), having markings that simulate their surroundings. Sparrows in a grassy field, marsh birds (bitterns, snipe, marsh wrens) in a marsh (Fig. 3.14), and ground nesters such as Whip-poor-wills in the woods and nighthawks on open ground (sand or gravel), resemble the backgrounds where they nest. Vireos and warblers, often with shades of green and yellow predominating, resemble the foliage of the trees they inhabit and are difficult to detect, especially when they "freeze" in times of danger. Larks, on a world-wide basis (75 species), show remarkable variations to blend with their environment—pale forms on desert sands, reddish forms on red clay, and dark forms on black volcanic ash.

Other cryptically colored birds illustrate the principle of *countershading* for concealment; that is, they are the darkest above where they receive the most light and lightest below where they receive the

FIG. 3.14. *Short-billed Marsh Wren at nest. Marsh birds often have brown-streaked plumages resembling the marsh grasses among which they live.* [Photo by L. H. Walkinshaw.]

most shadow. Shorebirds on a beach, with their light underparts and darker backs, illustrate countershading. *Disruptive* patterns, as in a Killdeer, may also serve to protect a bird by visually disrupting the birdlike shape. Notable also are the seasonal color changes in ptarmigan from the white of winter to the browns of summer.

Still other birds, not necessarily protectively colored, exhibit *deflective coloration,* with conspicious plumage features that are said to startle, confuse, or deflect the aim of a predator. Thus the flashing white outer tail feathers of birds like juncos, or the white or brightly colored rump patches of flickers or Killdeer, might divert the aim of a pursuing predator from a vital to a less vulnerable part of the intended victim's body. These also serve as *signal* colors to keep members of a pair or flock in contact.

Other theories purport to explain the strikingly marked or brilliantly colored plumages (*epigamic coloration*) that serve to attract mates in the breeding season. Darwin's theory of sexual selection, which presupposes that the females select the most strikingly colored males, has not stood the test of time, but there is no doubt that special markings on many male birds aid in sex recognition, and probably sex attraction, and that brilliant plumages often function meaningfully in mating season displays. In some cases secondary sexual characters, as in peacocks and birds of paradise, have seemingly evolved beyond the point of any apparent practical use, yet having attained this high development it is not likely that such birds could return to simpler patterns and still complete their sexual cycles. Many special markings on birds, however, such as the head patterns and wing marks of sea birds and ducks, are probably primarily for species recognition rather than sex attraction.

Another aspect of coloration is purely utilitarian from man's standpoint. Much of the popularity of birds stems from their pleasing variety of harmonious colors, and of course colors are the chief but not the only criteria for identifying birds either in the field or in the laboratory. Plumages of birds can be very confusing to the beginner, not only because of the many differently marked species but also because of the wide range of sexual, age, and seasonal variations within a species. Among the adults of some species, for instance, the sexes are marked alike (Blue Jay), or the males may possess some special marking (crown patches of kinglets or facial marks of flickers), or the males may be merely more brightly colored (Cardinal), or of an entirely different plumage (Bobolink). In some cases the female is the more brightly marked, throughout the year, as in the Belted Kingfisher, or only during the breeding season, as in the phalaropes. Young birds are commonly marked or colored differently than the adults, sometimes differing only from the male (as in the Rose-breasted Grosbeak), sometimes differing from both male and female (Red-headed Woodpecker), while in other cases (crows, jays, titmice) the young and the adults are essentially alike. And as already mentioned, some birds have different first-year, second-year, and third-year plumages, such factors notoriously increasing the difficulties of identification in the gulls. Add to all these such traits as melanism, albinism, and dichromatism, and the beginning ornithologist is truly beset with identification problems.

Other Integumentary Structures

A bird is so fully clothed with feathers that there is limited opportunity for development of other integumentary structures, but some do occur. Among these are the combs, wattles, and facial decorations on gallinaceous birds (poultry, turkeys, pheasants), the pouches and gular sacs of pelicans and cormorants, the spurs on the tarsi of pheasants or on the wings of jaçanas (Jacanidae) and screamers (Anhimidae), and bare areas, often highly colored, on the face (toucans), heads (Sandhill

Crane), or head and neck of certain vultures and guineafowl (Numid-idae). Such structures are usually derivatives of the epidermis, though some of them have a blood supply in the underlying dermis. In most cases they serve a secondary sexual or defensive function, and are under hormonal control. The frontal shield of the American Coot, which is a continuation of the covering of the maxilla up over the forehead, develops a highly pigmented *callus* in the breeding season and is used for sex recognition and for territorial threats and displays (Gullion, 1951). Similarly, the inflatable sacs on the neck of the Prairie Chicken (see Fig. 9.7) are composed of pigmented integument, but are connected (indirectly) with the lungs to act as a bellows in producing the booming sound heard on the bird's dancing grounds. The pouch of frigatebirds (Fig. 3.15) is a similar structure, but bright red in color, so that a colony of brooding males with inflated sacs is truly a spectacular sight.

Feather Maintenance by the Oil Gland

Feathers and other integumentary structures are maintained in good condition by various grooming activities, such as preening, bathing,

Fig. 3.15. *In frigatebirds the males possess a bright red inflatable gular sac used for courtship and intimidation displays. The exact inflating mechanism is not well known.* [Courtesy of Amer. Mus. Nat. Hist.]

sunning, and dust-bathing, but secretions from the oil gland supplement these other activities in care of feathers. The oil or preen gland (*uropygium*) is a conical, bilobed structure, often with a tuft of tiny feathers that serve as a wick, located immediately in front of the tail. It secretes an oily substance that serves as a "dressing" for the feathers during preening. Some birds—the ostrichlike birds, some parrots and pigeons—have no oil gland; in others, such as caprimulgids, it is probably largely functionless. Many other species, including flamingos, herons, toucans, and skimmers, would have an awkward time trying to distribute the oil among the feathers by means of the bill, although they could do so by rubbing their head feathers over the oil gland and thus transferring the secretion to the rest of the body.

Law (1929), using a variety of tests on the feathers of a large number of species, concluded, in disagreement with others, that birds do not waterproof their feathers, but that in preening they rub the beak over the oil gland to keep the bill in good condition. Later, Elder (1954) reviewed the whole situation and showed by extensive experiments and observations that ducks, which have large oil glands, depend on the secretion to keep their plumage in good condition; without it the feathers become brittle and frayed and lack waterproof structure. The secretion also keeps the bill and legs from chafing and peeling. Whitaker (1957) observed a captive Lark Sparrow deliberately and repeatedly oiling its tarsi but *not* its plumage.

A series of studies by a Chinese investigator (Hou, 1929–1931) suggests that the oil gland also has an antirachitic function; experimental extirpation of the gland in some but not all birds is followed by rickets. In preening, oil is distributed over the feathers where it is activated by the rays of the sun and the irradiated material absorbed through the skin or obtained by the ingestion of feather particles. Removal of the oil gland does not cause rickets in the House Sparrow, however, suggesting a differential degree of need for vitamin D in different species, or that different species get their vitamin D in different ways (Friedmann, 1925). Some authors maintain that there is no vitamin D in the secretion of the oil gland.

Thus the oil gland seems to serve at least three important functions, though apparently both usage and need vary greatly in different species: (1) it helps keep the plumage water-repellent, particularly in waterbirds, which have the largest oil glands; (2) it lubricates the beak and tarsi, thus preventing chafing; and (3) it may provide a source of vitamin D in some species through irradiation of a provitamin when exposed to sunlight. In addition, the strong musty odor emanating from the oil gland of female hoopoes (Upupidae) may have a nest-protecting function.

SELECTED REFERENCES

DWIGHT, J., JR. 1900. The sequence of plumages and moults of the passerine birds of New York. Annals N. Y. Acad. Sci., 13:73–360.

ELDER, W. H. 1954. The oil glands of birds. Wilson Bull., 66:6–31.

FOX, D. L. 1953. *Animal Biochromes and Structural Colors.* New York: Cambridge Univ. Press.

HUMPHREY, P. S., and K. C. PARKES. 1959. An approach to the study of molts and plumages. Auk, 76:1–31.

LILLIE, F. R., et al. 1932–1944. Physiology of development of feathers. I–VII (A series of 7 papers by Lillie and Juhn, and Lillie and Wang). Physiol. Zool., 5–17.

PETTINGILL, O. S., JR. 1970. *Ornithology in Laboratory and Field.* Minneapolis, Minn: Burgess. Parts on topography, feathers and feather tracts, plumage and plumage coloration.

RAWLES, M. E. 1960. The integumentary system. Chapter VI in A. J. Marshall's *Biology and Comparative Physiology of Birds.* New York: Academic.

THAYER, G. H. 1909. *Concealing Coloration in the Animal Kingdon.* New York: Macmillan.

THOMSON, A. L. 1964. *A New Dictionary of Birds.* New York: McGraw-Hill. Feathers and feather maintenance.

VAN TYNE, J., and A. J. BERGER. 1959. *Fundamentals of Ornithology.* New York: Wiley. Chapter 3.

WELTY, J. C. 1962. *The Life of Birds.* Philadelphia and London: Saunders. Chapter 3.

WETHERBEE, D. K. 1957. Natal plumages and downy pterylosis of passerine birds of North America. Bull. Am. Mus. Nat. Hist., 113:339–436.

4

Bones, Muscles, and Locomotion

The Skeleton

IN DESCRIPTIONS of internal anatomy it seems logical to start with the skeleton, which provides anchorage for the muscles and houses the internal organs. Figure 4.1 illustrates and identifies the principal bones of a bird. The following paragraphs point out, briefly, the use that skeletal features serve in ornithological studies and, at more length, the many serviceable adaptations exhibited by the skeleton.

Skeletal features are important in classification. The higher categories, of ordinal rank or higher, are based largely on the structure and arrangement of bones, and skeletons are used, in conjunction with other features, in determining phylogenetic relationships. Bird skeletons are becoming an increasingly important part of museum collections.

There are pitfalls as well as advantages in using skeletal structures in determining relationships, for though they become modified more slowly than external features and therefore have been extensively used in systematics, even bones respond in time to habit. The jaws, for example, reflect feeding habits, as well as phylogenetic relationships; the wings become modified for flight, even within the same family (e.g., the Rallidae have both flying and flightless forms); and the hind limbs show adaptations for locomotion on land or in the water. The former division of birds into a more primitive *ratite* and a more advanced *carinate* group, on the basis of raft-shaped and keeled sterna, respectively, has long since been abandoned in favor of divisions based on palatal structure, a primitive (*palaeognathous*—old jaw) type and a more modern (*neognathous*—new jaw) type. Later studies on the bony palate (Mc-Dowell, 1948) suggest that even this criterion may be no longer tenable,

FIG. 4.1. *Skeleton of a domestic fowl.* [From Hegner and Stiles, *College Zoology*, 6th ed., Macmillan Publishing Co., Inc., New York, 1951.]

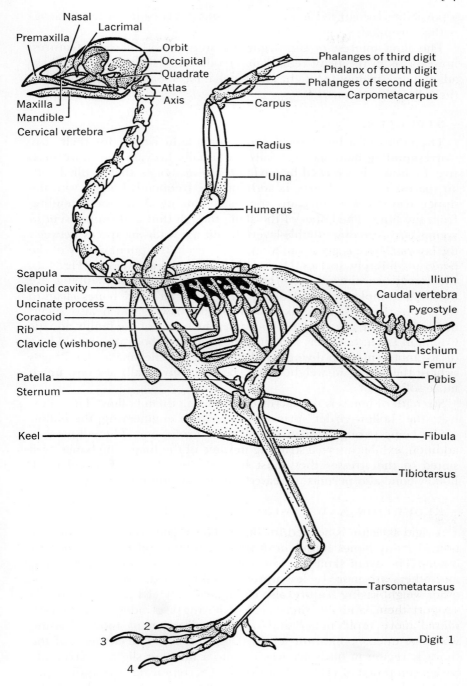

Nasal
Lacrimal
Premaxilla
Orbit
Occipital
Quadrate
Atlas
Axis
Maxilla
Mandible
Cervical vertebra

Phalanges of third digit
Phalanx of fourth digit
Phalanges of second digit
Carpometacarpus
Carpus

Radius

Ulna

Humerus

Scapula
Glenoid cavity
Uncinate process
Coracoid
Rib
Clavicle (wishbone)

Ilium
Caudal vertebra
Pygostyle

Ischium
Femur
Pubis

Patella
Sternum

Keel

Fibula

Tibiotarsus

Tarsometatarsus

2
3
4

Digit 1

a possibility recognized in current classifications (superorder Palaeognathae dropped).

The chief interest of the beginning student with skeletal features,
however, is in the truly remarkable modifications they exhibit. These
adaptations, mainly for flight and bipedal locomotion, are discussed in
the following sections.

STRUCTURE

The bones of a bird are of exceedingly light weight for their size.
Corresponding mammalian bones are usually heavier and more massive. Lightness is achieved by having *pneumatic* bones, that is, filled with
air spaces. The limb bones, in addition, are frequently hollow, with the
upper arm (*humerus*) in many species containing an air sac extending
from the lung. The bones of the skull, though thin and transparent in
young birds, become double-layered with age with air spaces between
the layers. There is great variation, however, in the pneumaticity of the
bones of different species. Oceanic birds like the albatrosses (Fig. 4.2),
which soar over the waves for hours with little or no wing flapping, are
the most completely aerial of all birds, with air-filled bones clear to
their toes. Land soarers like vultures and hawks, which also spend
much time aloft, are similarly buoyant. The bones of the highly aquatic
Common Loon, however, lack this pneumaticity—bouyancy would be a
disadvantage in deep diving—and the bones of the smaller land birds
(where the saving in weight would be relatively small) also are much
more solid.

Strength of bones is not sacrificed by having them hollow. They illustrate the "hollow-girder" principle familiar in engineering, the hollow
girder being stronger for its weight than a solid one. The vultures, in
addition, exhibit internal strutting in some of the long wing bones, well
suited to the stresses they must bear (Fisher, 1946). The enlarged,
hollow bone also permits increased surface for muscle attachment.

REDUCTION AND FUSION

A rigid skeleton is needed for flight. This is achieved by the elimination of many bones found in other vertebrates and by the fusion of
others. The avian skull has lost some of the bones found in reptiles,
and others are more or less unrecognizably fused. Loss of teeth, another weight-saving feature, also eliminated the need for heavy jaws to
support them, and the alternative of having a grinding gizzard permitted more rapid intake and quicker utilization of food—chewing
would be uneconomical for a bird (Dilger, 1957). The elements of the
thoracic region in birds are welded into a firm but slightly expansible
basket composed of (1) tightly articulated vertebrae, (2) ribs with upper
(vertebral) and lower (sternal) segments that connect the back with the

FIG. 4.2. *Wandering Albatross in flight. The albatrosses, some with wingspread exceeding 10 feet, are the most completely aerial of all birds and have light pneumatic bones and a large sail area for efficient soaring. Wilson's Storm Petrels are shown below.* [Courtesy of Amer. Mus. Nat. Hist.]

sternum, and (3) the peculiarly avian *uncinate processes* that overlap successive ribs (Fig. 4.1). The pelvic or *innominate* bones (*ilium, ischium,* and *pubis*) fuse with the lumbar and sacral vertebrae. The solidly welded portion of the vertebral column separating the two innominate bones is called the *synsacrum;* it is composed of the last thoracic, the lumbars, sacrals, and anterior caudals. The caudal vertebrae are reduced and partially fused to form the *pygostyle,* which, with its surrounding flesh ("pope's nose"), supports the tail feathers. Richardson (1972) found a pair of accessory pygostyle bones in all the falconids he examined and believed that they served for the attachment of the well-developed depressor muscles (page 97) needed for the increased stress that the tails of kestrels and related falconids have to bear for maneuverability and braking power.

The forelimbs in birds, particularly the distal parts, are greatly modified. The *ulna,* which bears the secondaries, is enlarged, but the *carpals* (wrist bones) are reduced to two. Of the five *metacarpals* typically found in the quadruped hand, the first and fifth are gone, the second is a mere remnant fused to the third, while the third and fourth are united at their ends to form the hand (*manus*), which bears the primaries. The *phalanges* or segments forming the three digits are also much reduced, varying from about four to seven in different species. One or two phalanges of the second digit (see Fig. 4.1) form the *alula* (bastard wing); two or three phalanges form the third finger; and one or two the fourth. (Some authors number the digits and metacarpals I, II, and III instead of II, III, and IV.)

The leg bones likewise exhibit considerable fusion, particularly at the tarsal joint (ankle), where the proximal *tarsals* are completely fused with the *tibia* to form a *tibiotarsus,* and the distal tarsals are fused with the united *metatarsals* to form a single lower leg bone, the *tarsometatarsus.* Birds typically have four toes, which facilitates perching, but some cursorial and scansorial species have only three, and the Ostrich only two. The development of the leg bones shows some correlation with locomotor activities, such as walking, running, and swimming.

Special Features

The chief exception to the prevailing tendency toward reduction and fusion in the avian skeleton is the exceedingly flexible and often greatly elongated neck, which facilitates stretching, reaching, and preening and compensates for an otherwise rigid skeleton. It also offsets to some extent the loss of manual dexterity caused by having wings instead of forepaws, and provides for rapid directional peering with eyes that are quite firmly fixed in sockets. Unlike mammals, where the number of cervical vertebrae in a giraffe or a bat is the same, in birds a long neck is due to an increase in cervicals as well as to a lengthening of individual bones; the number of cervicals varies from 13 in some hawks, parrots, swifts, and a few other birds to 24 or 25 in swans. Often there is a variation of one vertebra in different individuals of the same species (Fig. 4.3).

Among other special features perhaps the most important is the *sternum.* In flightless birds such as the Ostrich the sternum is flat or raft-shaped (ratite), but in all flying forms (and penguins) it has a median ridge or keel for the attachment of muscles. Birds with rapid or powerful wing strokes, such as hummingbirds, swallows, pigeons, and falcons, have greatly deepened or expansive keels on their sterna, whereas soaring birds like albatrosses, which depend on buoyancy and a large sail area (Fig. 4.2), have less expansive keels and smaller flight muscles. In flying birds the sternum is prolonged posteriorly to support the internal organs in rapid flight, and in woodpeckers there is increased ossification of the posterior border to give additional support for climbing (Feduccia, 1972). At its anterior end the sternum joins the pillarlike *coracoids,* which in turn unite with the other elements of the

FIG. 4.3. *Extremely long necks in flamingos (Phoenicopteridae) are due to an increase in the number of cervicals (18 or 19) and to lengthening of the individual bones. Flamingos have 5 to 7 fewer cervicals than a swan and only 4 or 5 more than an owl. Other adaptations include long legs for wading and a lamellate bill for straining food items out of bottom ooze, which they obtain by scooping with the bill in an inverted positon between the legs.* [From a Kodachrome by Roger T. Peterson.]

shoulder girdle (*clavicles* and *scapulae*) to provide the strong support needed for powerful wingbeats. The clavicles (*furcula*) are reduced or absent in flightless birds.

CENTRALIZATION

The parts of the skeleton, as well as the investing muscles and enclosed viscera, are organized in such a way as to shorten the body axis and to centralize body weight. The head is small and light; a long neck can, if necessary, be retracted in flight; and the feathered tail is a flying aid rather than an encumbrance. Thus there is little or no anterior or posterior drag in flight; the body is streamlined and well balanced with the center of gravity below the supporting wings. The center of gravity, of course, is an imaginary entity and can be shifted forward, backward, or laterally by movements of the head, wings, and tail.

The Muscular System

There are many standard references on musculature, from the early classical treatises of Gadow and Fürbringer (early German works), Shufeldt (1890) on the myology of the raven, and Hudson and co-workers on the pectoral and pelvic appendages (many papers from 1937 to 1969). Then there are works on functional anatomy, such as those of Fisher (1946) on the locomotor apparatus of vultures and Beecher (1951) on jaw musculature in icterids. Now a standard text on Avian Myology (George and Berger, 1966) is available, to which the student can refer for the many papers on musculature (eight pages of references). Unfortunately, muscle terminology is rather cumbersome and not always uniform; this brief section therefore merely calls attention to some of the muscle groups that perform special functions in birds.

Feather muscles in the skin for moving feathers are involuntary non-striated muscles under the control of the sympathetic (autonomic) nervous system, but the cutaneous components of striated body muscles that insert on the skin also contribute to the movement of feathers. Each feather follicle has an *erector* and *depressor* for raising (fluffing) and lowering (depressing) the feather (Stettenheim et al., 1963; Osborne, 1968). Feather muscles are arranged in groups corresponding to the feather tracts to which they are distributed. Their action includes such functions as shaking the water from the feathers of aquatic birds when they come ashore, or drying the feathers of land birds after a bath; fluffing out or depressing the feathers in response to temperature changes; erecting crests, pinnae, or breeding plumes in sexual displays; and parting the breast feathers for exposing the brood patch in incubating birds (see Fig. 10.15).

The *jaw muscles* of birds show some unique modifications for feeding (Beecher, 1951). In birds of prey, and in seed-eaters that actually crush seeds, the adductor muscles (*M. adductor mandibulae*) that close the jaws are strongly developed, whereas in insectivorous birds, or in birds that merely pick up and swallow seeds (pigeons), the muscles are generally reduced. In the Common Grackle, which saws open acorns by means of a sharp keel on the roof of the mouth, the aductor muscles are powerfully developed to perform this special function (Fig. 4.4a). Conversely, the protractor or gaping muscles (mainly *M. depressor mandibulae*) are less massive in the seed-eaters, but strongly developed in those fruiteaters that puncture fruit with a closed bill and then open the bill to extract pulp or juice from the fruit. The meadowlarks likewise have well-developed protractors (Fig. 4.4b) for prying under grasses with closed bill and then gaping to obtain the food. Several other muscles and ligaments help coordinate the complicated motions of the jaws (Zusi, 1967).

Better known are the large *flight muscles* on the breast of a bird. These consist of the usually enormous superficial pectoral muscle (*M. pectoralis major*), which has its origin along the whole ventral border of the sternal crest and clavicle and inserts on the ventrolateral surface of the humerus for depressing the wing, and the deep pectoral muscle (*M. pectoralis minor* or *supracoracoideus*), which also originates on the sternum, beneath the major, and inserts on the dorsomedial aspect of the humerus. Insertion of the minor on the *upper* surface of the arm

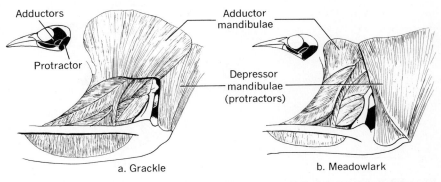

F I G. 4.4. *Jaw musculature of two icterids showing adaptations for food-getting.* a. *Grackle, with powerful adductors for cracking acorns and grain, and reduced protractors.* b. *Meadowlark, with reduced adductors and powerful protractors for gaping.* [Redrawn by Vito Cangemi from W. J. Beecher, *Auk*, 68, 1951.]

bone is achieved by passage of the tendinous portion of the muscle through an opening, the *foramen triosseum*, between the shoulder bones in a unique rope and pulley arrangement, so that contraction of a *ventral* muscle on the breast *elevates* the wing. Figure 4.5 illustrates this unique arrangement.

Birds with a deep keel have large and powerful breast muscles whereas soaring birds do not. Nair (1954) points out that the smaller flight muscles in soaring birds may have two origins and two bellies and thus act independently and alternately, preventing fatigue. In most birds the depressor muscles, which control the powered downstroke, are much the larger, often ten times the size of the minor, which only elevates the wing, but a hummingbird's unique manner of flight also utilizes a powered upstroke so that the minor is one half the weight of the pectoralis major (Grinyer and George, 1969). A grouse in rapid flight and underwater swimmers like murres also use a powered upstroke.

Perhaps understandably, the flight of birds (pages 106–17) is not

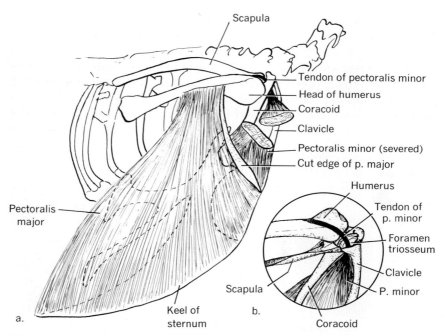

FIG. 4.5. a. *Origin and insertion of the pectoral muscles. The clavicular portion of pectoralis major cut away to show the underlying p. minor. Origin of p. major along the keel of the sternum and insertion on ventrolateral aspect of the humerus depresses the wing. b. Inset showing medial side of left shoulder elements with passage of the tendon of p. minor through the foramen and insertion on the dorsoposterior surface of the humerus to elevate the wing.* [Drawings by Vito Cangemi.]

achieved by the action of these two wing muscles alone; things are not that simple. Six different muscles supplement the action of the minor in elevating and tilting the wing, several others assist the major in depressing it, and several dozen others (not including the muscles to individual feathers) play a role in the complicated motions of the wings.

The breast muscles of birds usually consist of a mixture of red and white fibers (dark meat and/or white meat), but domestic fowl have only white fibers and hummingbirds have only red fibers. In strong fliers dark fibers predominate. The broad white fibers are glycogen-loaded (for quick but not sustained energy) and are relatively free of fat, whereas the narrow red fibers have abundant fat storage for sustained muscular activity. The smaller dark fibers have more capillaries per unit area and thus a richer blood supply (Fig. 4.6). A series of papers by George and co-workers in India, which describe these histological features in detail, are summarized by George and Berger (1966). Storage of lipids in the dark breast muscles of migrating birds (page 339) is of great importance in sustained flight.

The muscles of the pelvic appendage are similar in name and function to those of other land vertebrates, and only those involved in the "perching mechanism" need be mentioned here. The perching function has often been assigned to the *ambiens,* a slender muscle that arises on the pectineal process of the ilium and is continued over the kneecap into a tripartite flexor of the fore toes, but actually the ambiens plays only a minor role in perching (mainly flexion of the second toe) and can be severed without loss of perching ability. Several leg muscles, no-

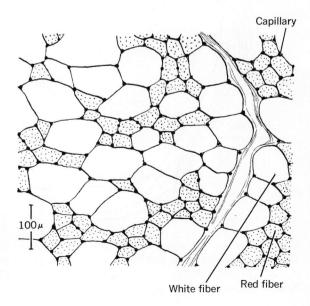

Fig. 4.6. *Cross section of pectoralis of Common Pigeon showing the narrow red fibers, the broad white fibers, and the distribution of the capillaries supplying the muscles.* [Drawing by R. M. Naik for J. C. George, *Auk,* 77, 1960.]

tably the big "calf" muscle (*gastrocnemius*) and the *peroneus longus* on the anterior aspect of the lower leg, assist in perching, but it is chiefly the action of the flexors of the digits (eight separate but coordinated muscles, six to the fore toes, and two to the hallux or hind toe) that accomplishes the tight grip of the foot on a perch (Fig. 4.7a). Strongly developed flexors are of obvious importance to raptors in grasping prey, but oddly perhaps, falcons and owls use different flexors in dif-

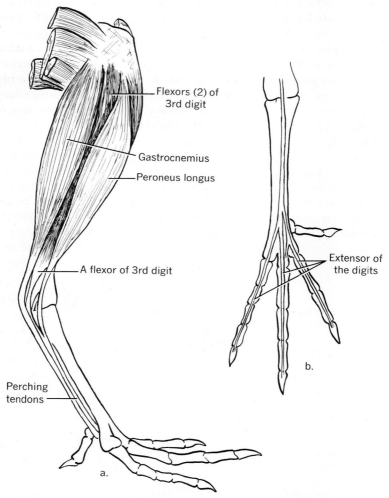

FIG. 4.7. a. *Lateral aspect of right leg of fowl, illustrating a part of the complex perching mechanism of the foot (underlying flexors not shown).* b. *Frontal (anterior) aspect of right foot, showing extensors.* [From Chamberlain's Avian Anatomy, Memoir Bull. 5, Mich. Agr. Exp. Sta.; drawings by Vito Cangemi.]

ferent ways to catch their prey (Goslow, 1972). Madura (1952) demonstrated a nerve center on the plantar surface of the foot by stimulating the muscle with tweezers, apparently initiating the perching reflex. Several other muscles act as extensors to open the toes (Fig. 4.7b). It should be noted that only tendinous parts of muscles reach the toes, a feature that is important for conserving body heat (page 157).

Birds have a variety of muscles for raising, lowering, and fanning out the tail feathers. Of particular importance are the *M. depressor caudae* muscles, which depress the tail to serve as a brake in landing. Pigeons with excised depressors show a 40–60 percent increase in landing force because of loss in braking power (Fisher, 1957).

The "hatching" muscle (*M. complexus*) in chicks, grebes, ducks, and gulls (see Fisher, 1966a, for review of his earlier papers) originates on the vertebral column, inserts on the dorsal surface of the head, and assists in the pipping of the egg. It reaches its maximum development at hatching time; then a portion of it atrophies, but another part persists and functions in postnatal life for raising the head.

Locomotion on Land and Water

Birds, perhaps more than any other animal group except the insects, have mastered the art of navigation in nearly all media. Though mastery of the air is their most spectacular accomplishment, birds also climb, walk, run, jump, burrow in the ground, swim on and under water, and dive. True saltatorial (jumping) forms and fossorial forms (excluding nest burrowers) are not characteristic of birds and are not discussed here, but climbing, running, navigation in the water and in the air, and the adaptations associated with each are discussed.

Scansorial (Climbing) Birds

From fossil evidence it is believed that the first birds were arboreal and climbed trees to gain altitude before launching out on short gliding flights. The arboreal habit necessitated, or possibly followed, the development of clawed toes and a perching type of foot for roosting and for hopping from branch to branch. In *Archaeopteryx* the three fingers were also separate, not fused as in modern birds, and terminated in claws. These claws on the wing, as well as the clutching foot, supposedly enabled *Archaeopteryx* to clamber among the branches of trees before taking flight.

Although quite a few modern birds—about 17 species in 4 families according to Rand (1954)—have well-developed spurs on the wings, these are mainly offensive or defensive weapons, not used for climbing. The classic example of functional claws on the wing among living birds

FIG. 4.8. *Odd in many ways, the Hoatzin (Opisthocomidae) of South America uses its crop as a gizzard, and the young have claws on the wings for clambering about in trees. Here a nestling uses its claws for climbing.* [From Stiles, Hegner, and Boolootian, *College Zoology*, 8th ed., © copyright 1969 by Macmillan Publishing Co., Inc., New York; photo by Beebe.]

is the Hoatzin (Fig. 4.8). The young of this species have well-developed claws on the second and third digits that aid them in climbing about the branches in their nest trees; if lost from the nest, they scramble back to it like quadrupeds, but in a few days the claws are shed and the young Hoatzins become bipedal. Percy (1963) reports that an African finfoot (Heliornithidae) has a 12–18 mm claw on its wrist (carpal joint) and uses it like a young Hoatzin to climb trees.

In other scansorial birds, modifications in the pelvic appendages and tail are the chief structural features associated with climbing. These include well-developed opposable toes with sharp claws, a shortened tarsus, and (except in the nuthatches, piculets, and wrynecks) a strong tail whose stiffened shafts often terminate in sharp barbless spines. In the woodpeckers (Picidae), the zygodactylous foot (see Fig. 5.10) enables them to get a vicelike grip on the bark of trees. Different species of woodpeckers show slightly different anatomical modifications that govern their "stance" or position on the trunk of a tree when climbing or pecking; sapsuckers, for instance, are the most efficient climbers but deliver the weakest blows in pecking (Spring, 1965). The woodhewers (Dendrocolaptidae), true creepers (Certhiidae), and nuthatches (Sittidae and allies) have the typical passerine foot structure (three toes in front and one in back), but long sharp claws enable them to cling to vertical surfaces. The stiff spiny tail in most species gives additional support, but largely limits climbing to ascending tree trunks and makes descending awkward. A nuthatch, by contrast, can clamber up or down or around a tree trunk without using its tail for support.

Parrots (Psittacidae) are also expert climbers and have zygodactylous feet with a reversible outer toe to aid them in maneuvers, but the strong bill also is used to pull them from one perch to another. In

India large numbers of parakeets (which are economic pests) elec-
trocute themselves by standing on one wire and grasping an adjacent
wire with the bill (Dilger, 1954). Swifts (Apodidae) have small feet and
are not good climbers, but use their strong curved claws and conspicuously spine-tipped tail (in some species) for support in chimneys,
mine shafts, and on the sides of cliffs and caves.

Cursorial Birds

A less generally accepted view of the origin of flying birds is that
bipedal dinosaurs achieved flight by fast running and flapping of their
free forelimbs, which developed frayed scales (feathers) on their posterior margins, thus facilitating flight. Later some flying birds lost their
power of flight and became strictly cursorial (e.g., the ostrichlike birds)
or aquatic (e.g., penguins and some auks).

Though the origin and subsequent evolutionary history of running
birds is still somewhat nebulous, many present-day forms, flying and
flightless, illustrate cursorial adaptations. These usually include an
elongation of the hind limb, a reduction in the surface area of the foot
coming in contact with the ground, and a reduction in the number of
toes. Such features are well exemplified by the Ostrich, rheas, emus,
and cassowaries. All are heavy-bodied, flightless forms with fluffy
feathers ill-adapted for flight; show a reduction in toes; and, except for
the forest-dwelling cassowaries, inhabit open areas where their speed is
unimpeded. The Ostrich has been credited with a speed of 60 miles per
hour, measuring 25 feet at a stride when full momentum is attained,
but perhaps such speed is an exaggeration. One paced by a car in an
African park ran at 30 miles per hour for more than 20 minutes
(Grzimek, 1961), but this may not have been maximum speed. Men on
horses can run down Ostriches, which run in wide circles, by taking an
inside track and bringing in fresh steeds occasionally. Rheas in South
America are sometimes hunted by a more picturesque method;
"gauchos" on horseback pursue and surround groups of the big birds
and lasso them with bolas—three-pronged ropes with weights at the
ends—which trip up or entangle the helpless birds.

Somewhat transitional in nature are the neotropical tinamous (Tinam-
idae), which can fly, at least to a limited extent, but are primarily terrestrial and depend on running, often in dense forest cover (or among
tall grasses and shrubs in the upland species), for escape. Flights, when
attempted, are of short duration, in a straight line, and result in quick
exhaustion after a few take-offs and crash landings.

Many other birds among the nonpasserines are true runners, as opposed to most terrestrial passerines, which walk (larks, pipits, blackbirds) or hop (thrushes, sparrows). Nonmigratory galliform birds, such
as pheasants and quail, take short rapid flights to escape terrestrial

predators, but otherwise depend on fast running and concealment for escape. Rails, though capable of sustained flight during migration, are difficult to flush in their breeding marshes; they depend on maneuverability of their laterally compressed bodies (hence, "thin as a rail") in running through dense vegetation. Among the shorebirds, some of the sandpipers are particularly fleet-footed, apparently as an adaptation for rushing ahead of incoming waves and then back-tracking rapidly for food items left stranded by the receding waters.

Cuckoos (Cuculidae), a heterogeneous family of some 127 species, are primarily aboreal, but one aberrant subfamily group of about 13 species has taken up a terrestrial mode of life. Best known of these is the Roadrunner of our southwestern deserts. These birds have been credited with amazing speed and dexterity in pursuing swift lizards, outmaneuvering hounds, and outdistancing human pursuers (Sutton, in Bent, 1940); two actual checks with car speedometers set their speeds at 10 and about 15 miles per hour, respectively. Cottam et al. (1942), timed two Roadrunners at 12 and 15 miles per hour, respectively (in soft sand and upgrade), but found two Ring-necked Pheasants (at 15 and 21 miles per hour) and a Chukar (at 18 miles per hour) equalling or exceeding the speed of the Roadrunner. Dissection of the pelvic musculature of both aboreal and terrestrial cuckoos indicates that while muscle formulas are similar in both groups, the Roadrunner's cursorial habit is reflected in the elongation of the hind limbs and a corresponding increase in the length of the pelvic muscles (Berger, 1952).

WADING BIRDS

Some nonswimming birds have developed useful adaptations for wading in shallow water for prey not usually available on shore. Extremely long legs, counterbalanced by a long neck and long bill for making sudden thrusts or for probing, are characteristic of herons (Fig. 4.9), cranes, flamingos (see Fig. 4.3), and some shorebirds. The kink in the neck of herons is due to a lengthened sixth cervical vertebra that articulates at an angle with the fifth and seventh cervicals and provides increased surface for muscle attachments. The cervical muscles are used in making quick jabs for prey. If walking on soft surfaces is necessary, the toes are elongated (a handicap for running or perching) as in the case of herons, or webbed as in flamingos, to give additional support. The elongation of the toes is carried to extremes in jaçanas and gallinules, which walk on floating vegetation. Among the shorebirds, curlews, yellowlegs, avocets, and stilts are particularly noted for long legs and bills for foraging in shallow waters; the stilts, in fact, are believed to have longer legs in proportion to their size than those of any other bird.

FIG. 4.9. *Long sharp beaks, long flexible necks, long legs, and widespreading toes are among the adaptations exhibited by many wading birds for capturing prey in shallow water. Here a Great Egret is presumably stalking a fish or frog. The kink in the neck is due to the lengthened sixth servical.* [Photo by Allan D. Cruickshank.]

SWIMMING BIRDS

Most land birds can swim for short distances under duress. Downs (1963) watched a Blue Jay that fell into a farm pond swim 25 feet to shore, using a propelling motion of wings and feet and splashing water over itself. A covey of quail that fell into a pond has been reported swimming a greater distance to shore, heads held high out of the water. Such land birds have no specializations for living in water, but true swimmers show special adaptations for aquatic existence. These usually involve an expanded foot with the toes fully webbed or lobed (see Fig. 5.9), a dense compact waterproof plumage (except in some pelecaniform birds), and an enlarged oil gland. Some groups show additional adaptations (see the discussion of loons and grebes, for example) but aquatic birds are numerous, many of them apparently independently evolved, and it is not possible to discuss them all here. Many oceanic birds, such as petrels, shearwaters, gulls, and terns, are primarily aerial and merely rest or float on the water rather than doing much active swimming.

Hesperornis (see Fig. 2.2), a flightless bird that inhabited the inland seas of North America during the Cretaceous period, is proof that birds took to the water at an early date. *Hesperornis* had powerfully developed hind limbs with large feet and broadly webbed or lobed toes. The legs were set far back on the body, useless for locomotion on land, and were apparently laterally disposed for sidewise strokes for swimming both on and under water, presumably in pursuit of fish.

Among modern birds the loons and grebes are expert swimmers both on and under water. The loons in particular exhibit a streamlined body, have their legs set far back (making walking on land difficult and awkward), have their tarsi flattened and bladelike to lessen the resistance of the water when the foot is drawn forward, and show various modifications in pelvic musculature and in the skeleton for life in the water (Wilcox, 1952). They bring both feet forward simultaneously, closing the toes when the foot is brought forward and opening them for the backward stroke. Grebes (Fig. 4.10), which have lobed rather than webbed toes, bring the foot forward in a sidewise stroke, with the top of the foot parallel to the surface of the water (Storer, 1956).

FIG. 4.10. *This Western Grebe is highly specialized for aquatic life and rarely ventures out on the land. It builds a floating nest, performs spectacular displays by "dancing" on the water with its mate, and is an expert swimmer both on and under water.* [Photo by Allan D. Cruickshank.]

Grebes also have a dense hairlike plumage that encloses little air and thus facilitates sudden submergence and underwater swimming.

Among the Anatidae, the geese and swans are surface swimmers and feed by "tipping up" in shallow water, or by grazing on land. The ducks include both "tip-up" types and diving forms—pochards (Fig. 4.11), sea ducks, and mergansers. Ducks of both types usually swim with alternate strokes of the feet, one forward and one backward, and underwater swimmers often use their wings, but there appears to be considerable variability among different species, as well as intraspecific differences under different conditions, such as depth of water, fast pursuit, or leisurely foraging. Humphrey (1958), observing the activities of a captive Common Eider in a tank, says that it used both feet and wings during submergence (for quick descent), but used only its feet when foraging on the bottom (to conserve energy) and rose to the surface by buoyancy without noticeable movements of either feet or wings. Most ducks, as well as penguins and alcids, use their wings for underwater swimming, but submergence in diving ducks (pochards, scoters, and mergansers) is usually accomplished without use of the wings.

Swimming speeds are available for a few birds. Hochbaum (1944) gives the rate of flightless adult Canvasbacks as 2–3 miles per hour and

FIG. 4.11. *This Ring-necked Duck, a diving pochard, has presumably just emerged from an underwater swim; its dense waterproof plumage enables it to shed water effectively.* [Photo by Allan D. Cruickshank.]

the "skittering" speed as 8-10 miles per hour. Stewart (1958) timed day-old ducklings of the Wood Duck swimming at the average speed of 0.6 mile per hour and skittering at 5.8 miles per hour (for short distances). An adult male in the flightless stage skittered at the rate of 9.5 miles per hour. Penguins, however, can attain speeds of 25 miles per hour under water (Austin, 1961).

A remarkable aquatic specialization is found in the dippers or water ouzels (Cinclidae), small passerine birds that inhabit swift-flowing mountain streams in the Americas and in the northern parts of the Old World (Fig. 4.12). Though primarily land birds, they feed almost exclusively on aquatic organisms gleaned from streams. They plunge into and swim or walk under the water, usually progressing upstream. Goodge (1959), both from observations in their natural haunts and from the analysis of motion pictures of captive birds in tanks, reports that they use their wings almost exclusively for underwater swimming, but employ a paddling motion of their unwebbed feet (webs would be a handicap in clinging to rocks) when swimming on the surface. They dive into a stream either from the air or from the surface of the water and can emerge from the water flying.

FIG. 4.12. *This water-loving songster, the Dipper, inhabits Rocky Mountain cascades, builds its mossy nest beside a stream, and forages under water for insect prey.* [Drawing by Dale A. Zimmerman.]

FIG. 4.13. *King Penguins* (Aptenodytes patagonica) *in the Antarctic are most at home in the water, swimming and diving, but they shuffle around awkwardly on the land during the prolonged breeding season.* [N.Y. Zool. Soc. photo.]

Perhaps the most specialized of aquatic birds are the penguins (Fig. 4.13), which have forsaken flight entirely and eke out an existence in some of the bleakest and most inhospitable regions on earth (pages 250–51). The forelimbs in penguins are scale-covered flippers with no remiges. The bony elements in the wing are flattened, the dense plumage covers the entire body (no apteria), and a heavy layer of fat insulates the body against the cold. Penguins use their flippers almost entirely for underwater swimming. By building up momentum they can leap up out of the water for six feet or more to land on rocks or floating ice cakes. On shore they progress by an awkward upright gait, or occasionally drop onto their ventral surface and "toboggan" over the ice and snow by using their flippers and kicking with their feet. On the long inland march of certain species to distant breeding rookeries they alternate walking and tobogganing and may take two or more months to reach their destination (Rivolier, 1959).

DIVING BIRDS

Diving is performed by two methods: (1) surface diving, such as employed by penguins, loons, grebes, cormorants, and most ducks, and (2) plunging into the water from a height, as illustrated by the Gannet,

Brown Pelican, kingfishers, and Osprey. Some birds use both methods. Many others merely dip into the water for surface prey with or without submerging.

Land birds that dive from considerable heights include the Osprey and many kingfishers. The former hovers over the water, then folds its wings and plunges down with great speed, but breaks before hitting the water and grasps fish with specially equipped talons (see Fig. 5.9c), often without submerging completely. The kingfishers, by contrast, dive head first into the water, either from a perch or a hovering position. Fish are speared (punctured) with the beak closed, then taken to a convenient perch to be disengaged and swallowed.

The North Atlantic Gannet is a spectacular diver. It plunges headlong from heights up to 150 feet and hits the water with a tremendous impact that carries it deep beneath the surface. Subcutaneous air cells act as shock absorbers to cushion the blow. Brown Pelicans (Fig. 4.14) use a similar technique, but, curiously, turn under water and emerge facing the opposite direction from which they entered, or sometimes merely bounce back from the water without submerging completely.

Deep divers such as the Common Loon and Oldsquaw, which have been recorded at depths up to 200 feet, face the problem of reducing their buoyancy before submergence (Schorger, 1947). Apparently they expel the air from their air sacs as well as the air trapped in the plumage and draw on the oxygen converted from the oxyhemoglobin and oxymyoglobin stored in the muscles. Such divers also are capable of tolerating a high carbon dioxide level and of reducing the flow of blood to the muscles while under water. Voluntary submergence in water birds is of short duration, usually less than a minute, often only a few seconds, but domestic ducks have been held under water in experimental tests up to a maximum of 16 minutes before asphyxiation, or 27 minutes when the trachea was tied off. Presumably in such cases, both air from the air sacs and "converted" oxygen are utilized. But wild birds active under water use up their oxygen supply rapidly; even Oldsquaws drown when trapped and struggling in a net. Emperor Penguins, the real champions among divers, were recorded by Kooyman et al. (1971) at a maximum depth of 870 feet, and one stayed down for 18 minutes (maximum duration of 238 dives), but most dives were for less than a minute.

Bird Flight

In preceding sections reference has been made repeatedly to a bird's adaptations for flight: (1) the body is streamlined, thus offering the least resistance to air; (2) feathers are lightweight and arranged to aid

FIG. 4.14. *The Brown Pelican, though grotesque in appearance when perched, is a skillful aerialist and expert diver, plunging into the water for fish. Often, however, the pelicans congregate on wharves and pilings and wait for fishermen to toss them handouts.* [Photo by Allan D. Cruickshank.]

flight rather than hinder it; (3) bones are pneumatic and the skeleton is largely remodeled from the original vertebrate plan to facilitate flying; (4) the internal organs are well centralized, with certain structures reduced and others eliminated; (5) respiration, unique in itself, is supplemented by a remarkable system of air sacs; (6) rapid metabolism provides for high intake, quick consumption, and economic utilization of fuel; and (7) a marvelously designed wing, blunt at the leading edge (bones and muscles of forearm) and tapering at the rear (feathertips only)—the ultimate in engineering design—has been developed as an instrument of propulsion through air.

DEFINITIONS

An analysis of flight involves engineering technicalities beyond the scope of this course—some methods of flight defy description. Some of the terms commonly used in describing flight are explained below:

Angle of attack—The amount of tilt of the leading edge of the wing from the horizontal (Fig. 4.15b); that is, the edge of the wing needs to be tilted upward at the proper angle to permit smooth flow of air over the upper surface.

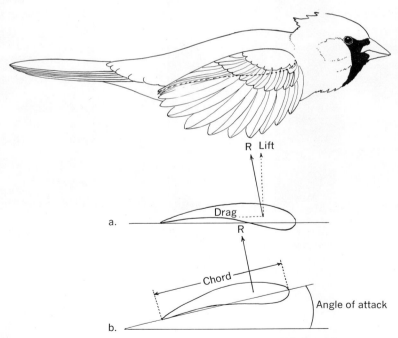

FIG. 4.15. *Mechanics of lift (angle of attack) in a bird's wing. See text for explanation.*
R = reaction. [Drawings by Diane Pierce.]

Aspect ratio—The ratio of the length of the wing to the width. It varies greatly in different wing types (see page 141), from low (about 5) in gallinaceous birds to high (20 or more) in albatrosses.

Camber—The convex upper surface of a "rounded" wing is spoken of as camber (Fig. 4.16f). Some wing designs are flat with little or no camber, others are quite convex above and concave below.

Drag—". . . the force of air opposing the wing in its movement through the air" (Van Tyne and Berger, 1959). Efficient flight has to minimize drag as much as possible (Fig. 4.15a).

Lift—Decreased or negative pressure (vacuum) on the upper surface of the wing produces lift, which minimizes the effort of raising the wing and keeping the bird aloft.

Stalling—If the angle of attack becomes too great, smooth flow of air over the wing is prevented and stalling occurs. Some birds (e.g., swifts) can "stall" with one wing to effect turns and aerial maneuvers.

Slotted wing—The tips of the primaries can be opened to produce "slots" (Fig. 4.16b) to decrease resistance of air on the upstroke or to lessen turbulence (see next item). The primaries in some birds, especially raptors, are emarginate or notched (Fig. 4.16e) for greater efficiency. The tips of the primaries are also very flexible and serve as individual propeller blades. The alula on the forearm provides an additional slot that is useful for directing the flow of air over the wing.

Turbulence—When the angle of attack becomes too great, smooth air flow over the upper surface of the wing is interfered with and turbulence is produced at the trailing edge of the wing.

Wing loading—The weight of the bird divided by the wing area. Birds show many adaptations for reducing the amount of weight to be lifted by the wings. Large-bodied birds like vultures have a tremendous wing span, thus reducing the wing loading (weight per unit of area).

GLIDING FLIGHT

Four different types of flight are commonly recognized, although all four kinds may be used at different times by the same bird. The first and simplest of these is gliding flight (Fig. 4.17a), presumably worked out by the earliest birds, which climbed trees and launched out on set wings to coast as far as the prevailing conditions of air and height of take-off permitted. We see a similar performance in many birds today, as when a pheasant works up considerable initial velocity by rapid wing flapping and then goes into a long glide. Air conditions, wind direction, and speed of take-off determine how far the bird can go before the pull of gravity and friction (decreasing lift and increasing drag) brings it down. Shorebirds coming in for a landing, ducks descending over

F IG. 4.16. *Flapping flight of Red-tailed Hawk based on analysis of motion pictures. Shown are a complete downstroke, from the highest position of wing* (a) *to lowest* (e), *and beginning of next upstroke* (f and g). *Note slotting, camber, flexibility of tips of primaries.* [Drawings by Diane Pierce.]

a.   b.

FIG. 4.17. a. *Gliding flight, illustrated by a Gambel's Quail.* b. *Flapping flight of Arctic Tern.* [Drawings by Dale A. Zimmerman.]

water, swallows and swifts in the air, are among the birds commonly seen to utilize gliding flight.

FLAPPING FLIGHT

The most common type of flying is ordinary flapping or cruising flight (Figs. 4.16 and 4.17b), perhaps originally developed by arboreal or cursorial reptiles and birds to increase the distance or effectiveness of the glide. Flapping flight, as the name implies, consists of up-and-down motions of the wings, which are so skillfully executed that the effort of keeping the bird aloft is minimized and forward velocity achieved at the same time.

Keeping the bird aloft involves economy on the upstroke in several ways: (1) the convexity (camber) of the upper surface of the wing in most birds permits ready air flow off the feathers and increases lift; (2) separation of the tips of the primaries (wing slots) allows free passage of air through them and reduces turbulence at the trailing edge of the wing; (3) the wing is partially folded at the wrist and elbow and drawn back in toward the body, thus reducing friction; and finally (4) the upstroke is often accomplished by a quick flip, especially at the end of the stroke, the air itself helping to throw the wing backward and upward. Conversely, the effective power of the downstroke is increased because: (1) the concave under surface of the wing in most birds grips more air, (2) the flight feathers are held firmly together to close the slots (Fig. 4.16d), (3) the wing is fully expanded and extended, and (4) the relatively slow (compared to upstroke) and powerful motion of the wing is downward and forward. The vacuum created on top of the wing by the downstroke provides additional lift. We have already seen (page 94) that the pectoralis major muscle is especially well developed in strong fliers, less so in soarers, and that the deep pectoral (supracoracoideus), except in swifts and hummingbirds, is relatively small.

Up-and-down motions of the wings serve to keep the bird aloft, but this does not explain forward progress. Of course, as motion pictures show (Fig. 4.16), the movements of the wings are not strictly up and down, but upward and backward, then powerfully downward and forward. On the downward-forward motion, the leading edge of the wing is depressed, with the secondaries providing the main lift and the flexible primary wing tips providing the chief propelling action drawing the bird forward. Then, as the wing is moved posteriorly at the end of the stroke, further forward motion is imparted to the body.

Though this is the pattern of strokes characteristic of most birds, there are many variations. The aerial gymnastics of the swifts involve many different wing motions—banking and veering and stalling with one wing in a jerky sort of flight that gives the optical illusion of alternate wing beats (Savile, 1950). Hummingbirds and swifts, as previously mentioned, utilize a powered upstroke as well as downstroke. Hummingbirds flap their wings from 55 to 75 or more times per second compared to the slow flapping of a vulture at one stroke per second. The flight feathers, particularly the first primary, as well as the various parts of the wing (alula, manus, forearm, and so on) can be moved more or less independently, thus allowing for all sorts of adjustments in position, balance, and maneuvers in correlation with wind and air conditions. Other variables are that the requirements for the take-off may be quite different than those for leisurely flying, that a sudden burst of speed imposes new demands,and that landing involves new techniques and the use of other muscles.

Soaring

Soaring (Fig. 4.18) is a highly specialized flying skill that employs alternate flapping and gliding, but, in its highest development, utilizes updrafts and air currents so expertly that flapping is often dispensed with for long periods. Soaring is characteristic of birds with a large sail area per body weight, but two quite different wing types seem to be equally effective for this kind of flight: long, narrow wings in *dynamic soarers* like albatrosses and frigatebirds and short, broad wings in *static soarers* like hawks. Both types require a large supporting sail area; the broad tail fans of the hawks help materially in this respect. Oceanic birds may drift along on nearly motionless wings for long periods, riding on turbulent air currents or on updrafts created by ships. Updrafts on the open sea, however, are usually inadequate for keeping an albatross aloft for long periods; it also makes skillful use of *variations* in wind speed at different levels. The surface winds are slowed down by friction against the waves, whereas higher winds (above 50 feet) are unimpeded, and the resulting *wind gradient* is much utilized by oceanic birds.

FIG. 4.18. *Frigatebirds have long narrow wings (high aspect ratio), providing a large sail area. An extremely light, pneumatic skeleton gives them additional buoyancy, and they remain aloft for long periods, often without visible movement of the wings.* [Photo by Allan D. Cruickshank.]

Land birds in mountainous regions make use of updrafts from adjacent valleys; hawks at Hawk Mountain, Pennsylvania, move long distances along a famed migration route by taking advantage of updrafts from precipitous Appalachian Mountain slopes. An Osprey has been observed to attain the amazing speed, while still coasting (gliding), of 80 miles per hour (Broun and Goodwin, 1943). Hawks can also be observed circling on sunny spring days over warming patches of meadows, taking advantage of the rising air (thermals) associated with the earth's radiation. Again a hawk may execute wide circles by riding downwind to pick up velocity, then turning and riding upwind until it begins to lose altitude, when it will again ride downwind to regain speed for the upwind segment of the circle. Falconiform birds, in fact, admirably illustrate all methods of flight—gliding, soaring, cruising, and hovering. Brown and Amadon (1968) cite dramatic examples of these from around the world.

Savile (1957b), elaborating on the traditional division into low and high aspect ratios (wing length:breadth ratios), recognizes four different though often overlapping wing types: (1) the elliptical wing with low aspect ratio, characteristic of birds flying across small treeless areas to nearby cover (gallinaceous birds, doves, woodpeckers, most passerines); (2) a high-speed wing, with moderately high aspect ratio, characteristic of birds with narrow tapered wings for fast flight in open spaces (falcons, swifts, hummingbirds, swallows); (3) the high-aspect-ratio wing found in oceanic birds (albatrosses, frigatebirds); and (4) the high-lift or slotted soaring wing that combines a moderate aspect ratio with pronounced slotting and camber, found in hawks and owls that inhabit wooded areas (Van Tyne and Berger, 1959).

HOVERING

When a bird reduces its speed to that of the wind, it is said to be hovering. When hovering, a bird can be suspended on nearly motionless wings in the air, or it can be fluttering its wings rapidly to maintain position. Gulls are experts at riding or resting on winds that are deflected upward from a ship, a cliff, or waves (Fig. 4.19). But when a bird hovers in still air, as is often observed in kestrels and humming-birds, it assumes a nearly vertical body position and fans its wings backward and forward rapidly to maintain position. Most birds tire quickly when hovering in this fashion, since it puts a severe strain on the relatively small supracoracoideus muscle, but hummingbirds, as previously noted, have a well-developed supracoracoideus and can hover in front of flowers for extended periods. Kingfishers utilize hovering for short periods before dropping or plunging to the water for prey.

TAIL, WIND, AND SPEED OF FLIGHT

The tail of a bird is an important instrument in flight. Though a bird temporarily deprived of its tail is not necessarily flightless, its progress is more labored, and its balance in the air more precarious. The tail serves a variety of balancing and steering functions and in long-tailed birds is a decided asset in aerial maneuvers. Closing and tilting of the tail fan adjust a bird to surrounding air conditions. In horizontal flight, spreading the tail fan would give additional lift toward the rear, the

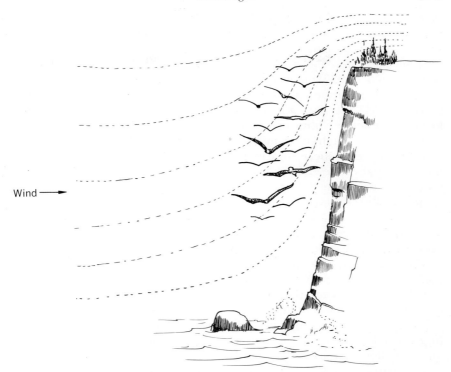

Wind →

FIG. 4.19. *Updrafts from boats or vertical cliffs and thermals over open water enable gulls and other sea birds to soar and hover with little expenditure of energy.* [Drawing by Diane Pierce.]

center of gravity would shift anteriorly, and the head of the bird dip for downward descent; conversely, closing the tail fan would shift the center of gravity backward and the bird would rise. Turns may be initiated by tilting the tail sideways, but of course merely turning the head in the desired direction and stalling with one wing also help execute a turn. Tails can be used as an air brake coming in for a landing, as is observable in ducks descending on a pond or a bird alighting upon a perch. The short tail of the Chimney Swift is offset by the versatility of its wings. Ducks are somewhat handicapped by an extremely short tail, and straight, fast flights without quick turns are characteristic of this group of birds.

Wind as a factor in modifying flight is important. Birds have two speeds, an air speed due to their progress irrespective of the accelerating or retarding effect of the wind and a ground speed, which is measured in terms of actual progress through space. That is, a duck flying with an air speed of 40 miles per hour in a tail wind of 10 miles per hour would have a ground speed of 50 miles per hour; conversely, in flying into the same wind the bird's ground speed or actual progress would be slowed down to 30 miles per hour.

Many measured flight speeds are now available for many different birds. Cooke (1937) tabulated the early records for more than 100 species and Cottam et al. (1942), measured (or listed) flight speeds for 91 birds of 57 species. Cruising speeds among the latter varied from a low of 12 miles per hour in a Burrowing Owl and 18 miles per hour in two herons to a high of 55 miles per hour in a Redhead when pressed.

Most passerine species normally fly at about 15–25 miles per hour, but can usually accelerate this to about 35 miles per hour for short distances when pursued. Swallows and Starlings, among passerines, are somewhat faster; Starlings attain speeds of 40–48 miles per hour. Many of the larger nonpasserine birds are capable of higher speeds; records for most ducks and geese range from 40 to 60 miles per hour, but a Canvasback and an Oldsquaw have been clocked at about 72 (Speirs, 1945). Aviators claim that they can overtake most ducks when their planes are traveling 65 miles per hour. Some of the shorebirds are also fast fliers, with speeds ranging from 40 to 60 miles per hour. McCabe (1942) estimated the flight speed of two flocks of sandpipers (which *passed* his plane) at 110 miles per hour. Pittman (1953) clocked the flight speed of a Common Loon at about 90 miles per hour as it went into a shallow dive in front of a plane.

The swifts and falcons are believed to be the fastest-flying birds. E. C. Stuart Baker's oft-quoted record of a swift in Asia achieving a speed of 171.4–200 miles per hour has sometimes been questioned, but Lincoln (1939) estimated that a large neotropical swift (*Streptoprocne zonaris*) might approach that speed (150 miles per hour estimated). Two well-publicized records (from Cooke 1937) for two Peregrine Falcons give a speed of 165–180 miles per hour for one and 175 for another, both apparently attained in the sudden downward plunge or "stoop."

Birds fly readily into or with the wind; the view that a bird oriented with the wind is handicapped by having its feathers ruffled applies only to a perched bird, as (obviously) a bird in flight is traveling faster than the wind (its own speed *plus* that of the wind). A bird on the ground or in a tree, however, nearly always faces into the wind, not only to avoid ruffled feathers, but also to facilitate the take-off. Many birds cannot rise in a tail wind but have to face into it. A flock of sparrows on a highway, when approached by a car traveling with the wind, may be observed to fly *toward* the car momentarily before swerving out of its path. Large-winged birds like vultures and albatrosses, so graceful in soaring flight once they are launched, are very awkward on the take-off and have to get a running start into the wind before they can rise. South Americans have capitalized on this in capturing condors in baited enclosures, from which the birds cannot escape because of the lack of an adequate runway.

Crosswinds affect a bird's flight more seriously. To maintain a straight course in a strong crosswind, a bird has to head slightly into the wind, as an experienced boatman heads his vessel upstream to cross a river in a straight line. It is not clear whether a bird in sustained flight, as over a body of water, consciously pursues a straight course by constantly tacking slightly into the wind or allows itself to be blown out of its direct course and then regains its position by flying into the wind. Limited observations suggest that both methods may be used: Starlings and other birds are often observed, over short courses at least, to fly sideways, and migrating birds have been observed to come ashore *into* the wind, though a direct line of flight would have brought them in on a crosswind.

SELECTED REFERENCES

AYMAR, G. 1935. *Bird Flight*. New York: Dodd.

BEECHER, W. J. 1951. Adaptations for food-getting in the American black-birds. Auk, 68:411–440.

BERLEE, J. 1964. Flight. In Thompson's *A New Dictionary of Birds*. New York: McGraw-Hill.

BROWN, L., and D. AMADON. 1968. *Eagles, Hawks, and Falcons of the World*. New York: McGraw-Hill. Chapter 5 on Flight.

CHAMBERLAIN, F. W. 1943. Atlas of avian anatomy: osteology, arthrology, myology. Mich. State Coll. Agr. Exp. Sta. Memoir Bull., 5:1–213.

COOKE, M. T. 1937. Flight speed of birds. U.S.D.A. Circ. No. 428.

FISHER, H. I. 1946. Adaptations and comparative anatomy of the locomotor apparatus of New World vultures. Amer. Midl. Nat., 35:545–727.

GEORGE, J. C., and A. J. BERGER. 1966. *Avian Myology*. New York and London: Academic.

HUDSON, G. E. 1937, 1948. Studies on the muscles of the pelvic appendage in birds. Amer. Midland Nat., 18:1–108; 39:102–127. (Many later papers on myology with co-workers.)

JAMESON, W. 1958. *The Wandering Albatross*. London: Rupert Hart-Davis.

SHUFELDT, R. W. 1890. *Myology of the Raven (Corvus corax sinuatus)*. London: Macmillan.

STORER, JOHN H. 1948. The flight of birds. Cranbrook Inst. Sci. Bull. No. 28.

SUTTON, O. G. 1949. *The Science of Flight*. Harmondsworth, Middlesex (England): Penguin Books.

VAN TYNE, J., and A. J. BERGER. 1959. *Fundamentals of Ornithology*. New York: Wiley, 1959. Chapters 2 and 8.

WELTY, J. C. 1962. *The Life of Birds*. Philadelphia and London: Saunders. Chapters 4 and 21.

5

Digestion, Food, and Feeding Habits

The Digestive System

THE DIGESTIVE system provides for the intake and breakdown of food products, which are then transported by the blood to other parts of the body. Because of the high rate of metabolism in birds, food requirements are great and digestion is rapid, but in the interests of economy for flight, most species are highly selective in their diet, not ordinarily taking in items that cannot be promptly and fairly completely utilized. Digestive structures in birds (Fig. 5.1) exhibit a number of unique features that are peculiar to birds. These features are described and their functions explained in the following paragraphs.

The interior of a bird's mouth is comparatively featureless. There are no teeth in any modern forms, and mucous-secreting salivary glands for lubricating the hard dry *palate* or roof of the mouth are, with some exceptions, largely absent or poorly developed. In general, salivary glands seem to be absent or poorly developed in aquatic birds (except waterfowl) and "undeveloped" in semiaquatic birds (Tucker, 1958), but the full complement is present (see Tucker for complete classification) among certain seed-eaters, insectivores, and omnivores (Farner, 1960). Use of a gluelike salivary secretion for nest building (page 260), is well known in the swifts. Apparently, the salivary glands in the Chimney Swift have a cycle of recrudescence and regression like the gonads (Johnston, 1958). Gray Jays, unlike other corvids, have well-developed salivary glands, presumably enabling them to probe into bark crevices and cones to extract food items on a mucous-coated tongue (Bock, 1961) or to form compact food balls for storage in crevices in conifers (Dow, 1965).

Taste buds are scanty in birds; a few scattered over the palate or on the usually undifferentiated tongue permit some taste discrimination, but most birds are accustomed to bolt food items quickly without tasting or moistening them before swallowing. Indeed, in young birds,

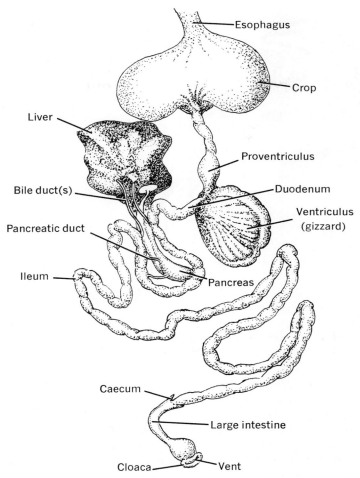

FIG. 5.1. *Digestive system of a Common Pigeon.* [Drawing by Diane Pierce.]

food is placed far back in the throat, so that taste buds, salivary glands, and even a much-differentiated tongue may be superfluous.

The tongues of birds, though usually relatively undifferentiated or even degenerate in certain fish-eating species, exhibit many useful modifications, primarily for food gathering. In typical woodpeckers the tongue is barbed at the tip to facilitate extracting grubs from bark crevices and excavations in trees, but in sapsuckers it is brushlike for licking up sap that fills the little well holes they drill into the cambium. In woodpeckers in general, but notably in the flickers, which are largely ant-eating ground-feeders, the tongue is conveniently extensible by means of the extremely prolonged *hyoids,* which extend backward and

FIG. 5.2. *Hyoid apparatus and tongue of a flicker, showing extensions of hyoids over the skull and insertion in nostril.* [Drawing by Diane Pierce.]

curl up and over the base of the skull and forehead and attach near the nostrils (Fig. 5.2). In nectar-feeders — hummingbirds, sunbirds (Nectariniidae), and various honeycreepers (Fig. 5.3) — the tongue is usually long and forked, and the two branches are often rolled into tubes for sipping nectar, not by suction (there is no vacuum at the base of the tongue) but by a sort of capillary action. In most sunbirds the tongue is a closed tube formed by inward rolling and meeting of the edges (Gill, 1971), but in many other nectar-feeders it is semitubular rather than completely closed. The tongues of parrots are thick and fleshy, presumably correlated with their fruit-eating habits but perhaps also as an aid in modulating the sounds made by this notoriously articulate group of birds. The tongues of waterfowl, except for the fish-eating mergansers, are also fleshy; in addition, they are fringed along the lateral margins for sifting the fine materials gathered up in bottom ooze.

FIG. 5.3. *Tongue of honeycreeper* (Coerebidae), *showing tubular structure with fringed tips.* [Drawing by Diane Pierce.]

The part of the digestive tube leading from the pharynx, or posterior part of the mouth, to the stomach is the esophagus or gullet. In many birds, particularly in galliform and columbiform birds, a diverticulum known as the *crop* develops midway down the tube. This is chiefly a storage receptacle for birds that need to take in large quantities of food at a faster rate than the small stomach can accommodate. Thus gallinaceous birds fill their crops quickly and then pass on the stored items a few at a time to be digested while the bird is resting or roosting. Pheasants and grouse, for example, have two main feeding periods, in the morning and in the late afternoon or the evening. No doubt the usually heavier evening meal helps sustain them overnight. By contrast, insectivorous birds feed much more frequently and have less need for a specialized storage place. Hummingbirds, with the highest food consumption of any animal, feed almost continuously throughout the day; yet even then the males often cannot take in enough nourishment to sustain themselves overnight at the daytime rate of metabolism and they become torpid (see page 159), reducing body temperature and breathing rate until they are in a state of dormancy (Pearson, 1953). Lasiewski (1963) concludes that this is not a nightly occurrence, but an "emergency measure," when food reserves are low.

The crop, or at least a temporarily distended esophagus (false crop), also serves as a market basket for fish-eating species that go far afield on fishing trips or transport food long distances for their young. It is used as a fruit basket for birds like the Cedar Waxwing, which carries a cargo of berries to its young. Birds of prey commonly fill their gullet to the maximum capacity when food is available, and then may not feed again for hours or even a day or more, while the body gradually assimilates the bounteous meal. Vultures, in fact, often stuff themselves so full of carrion that they are unable to rise from the ground without disgorging (Fig. 5.4).

Pigeons are noted for their unique ability to produce "pigeon's milk," a caseous (cheesy) secretion formed by proliferation and sloughing off of cells lining the crop. It is used to nourish the squabs until they are old enough to eat grain. The "milk" is rich in proteins, fats, and vitamins and is produced by both sexes during the breeding season. Another unusual function of a crop is found in the Hoatzin, which uses the crop as a grinding organ for crushing masses of thick foliage (largely arum leaves) on which it feeds.

Other modifications of the anterior digestive tract for special purposes include (1) gular sacs in the Pine Grosbeak (French, 1954) used during the breeding season for carrying seeds to their young, (2) an esophageal pouch (false crop) in redpolls for storing seeds in the winter (Fisher and Dater, 1961), and (3) a bellowslike distention in certain

F IG. 5.4. *The Turkey Vulture, like other scavengers and birds of prey, is an opportunist and gorges itself when food is available, storing the surplus in an expansible gullet.* [Photo by Allan D. Cruickshank.]

grouse, pigeons, and bustards that serves as a resonance chamber for producing sound during courtship.

The stomach of a bird is usually divided into two parts, an anterior small but thick-walled glandular *proventriculus* and a posterior muscular *ventriculus* or *gizzard*. The former secretes gastric juice and initiates the breakdown of food before it is passed on to the gizzard, whose muscular walls abrade and crush hard materials, often with the aid of grit, for subsequent assimilation in the intestine. Insect-feeders have a small gizzard and a small but well-developed proventriculus, whereas birds subsisting on hard items usually have a large powerful gizzard but a hardly noticeable proventriculus. There is great variation in the degree of development of these structures in different birds; in fact, they may vary somewhat in the *same* species at *different* seasons, depending on their seasonal food habits. Some tropical fruit-eaters have evolved a nearly straight and relatively featureless digestive tube and pass incompletely digested berries more or less continuously through the whole tract. The grinding power of the gizzard is well illustrated by birds that feed on acorns, beechnuts, and hickory nuts (Wood Duck, Ruffed Grouse, Turkey). The hard nuts are crushed and pulverized by muscular contractions of the gizzard. Schorger (1960) found that penned Turkeys required about an hour to crush pecans, but that hickory nuts (perhaps not a part of their natural diet) required 30 to 32 hours.

Some disagreement exists regarding the need for grit in seed-eating birds. Undoubtedly grit aids materially in the mechanical mastication of hard materials, for gizzard stones soon become smooth and polished, but in the absence of grit gallinaceous birds at least can utilize hard seeds that abrade against each other and the hard lining of the organ (Beer and Tidyman, 1942). Grit of the right kind is also a source of valuable minerals, such as calcium and phosphorus, needed especially for bone and feather development. Sadler (1961) noted that female pheasants selected calcium-bearing grit (limestone) over noncalcareous material during egg laying and suggested that pheasants might be limited to areas where calcium is available. Verbeek (1971) found female Anna's Hummingbirds feeding on sand repeatedly while laying eggs, then eating more sand after laying, presumably to replace the calcium lost from bones during egg production. He had the sand analyzed and it contained calcium. In other cases sand, which is largely quartz, serves mainly as an abrasive (Dennis, 1951, 1952). Jenkinson and Mengel (1970) reported that many, but not all, caprimulgids ingested stones, apparently to help grind the chitinous bodies of insects. Some grit is taken in accidentally in normal feeding operations by flickers in gleaning ants from an anthill and by the American Robin and American Woodcock in probing for earthworms. Beach-combing shorebirds may have 10–60 percent of their stomach contents composed of incidental sand (Reeder, 1951).

Digestion in birds, as in other vertebrates, is mainly consummated in the small intestine whose duodenal portion receives bile from the liver and pancreatic enzymes from the pancreas. The bile assists in the chemical preparation of food for further digestion (neutralizes acids and emulsifies fats), and the enzymes from the pancreas break down carbohydrates, fats, and proteins. The large liver has many other important functions in birds, but most of them, such as the storage of lipids and glycogen, are nondigestive (Farner, 1960). A gall bladder, for storage of bile, is present in some birds, but absent in many; when it is absent, bile goes directly via the hepatic ducts to the duodenum.

The small intestine is comparatively short and moderately coiled in most animal-feeders, but is long and extensively coiled in omnivorous and herbivorous species. The intestine of the largely vegetarian Ostrich is 46 feet long, but in the insect- and nectar-feeding Ruby-throated Hummingbird it is only 2 inches long. This is associated with the greater needs for intestinal space for digestive processes in the herbivorous forms, which in general require a greater bulk of food than the animal-feeders.

The large intestine in birds is much reduced, as digestion is relatively complete in both vegetarian- and animal-feeders, particularly the latter, and there is a minimum of nondigestible wastes. Many birds have economized further by evolving other methods of voiding indigestible items. Birds of prey, particularly the owls, form pellets (Fig. 5.5) of the bones, fur, and feathers of their prey in the ventriculus (gizzard), and periodically eject them through the mouth so that these parts do not pass through the entire digestive tract. Pellets may be retained in the gizzard six hours or more in Great Horned Owls (Grimm and Whitehouse 1963), or, in the Barn Owl, be passed forward to remain in the proventriculus until the sight of new food triggers ejection (Wallace, 1948; Smith and Richmond, 1972). In fruit-eaters a variety of methods is found. Evening Grosbeaks, with their powerful beaks, crack the pits of cherries and feed the kernels to their young; corvids form pellets of small seeds and spit them up; thrushes strip the flesh from berries in their crop or stomach and eject the pits. On the other hand, Cedar Waxwings and many tropical species pass even large pits through the digestive tract.

Paired *caeca,* usually rudimentary but well developed in gallinaceous birds, anatids, and the Ostrich, mark the transition between the small and large intestine. In gallinaceous birds the length of the caeca, and also of the intestines, is related to the diet, being longer in browsing forms like grouse than in seed-eating quail and pheasants, and longer in a coastal race of California Quail living on a low-grade winter diet than in a close relative living on richer seeds (Leopold, 1953). Studies on grouse in Finland indicate that the caeca, which harbor bacteria, function in the microbial decomposition of cellulose, which figures prominently in the diet of these birds (Suomalainen and Arhimo, 1945).

Passage of food through the digestive tract is rapid. Waves of contractions (*peristalsis*) move the food mass (*bolus*) along the alimentary canal toward the cloaca. Black Ducks digest and pass blue mussels, and probably other molluscs, in 30–40 minutes (Grandy, 1972), and Mallards pass crayfish in about 45 minutes (Malone, in Grandy). Pheasants fed chromium-treated tracer pellets required considerably longer for passage of food, a minimum average time of 1–2 hours, and maximum average time of 8.5 hours; if the food was stored in the caeca for bacterial decomposition 35 hours were required for passage (Duke et al.,

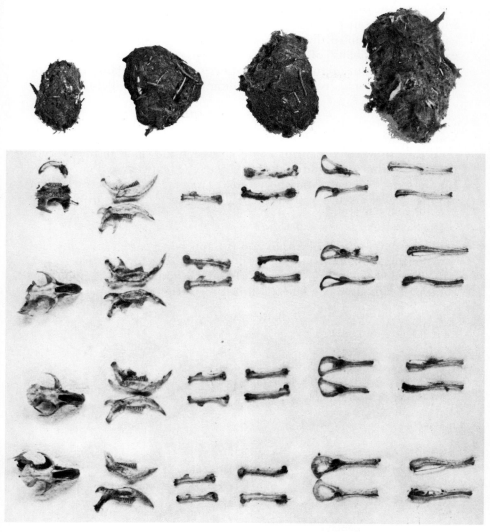

FIG. 5.5. TOP ROW *Sample pellets from a Barn Owl. The smallest probably contains the remains of only one mouse. The largest, when dissected, was found to contain the bones of four mice as shown in the four lower rows—skulls, lower mandibles, humeri, femora, pelvic bones, and tibia. One humerus was missing.* [Photos by Philip G. Coleman; courtesy of Mich. Agr. Exp. Sta.]

1968). On the other hand, fruit-eating species often pass pits and seeds within a half hour.

A final function of digestion not related to any specific organ is the extraordinary ability of birds to store up reserves of fat to utilize during migration. Migratory birds often increase their body weight by 15–40 percent or more by heavy feeding before a long flight.

Feeding Techniques

Some of the feeding methods employed by birds and the many adaptations they show for procuring food have been indicated in the preceding pages; others are scattered throughout the text (see particularly the chapters on Behavior and the Annual Cycle). Particularly significant here, however, is the use of the bills and feet for procuring food.

Use of Bills and Feet

Figures 5.6 and 5.7 illustrate and the following paragraphs describe some of the adaptational modifications of the beaks for procuring food. The flattened bills of most waterfowl and the odd bills of flamingos are equipped with strainer plates (*lamellae*) for sifting out submerged food

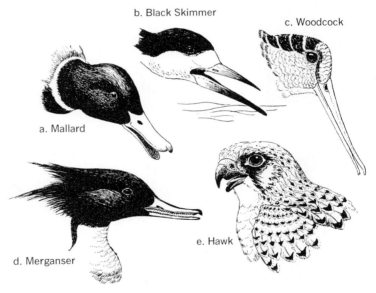

FIG. 5.6. *Adaptations of bills of aquatic, marsh, and predatory birds. a. Mallard, with strainer plates for sifting bottom debris. b. Black Skimmer, with elongated lower mandible for skimming and cutting surface of water. c. Woodcock, with elongated bill and flexible tip on upper mandible for probing in mud. d. Red-breasted Merganser, with saw-toothed (serrate) margins for holding fish. e. Hawk, strongly hooked for tearing up prey.* [Drawings by Diane Pierce.]

a. Downy Woodpecker   b. Evening Grosbeak

c. Kingbird

FIG. 5.7. *Adaptations of bills of insectivorous and seed-eating birds.* a. *Downy Wood-pecker, strong and chisel-like for drumming and drilling in trees.* b. *Evening Grosbeak, powerfully developed for cracking seeds.* c. *Kingbird, strong and depressed (flattened) with notch at tip and rictal bristles at base (tyrant flycatchers snap up insects in the air with a loud click of the bill).* [Drawings by Diane Pierce.]

items, but mergansers have *serrate* mandibles to hold onto slippery fish. Hawks, owls, and shrikes possess powerful hooked beaks, operated by well-developed mandibular muscles, for tearing up prey. The Corvidae have somewhat generalized but strong bills for their omnivorous habits. Semiaquatic shorebirds have long bills, some extremely long for probing into mud and water; the probing bill of the American Wood-cock, in addition, has a muscle-controlled, flexible tip on the upper mandible to facilitate seizing an earthworm deep underground. Wood-peckers have strong chisel-like beaks for drilling into wood for food or for nest construction, and have thickened, shock-absorbent skull bones and strong neck muscles to make such poundings feasible. The effectiveness of the rapierlike probing bill of hummingbirds, modi-fied to serve special needs in many tropical forms, is obvious. Swifts and swallows, which scoop up their prey into open mouths while on the wing, have small bills and a large gape, while flycatchers, which snap onto and crush insects, have strong, flattened, notched mandibles equipped with rictal bristles at the base. Insectivorous foliage-gleaners like warblers and vireos have small slender bills, whereas short stout bills for crushing seeds are characteristic of seed-eaters.

Like some fantastic plumage decorations, some peculiar bills have developed to a stage where they fail to show any readily apparent adap-tation, as in the enormous spongy beak of the toucans (Fig. 5.8), which may possibly serve a useful purpose in securing fruit, but is more likely for sexual displays and species recognition. On the other hand, the dextrally curved beak of the Wrybill Plover (*Anarhynchus frontalis*) of New Zealand is said to function admirably for extracting prey from beneath the edges of stones, which the bird circles in a clockwise direc-tion, although the bills are not used exclusively for this unique method of feeding. The peculiar bill of the Black Skimmer (Fig. 5.6b) serves the skimming purpose implied in the name; the thin, bladelike lower man-

F IG. 5.8. *The large beak of a toucan seems unwieldy, but it is spongy in character and exceedingly light. Though its length may serve a useful purpose in getting fruit and its bright colors may serve for sexual displays or species recognition, there is no known advantage for so large a bill.* [N. Y. Zool. Soc. photo.]

dible, which has a faster growth rate than the upper mandible, cuts the surface of the water, and fish are seized and swallowed. The bills of the extinct Huias (Callaeidae) of New Zealand were even more remarkable; that of the male was straight and sharp for tunneling into dead wood for grubs, that of the female was slender and decurved for extracting prey the male could not reach. Then they shared the spoils in an unusual example of cooperative hunting.

Sexual dimorphism in bill size, as well as in body size, tarsal length, and tail length, is important in foraging operations; it enables the male and female on the same territory to utilize slightly different prey and not compete with each other, thus permitting more efficient coverage of the foraging area. Interspecific differences between closely related species are even more important. In several genera of woodpeckers

(Kilham, 1970; Ligon, 1968; Selander and Giller, 1959, 1963), in accipiters (Storer, 1966), and in many sandpipers these differences lessen competition and permit a more efficient utilization of feeding areas. A good example of the importance of bill differences is in the Sora and in the Virginia Rail; the former has a stout, chickenlike bill and feeds on a diet of 73 percent seeds and 27 percent animal matter; the latter has a long slender bill and takes 62 percent insects and 38 percent seeds (Horak, 1970). Both inhabit the same marshes and nest in close proximity (Berger, 1951a). Weeden (1969) found that competition between sympatric Rock and Willow Ptarmigans in Alaska was avoided to some extent because the smaller-billed Rock Ptarmigans took smaller items, even including grit size. Willson (see page 135) found the thinner-billed fringillid species selecting the smaller seed sizes at feeding stations.

Similarly, the feet of birds (Figs. 5.9 and 5.10) are often modified to serve specific needs, especially for feeding purposes: webbed or lobed for pursuing underwater prey in aquatic birds, long-toed for walking over aquatic vegetation in some marsh birds, strongly clawed for scratching in gallinaceous birds, strongly taloned for seizing prey in predatory birds, yoke-toed (*zygodactylous*) in woodpeckers for clinging to and climbing on vertical surfaces, and well adapted for perching,

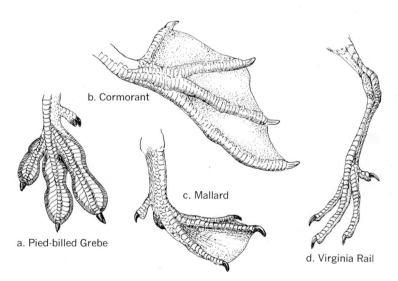

FIG. 5.9. *Adaptations of feet of aquatic birds. a. Pied-billed Grebe, toes broadly lobed with nails flattened and tarsus compressed, for swimming and diving. b. Cormorant, a "pelecaniform" foot, with all four toes united in a web, for swimming. c. Mallard, three toes united in a web, for swimming and paddling. d. Virginia Rail, long-toed for paddling over mud and marsh vegetation.* [Drawings by Diane Pierce.]

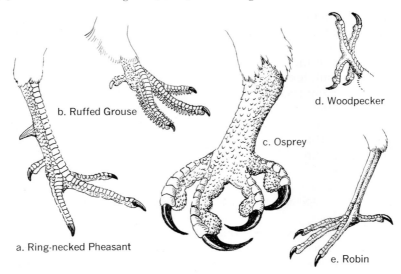

FIG. 5.10. *Adaptations of feet of land birds.* a. *Ring-necked Pheasant, with elongated tarsus and strong toes for running and scratching and, on the cock, a spur for fighting and for treading hens.* b. *Ruffed Grouse, with fleshy fringes (winter only) on toes for walking on snow.* c. *Osprey, with strong talons for grasping fish; outer toe reversible.* d. *Woodpecker, with strong claws; outer (fourth) toe permanently reversed (zygodactylous).* e. *Robin, "passerine" foot for perching, with long incumbent ("same-level") hindtoe or hallux.* [Drawings by Diane Pierce.]

hopping, and running in perching birds. The fleshy fringes that develop in winter on the sides of the toes of the snowshoelike feet of grouse also serve important adaptational needs; other developments, such as the long hind claws of the larks (Alaudidae) and certain fringillids, give extra support on snow or sand.

A unique foot structure for grasping prey is that of the Osprey (Fig. 5.10c). The sole of the foot is rough and scaly, for securely fastening onto slippery fish, the claws are long, sharp, and strong for securing a firm hold, and the outer (fourth) toe is reversible, making a more efficient grasping mechanism. The toes in some waders are also highly specialized; in jacanas the toes are so long they support the weight of the birds walking over lily pads and floating pond weeds, hence jaçanas can exploit a source of food (mainly insects) not readily available to other birds.

Food Habits

What birds eat and in what quantities are no longer conjecture. Ornithologists in this country have been making studies of food habits for nearly a century. Staff members of the former Biological Survey and its successor, the Fish and Wildlife Service, have now analyzed more than 100,000 stomachs of birds from nearly all parts of North America. Thus, in a general way at least, the food trends of most of our birds are already known. This does not mean that such studies are no longer needed. Analysis in Washington of sample stomachs from various parts of the country is not a good criterion of what Marsh Hawks eat in Georgia's quail country or North Dakota's waterfowl marshes, nor is the diet of a Great Horned Owl in an overpopulated pheasant area a true reflection of its prey at the peak of a rabbit cycle. More detailed data are needed on regional studies, with seasonal or year to year variations, before the role of a species in a community can be properly evaluated.

It is convenient, though not always logical, to describe the food habits of birds by dividing them into groups, such as seed-eating and insectivorous, but of course few birds belong exclusively in one group. The Common Crow, for instance, feeds to a considerable extent on seeds and grain, but it is also insectivorous, a scavenger, a predator, and even a fisherman at times. The following summary lists the prevailing food for the most prominent families of North American birds. In succeeding paragraphs the food-habit groups are discussed in more detail, concentrating on North American birds but bringing in additional data for certain foreign birds where pertinent. It should be borne in mind that there are great geographic and species variations and that the trends depicted here do not necessarily apply to all the species within a group on a world-wide basis.

FOOD HABITS OF SOME
NORTH AMERICAN FAMILIES OF BIRDS

Gaviidae—mostly fish, but consume some insects and other invertebrates.

Podicipedidae—fish, amphibians, aquatic invertebrates; ingest their own feathers regularly.

Pelecanidae—almost entirely fish.

Phalacrocoracidae—mostly fish, some amphibians and crustaceans.

Ardeidae—fish, frogs, mice, miscellaneous vertebrates and invertebrates.

Anatidae:

Swans—grazers, almost entirely vegetarian.

Geese—grazers, mostly vegetarian.

"Puddle" ducks—chiefly vegetarian; tubers, seeds, foliage; but ducklings feed on insects at first.

Fresh-water divers (Pochards)—mixed animal and vegetable; great species variations (see Bartonek and Hickey, 1969).

"Sea" ducks—predominantly animal, crustaceans and molluscs.

Mergansers—chiefly fish, sometimes depredatious in trout streams.

Cathartidae—scavengers, chiefly dead animals, but sometimes take nestling birds and newborn mammals.

Accipitridae—predators on animals, very variable (see discussion).

    Accipiters—chiefly birds.

    Buteos—chiefly mammals.

    Eagles—fish, mammals.

    Harriers—variable.

Pandionidae—almost exclusively fish.

Falconidae—large birds (Peregrine); insects, mice, and birds (American Kestrel and Merlin).

Tetraonidae—browsers; buds, foliage, berries, some insects.

Phasianidae—chiefly vegetarian, grain and weed seeds; some insects.

Gruidae—omnivorous; insects and small vertebrates, grain, fruit.

Rallidae—omnivorous; rails on a mixture of plant and animal matter (Sora 73 percent seeds, Virginia 62 percent insects, Horak, 1970); coots predominantly on vegetable.

Charadriidae—mainly animal matter.

Scolopacidae—mainly animal, beachcombers and scavengers.

Laridae—gulls omnivorous, chiefly scavengers on fish and refuse, insects in fields; terns feed on living fish and invertebrates.

Columbidae—grain, seeds, vegetable matter; feed young on "pigeon's milk."

Cuculidae—chiefly insectivorous (hairy caterpillars); Roadrunner on small vertebrates.

Strigidae—mostly mammalian prey (but see discussion).

Caprimulgidae—"Hawk" insects (mosquitos, moths) on the wing.

Apodidae—nearly 100 percent insects (aeroplankton) caught on the wing.

Trochilidae—insects, arachnids, and nectar from flowers.

Alcedinidae—mainly fish (in the New World); some amphibians and aquatic invertebrates.

Picidae—most species mainly insectivorous; some feed on cambium, sap, mast, and fruit.

Tyrannidae—almost exclusively insectivorous; phoebes sometimes catch fish.

Alaudidae—mainly seeds; some insects.

Hirundinidae—almost exclusively insectivorous, but Tree Swallow eats some fruit in winter.

Corvidae—omnivorous; predators (nest robbers), scavengers, and grain feeders.

Paridae—mainly insectivorous, but consume seeds, nuts, fruit, especially in winter.

Sittidae—mainly insectivorous, but consume nuts and seeds, especially in winter.

Certhiidae—almost exclusively insectivorous; a few seeds.

Troglodytidae—insectivorous, some spiders.

Mimidae—mainly insectivorous, but considerable fruit, especially in mockingbirds.

Turdidae—mainly insectivorous (or other invertebrates—earthworms, spiders); much fruit in season.

Sylviidae—almost entirely insectivorous (or other small invertebrates).

Bombycillidae—mainly fruit, some insects, buds, and flower parts.

Laniidae—predatory on small mammals, birds, and large insects.

Sturnidae—omnivorous, but mainly animal matter.

Vireonidae—almost exclusively insectivorous; some fruit.

Parulidae—mainly insectivorous; some fruit (puncture grapes) in fall.

Icteridae—omnivorous; orioles and meadowlarks predominantly insectivorous; others feed heavily on grain except in breeding season.

Fringillidae—very variable; some predominantly seed-eaters, especially in winter, some mainly insectivorous.

Insectivorous Birds

The insect-eating propensity of birds is one of their best-known traits (Fig. 5.11). Most passerine species, which comprise more than half of the birds of the world, are predominantly insectivorous; other passerines, even the seed-eaters, feed in part on insects, especially during the nesting season. Many nonpasserines (woodpeckers, swifts, goatsuckers, cuckoos, and most shorebirds) are also insectivorous; others, even ducks, herons, hawks, and owls, take some insects. Probably there is no family of North American birds that does not feed to some extent on insects, although pigeons and some finches (e.g., goldfinches) are almost exclusively granivorous.

It is probably apparent from preceding accounts that the great diversity in foraging habits of birds means that all habitats are thoroughly covered by insectivorous birds: bark-foragers on the trunks and branches of trees, foliage-gleaners on the twigs and smaller branches, ground-feeders on the forest floor and in pastures and meadows, shorebirds along beaches and mudflats, and a great variety of marsh and water birds in aquatic habitats. Diurnal aerial feeders comb the skies during the day, and when night falls caprimulgids take up the quest for nocturnal insects. Although there is much overlap in insect prey consumed, there are many areas of specialization, such as cuckoos feeding on hairy caterpillars that most other birds avoid.

Instances of insect-eating birds affecting man's interests adversely are rather rare. Kingbirds in this country and honeyguides in Africa sometimes annoy beekeepers, but stomach analyses of kingbirds show that

FIG. 5.11. *The insect-eating capacity of small birds in the nesting season, when insects are approaching their maximum abundance, is enormous. Many birds feed their young more than a hundred times a day. Here a Nashville Warbler in northern Michigan brings another load.* [Photo by L. H. Walkinshaw.]

the bees taken are mainly the almost useless drones while the incidental consumption of robber flies, which prey on honeybees, is presumably of material benefit to beekeepers. Many insects are beneficial, however, and of course some of these are taken by birds.

GRANIVOROUS BIRDS

This category, in the broad sense, includes not only the primarily seed-eating and grain-feeding forms, but also those that eat buds of deciduous and coniferous trees (budders), those that browse or graze in terrestrial or aquatic habitats, and those that feed on mast such as beechnuts and acorns.

Few birds feed exclusively on seeds or grain, since most of them feed their young on insects, but many fringillids (especially finches), Horned Larks, and some icterids (such as blackbirds) are predominantly seed-eaters much of the year. Most items taken are weed seeds or waste grains—Beal (1897) estimated that wintering Tree Sparrows in Iowa consumed 875 tons of weed seeds in one winter—but it is well known that many seed-eaters, especially blackbirds, raid grainfields in late summer, fall, and winter and do great damage to crops (see Chapter

15). Grazing ducks and geese are often a problem in this respect, especially on southwestern wintering grounds.

Perhaps the most strictly granivorous birds are some of the finches. Goldfinches, linnets, siskins, and redpolls are almost exclusively seed-eaters, even feeding their young in part on regurgitated (predigested) seeds. In winter they often resort to the catkins of birches and alders. Crossbills are specialists in extracting seeds from conifer cones, using their crossed bills to twist off the scales of cones and snip out the seed at the base. Pine Grosbeaks, though feeding heavily on fruit at times, also do considerable debudding; when they visit Christmas tree plantations in winter they often deform the trees by snipping off the terminal buds. Horned Larks sometimes raid nurseries for seeds or seedlings; bounties were once imposed on them in California to avert such damage.

Grosbeaks and Purple Finches do considerable debudding of trees; in commercial orchards this may be extensive enough to be an economic loss to orchardists, but often it constitutes natural pruning, possibly improving the quality of the developing fruit. Linnets in California do considerable debudding of apricot and almond trees (Biehn, 1951); woodpeckers and corvids take the mature almonds (Emlen, 1937).

Willson (1971, 1972), experimenting with seed selection and seed size preferences at feeding stands, found that selection was governed to some extent by bill size and ability to crack seeds. Cardinals could handle all sizes of seeds readily, but the thinner-billed sparrows chose smaller seeds. Caloric content—high in hemp—also seemed to be a factor. There was some advantage, she found, in taking the smaller seeds even if they had lower food value because they could be taken faster and handled more easily and might reduce the risk of predation. Proctor (1968) tested a variety of shorebirds, not normally seed-eaters, on seeds in gelatin capsules and found they could be retained in the digestive tract from a few hours up to 216 hours in a Least Sandpiper and 340 hours in a Killdeer. He concluded that migrating birds could carry viable seeds for long distances and regurgitate them in an entirely different habitat. Washing the feet of aquatic birds and testing the resulting solution shows that birds could carry seeds, as well as small invertebrates, from one pond to another.

Grouse and ptarmigan are budders and browsers of a different type (Fig. 5.12). They live on a low-grade diet of buds, bark, leaves, and twigs, especially in winter, and harbor bacteria in their intestinal caeca for the decomposition of cellulose. Ptarmigans, from Colorado to Alaska (three species), feed mainly on buds and catkins of willow (*Salix*), alder (*Alnus*), and birch (*Betula*) in winter, depending on availability, but change largely to leaves, flowers, and berries in spring and

FIG. 5.12. *Sharp-tailed Grouse on nest. Grouse eat buds, twigs, and a great variety of vegetable matter.* [Photo by Mich. Dept. Nat. Res.]

summer. Even in winter some fruit and herbage is available on wind-blown slopes (Weeden, 1967, 1969). Spruce Grouse subsist in part on conifer needles, of low nutritional value but unlimited in quantity, which impart a buddy flavor to their flesh, making them less popular with hunters.

Waterfowl—swans, geese, and many ducks (excluding those feeding on fish and aquatic invertebrates)—are primarily grazers, on land for grain and herbage or in water for seeds, tubers, stems, and succulent parts of plants. Celery-fed Canvasbacks are an epicurean's delight. Duck diets, however, vary greatly, even among closely related species, seasonally, geographically (whether on coastal or inland waters), and in different age and sex classes. Bartonek and Hickey (1969), for instance, found wide differences in three species of pochards at Manitoba in spring and summer. Male Canvasbacks ate 98 percent vegetable matter—mainly pondweeds (*Potamogeton*), not wild celery (*Valisneria*)—but the females and young ate mainly animal matter (92 and 87 percent respectively). In the Lesser Scaup both sexes and the juveniles took 98 or more percent animal matter; in the fall they are often largely vegetarian. The food habits of game ducks have been studied in great detail, primarily for management purposes (Cottam, 1939; Martin and Uhler, 1939).

Many vegetarian as well as omnivorous species resort to mast—acorns, beechnuts, and other dry fruits—when these are available. Wood ducks consume many beechnuts and acorns in the fall, omnivorous Blue Jays and grackles also feed on them, and some woodpeckers—notably Red-bellied, Red-headed, Red-cockaded, and Gila Woodpeckers—are mast feeders to a considerable extent. But perhaps the ultimate among mast feeders are the Acorn Woodpeckers. Long noted for their acorn-storing habits (page 211), these sociable woodpeckers live in groups, defend their oak trees vigorously against all intruders (except that fox squirrels introduced in California sometimes overpower them), and cache vast stores of acorns for winter (Mac-Roberts, 1970).

FRUIT-EATING BIRDS

In other parts of the world, particularly in the tropics, there are many fruit-eating species of birds, including parrots, fruit pigeons, toucans, hornbills, barbets, and tanagers. In this country, however, fruits and berries are usually merely supplementary items in the diet of seed-eating and insectivorous birds. Normally, such items are taken wholly or largely from Nature's vast storehouse of wild fruits, and thus have little or no economic significance except for the dissemination of fruit pits. Many wild plants, trees, shrubs, and vines, mostly highly desirable from the wildlife standpoint, are planted by birds, but of course some undesirable species, such as poison ivy, are also distributed.

Understandably, the presence of fruit-eating birds around commercial orchards and berry-producing farms poses problems for growers. Cedar Waxwings, the most frugivorous of our native birds, sometimes invade fruit farms, but the worst offender is probably the American Robin in cherry orchards, blueberry fields, strawberry and raspberry patches, and grape vineyards. Northern Orioles and transient Tennessee Warblers puncture grapes both for the pulp and the juice. Many other species—woodpeckers (e.g., Pileated), thrushes (Fig. 5.13), Starlings, vireos, and blackbirds—often supplement their otherwise insectivorous diet with wild or cultivated fruit.

In the American tropics many species, especially tanagers, are predominantly fruit-eaters. Often they assemble at particular trees, sometimes called "tanager trees"—a bonanza for museum collectors. Land (1963) listed 57 species of birds seen at a "feeding tree" between late

FIG. 5.13. *Hermit Thrush. Thrushes, highly insectivorous in spring, supplement their insect diet in late summer and fall with fruit.* [Courtesy of Cleveland Mus. Nat. Hist.]

June and early August in Guatemala. Although only 20 species were seen feeding on the fruit, others came for insects, nectar, or mere sociability. Willis (1966) observed 28 species, including 15 tanagers, feeding on the blue berries of a *Conostegia* tree in Colombia during short periods of observation on three in March. In these and other similar observations some competitive exclusion has been noted; unstable dominance relations exist, and there are temporary divisions in foraging levels and parts of the tree. Snow (1971) discusses the evolutionary aspects of fruit-tree dispersal by birds and the "strategies" used by trees (color, availability, food value, and abundance) to entice birds to eat and scatter the fruit.

PREDATORY BIRDS

Though in a broad sense any animal that preys on another is predatory, among birds the term has been applied chiefly to the hawks and owls, which feed largely on other vertebrates. The principal effect of predation is the suppression of rapidly reproducing prey populations, which, if unchecked, might become so abundant that they would devegetate their range. In performing their usually useful mission hawks and owls have fallen into disrepute because their prey sometimes includes game and poultry. As a result they have been needlessly persecuted, and some of the most useful species have been greatly reduced in numbers.

It is not feasible here to analyze food habits of hawks in detail. There are more than 200 species, world-wide in distribution and exhibiting great geographical variations in diet. Several neotropical hawks (kites) feed on snails, the Secretarybird in Africa is primarily a reptile hunter, an *Accipiter* in Korea lives largely on frogs (Wolfe, 1950), an eagle in the Philippines specializes on green monkeys, another in Africa takes young antelopes, and many other species have special diets. A few comments on the North American hawks follow.

The buteos (the Red-tailed, Red-shouldered, Broad-winged, Rough-legged, and Swainson's Hawks) are mainly rodent hunters; their wing structure and habit of soaring over open spaces favor such prey selection. But species diets differ: Red-shouldered Hawks take more amphibians and reptiles than Red-tails, and Broad-winged Hawks, smallest of the quintet, are quite insectivorous.

By contrast the accipiters or bird hawks—the fast-flying types that dart from concealment in a sudden sally to catch their prey—live largely on birds. The largest of these, the Goshawk, preys heavily on grouse; the medium-sized Cooper's Hawk takes medium-sized birds and some mammals; and the smaller Sharp-shinned Hawk lives mainly on small birds, perhaps 90 percent or more of its diet. By preying on the vulnerable surplus in populations, the accipiters tend to eliminate maladjusted individuals, such as the diseased, injured, or less alert, leaving survivors of a superior genetic strain and maintaining populations at a higher level of fitness.

The widely distributed Marsh Hawk shows great adaptability in prey selection. Quail hunters in Georgia formerly condemned the hawk, but Stoddard's classic work (page 322) on the Bobwhite disclosed that wintering Marsh Hawks destroy many cotton rats, which in turn are serious enemies of nesting quail. Because of a similar prejudice New Zealanders collected 200,000 harriers (*Circus approximans*) for bounties—supposedly to protect introduced California Quail, which the slow-flying harriers could not catch (Williams, 1952). Marsh Hawks in Manitoba, nesting in a marsh that included 21 duck nests, were found to be feeding on voles and blackbirds, with an occasional run of coots and young muskrats, but not on ducks or ducklings (Hecht, 1951). Marsh Hawks on a wildlife area in Michigan took young pheasants until they were nearly half grown, then ignored them; adult pheasants in open pens on a game farm in Vermont ignored Marsh Hawks flying over the pens but ran frantically for cover when a Goshawk appeared.

The falcons (Falconidae) are likewise quite variable in their diet. These include the far northern Gyrfalcon, feeding in part on ptarmigans; Peregrine and Prairie Falcons, feeding on a variety of birds and mammals; the small, largely insectivorous American Kestrel; and the bird-eating Merlin. Beebe (1960) calls attention to variations in the feeding habits of Peregrine Falcons in the Pacific Northwest; on certain islands they were living entirely on sea birds (alcids), in one case entirely on Cassin's Auklets. In some cities (New York, Hartford, and Montreal), Peregrines formerly nested on buildings and lived chiefly on pigeons.

The economic picture with respect to the owls is much simpler — and even more commendable. With the few exceptions mentioned below, the owls in eastern North America are predominantly mousers. In a three-year study of Barn Owls in Michigan, identification of 6,815 prey animals in 2,200 pellets yielded more than 90 percent mice, mostly meadow voles (Fig. 5.14), and nearly 99 percent small mammals (Wallace, 1948). No trace of poultry or game birds was found, and of the

FIG. 5.14. *Meadow vole* (Microtus) *skulls taken from pellets collected under a Barn Owl roost on the Michigan State University campus in the mid-1940's. The owls no longer occur on campus.* [Photo by Philip G. Coleman; courtesy of Mich. Agr. Exp. Sta.]

relatively small number of birds taken (1.07 percent) the majority (89 percent) were House Sparrows and Starlings. Studies of Barn Owls in other northern states have given similar figures, but the prey varies more in the southern and western states. In the south the catch runs more to cotton rats and rice rats; in a Louisiana marsh, where other small mammals were not available, the take was 97.5 percent rice rats (Jemison and Chabreck, 1962). In the Pacific Northwest Barn Owls have been recorded, in two cases, as living almost exclusively on an island population of petrels (Bonnot, 1928; Howell, 1920). In Europe, where Barn Owls also occur, the prey runs more heavily to insectivores, such as shrews.

Most other owls present a somewhat similar picture. The chief exception is the Great Horned Owl, a versatile opportunist that includes game birds, ducks, and poultry in its varied diet. Comprehensive studies by Errington and the Hamerstroms (1940), however, indicate that it is predominantly a rabbit hunter (at least in the midwest), a fact probably lauded by farmers but not by rabbit hunters. Sometimes the game birds taken may appear as serious inroads on local pheasant, grouse, or quail populations, but ordinarily only the vulnerable surplus is taken and the surviving nucleus is more or less impregnable. One item not disclosed in the Errington–Hamerstrom studies is the owls' frequent catch of skunks (owls have no sense of smell) and stray cats.

The Snowy Owl, which in its Arctic home feeds primarily on lemmings and some ptarmigans, takes any animal prey it can get on its periodic winter visits to the states. The Screech Owl, widely distributed throughout most of the country, feeds chiefly on insects, rodents, and sometimes small birds in the east, but in the southwest feeds largely on ground-dwelling arthropods. The Burrowing Owl of the prairies is mainly insectivorous. Ten other North American owls, not mentioned above, are primarily mousers, or, in a few cases, insectivorous.

FISH-EATING BIRDS

Piscivorous species are numerous the world over, though many oceanic birds, perhaps contrary to popular conception, live largely on marine invertebrates rather than on fish. Inland regions also have a considerable number of fish-eating birds inhabiting lakes and streams, including such well-known forms as loons, grebes, pelicans (Fig. 5.15), cormorants (mainly marine), herons, mergansers, Bald Eagle and Osprey, gulls and terns (partly scavengers), and kingfishers.

The fish-eating preferences of the Bald Eagle, as opposed to ducks or game, have been well documented in a study in New Brunswick (Wright, 1953) in which captive birds chose fish when both fish and ducks were provided. The diet of the wild birds was 90 percent fish, 9 percent birds, and 1 percent mammals. In Alaska, where eagles are still

F IG. 5.15. *The Brown Pelican is a fisherman par excellence and consumes large quantities of noncommercial fish. It inhabits the warmer coastal waters of both New World continents and formerly was a familiar sight along the Gulf Coast and Florida.* [Photo by Allan D. Cruickshank.]

quite abundant, they congregate on salmon spawning grounds, preying chiefly on spent and dead or dying salmon—a much-misinterpreted habit that has cost the lives of more than 100,000 eagles and payments of more than $100,000 in bounties in the past (Barnes, 1951). Laszlo (1970) found that Bald Eagles on San Juan Island, Washington, were not dependent on fish (although they apparently preferred fish when available) but fed heavily on road-killed rabbits and sheep carrion.

Sometimes fish-eating birds become a nuisance near hatcheries, rearing pools, and trout streams. Studies of American (Common) Mergansers wintering in the Great Lakes region (Salyer and Lagler, 1940) have shown that though they normally inhabit large bodies of water and subsist largely on noncommercial fish, in some years the freezing over of their usual wintering areas forces the birds onto open stretches of trout streams. Even there the catch of commercial or game fishes (32.79 percent) was lower than that of other, mostly undesirable, species. Fritsch and Buss (1958), checking on the food habits of mergansers on salmon streams in Alaska, found 48 salmonid eggs (less than 1 percent of the total diet) and 3 salmonid fry in the 60 stomachs examined, although some of the birds were feeding in salmon spawning grounds.

Abolishing the legal size limit of pan fish in some states reflects a changing attitude toward fish-eating birds, which wildlife biologists now think may actually improve fishing by the removal of undersized specimens, thus hastening the growth of the survivors. Reelfoot Lake in Tennessee, for instance, used to afford good fishing while fish-eating birds were removing 400,000 pounds of fish per year, whereas the killing of thousands of mergansers in reservoirs in New Mexico apparently resulted in waters teeming with stunted, undersized fish that yielded poor fishing (Elder and Kirkpatrick, 1952).

SCAVENGERS

Best known of the carrion-feeders are the vultures or buzzards, terms which in the New World apply solely to the falconiform family Cathartidae, but in the Old World refer to carrion-feeding hawks in the Accipitridae family. Scavengers have an important but unaesthetic function to fulfill, especially in southern regions where decaying dead animals might constitute a health hazard. The automobile has undoubtedly benefited vultures by providing an abundance of highway kills. Similarly, overstocked deer areas in such states as Michigan and Wisconsin have permitted the spread of the Turkey Vulture north of its former breeding range because of the availability of carcasses left from winter starvation.

Coincident with the spread of vultures in the north, however, sanitation on southern farms has probably decreased the food supply there. Perhaps because of this, Black Vultures in particular may have been forced to seek living prey; in Kentucky (Lovell, 1952) and Georgia (Hopkins, 1953) they sometimes attack newly born pigs, picking at the vent and pulling out the intestines. In other cases, Black Vultures have been known to harass skunks and opossums until they die, thus creating their own carrion, and to raid heron colonies containing helpless young (McIlhenny, 1939). In Texas farmers trap vultures extensively to prevent depredations on domestic animals (Parmalee, 1954); most of the damage is done by Black Vultures. Mrosovsky (1971) found Black Vultures preying on leatherback turtle hatchlings when they emerged from their nestholes.

Petrides (1959) describes an often-observed hierarchy among vultures in Africa. Small Hooded Vultures (*Necrosyrtes monachus*) locate carcasses but are soon displaced by larger numbers of White-backed Vultures (*Pseudogyps africanus*), after which three larger species successively force their way to the carcass. Petrides suggests that the larger species may not be independent hunters and may depend on the smaller species to locate prey for them.

Many other birds are partial scavengers, sometimes more or less incidentally. Nearly all predatory birds will capitalize on dead animals they find. A good example is the Bald Eagle (Fig. 5.16). Crows are quick to take advantage of any obtainable carrion, and often line heavily traveled highways in the early morning hours to feed on the kills of the night. Gulls are important scavengers about harbors, fishing ports, and inland lakes, and though normally useful in this respect, in large concentrations they may start plundering the eggs and young of other birds or even cooperate in overpowering larger birds like ducks.

A special and often overlooked group of scavengers are the sandpipers and plovers. Though they often wade into water and catch living prey, they also glean beaches of small animals, mainly invertebrates, that are washed ashore, sometimes in windrows. Thus they have an important clean-up mission to fulfill, perhaps making lakesides a pleasanter place for summer homes.

SPECIAL DIETS

There are many dietary oddities not covered in preceding accounts. Those of the snail-feeding kites and of nectar-feeding hummingbirds have already been mentioned. In the New World, honeycreepers (Coerebidae and Drepaniidae) and, in the Old World, the sunbirds (Nectariniidae), white-eyes (Zosteropidae), flowerpeckers (Dicaeidae), and honey-eaters (Meliphagidae) also feed to some extent on nectar. "Tick birds," or oxpeckers (Sturnidae), in Africa walk about on the backs of cattle and other ruminants, picking off ticks and other invertebrates. "Crocodile birds" on the same continent are alleged by historians to enter the mouths of crocodiles and clean the teeth of the huge reptiles, but both the habit and the species involved have been questioned. Recently, however, Guggisberg (1972), in a book on crocodiles, has documented some convincing evidence of his own or other observers, citing cases of at least six species of birds seen foraging in and around the jaws and teeth of crocodiles. Equally fantastic is his account of seeing Marabou Storks stealing fish out of the gullet of a living, open-mouthed crocodile.

Other dietary oddities include the peculiar habit of the Kea (*Nestor notabilis*) in New Zealand of feeding on the kidney fat of living sheep (page 210) and of Pigmy Parrots (*Micropsitta*) in the Solomon Islands living largely on fungi, which they glean from the bark of trees by creeping about in nuthatch fashion (Sibley, 1951). Even more unusual are the cerophagous (wax-eating) habits of the honeyguides (Indicatoridae, see Fig. 1.6) in Africa and Asia. Friedmann (1955) has shown that these birds, which consume considerable beeswax along with bee larvae and honey, can subsist for a time on the wax alone, though it is not an adequate diet and eventually they starve without other food. Apparently microbes or bacteria in the digestive tract enable the honeyguides to digest the wax; no other birds, or other vertebrates, are known to do this.

FIG. 5.16. *Bald Eagles are primarily scavengers, especially on fish.* [Courtesy Cleveland Mus. Nat. Hist.]

SELECTED   REFERENCES

Brown, L., and D. Amadon. 1968. *Eagles, Hawks and Falcons of the World.* New York: McGraw-Hill. 2 vols.

Cottam, C. 1939. Food habits of North American diving ducks. U.S.D.A. Tech. Bull., 643:1–140.

Errington, P. L., F. Hamerstrom, and F. N. Hamerstrom, Jr. 1940. The Great Horned Owl and its prey in the north-central United States. Iowa Agr. Exp. Sta. Res. Bull., 277:757–850.

Farner, D. S. Digestion and the digestive system. Chapter XI in A. J. Marshall's *Biology and Comparative Physiology of Birds.* New York: Academic.

Martin, A. C., and F. M. Uhler. 1939. Food of game ducks in the United States and Canada. U.S.D.A. Tech. Bull., 634:1–157.

Thomson, A. L. 1964. *A New Dictionary of Birds.* New York: McGraw-Hill. See alimentary system, feeding habits, food selection, and nutrition.

U.S.D.A. Farmer's Bulletins and Yearbooks. Numerous bulletins and reports on the detailed food habits of nearly all North American birds by F. E. L. Beal, A. K. Fisher, H. W. Henshaw, S. D. Judd, E. R. Kalmbach, W. L. McAtee, and others.

Van Tyne, J., and A. J. Berger. 1959. *Fundamentals of Ornithology.* New York: Wiley. Chapter 9.

Wallace, G. J. 1948. The Barn Owl in Michigan, its distribution, natural history, and food habits. Mich. State Coll. Exp. Sta. Tech. Bull., 208:1–61.

6

Other Metabolic and Physiological Functions

The Respiratory System

THE RESPIRATORY system provides for the various gaseous exchanges that take place between the outside medium and, via the lungs, the cells of the body. Fresh air (oxygen) is introduced into the lungs with each intake of breath (inspiration), and gaseous wastes (carbon dioxide and water) are expelled during expiration. The blood is the medium transporting fresh oxygen, in chemical combination with hemoglobin, to all the cells of the body and returning gaseous wastes to the lungs.

From the standpoint of gross anatomy the respiratory system in birds is fairly simple. Except for some pelecaniform birds in which the nostrils are rudimentary or absent, a pair of *external nares* opens into nasal cavities in the mouth. Typically these are mere slits or small oval-shaped openings in the horny beak, but in hawks they are surrounded by a soft membrane, the *cere,* and in pigeons by a swollen sensitive *operculum.* The nostrils may be *imperforate,* with a septum completely separating them, or *perforate,* without a nasal septum, as in the vultures. In the Procellariiformes the nostrils are tubular, with or without a septum, and in the kiwis they open near the tip of the bill. Figure 6.1 shows some of these variations.

The *internal nares* open directly into the pharynx from which a slitlike opening, the *glottis,* leads into the *trachea* or windpipe. The latter is usually long and flexible; it is 4 feet long in the Whooping Crane, 2 feet of which is looped within a concavity in the sternum. It is composed of a series of stiffened rings, bony on the ventral surface and cartilaginous on the dorsal. At the upper extremity of the trachea is the relatively undifferentiated *larynx,* largely lacking the complicated cartilages and vocal cords characteristic of mammals, but at the lower end of the trachea, at or near the juncture where it divides into short *bronchi* leading to the lungs, is a special structure, the *syrinx* or voice box.

a. Cory's Shearwater   b. Turkey Vulture

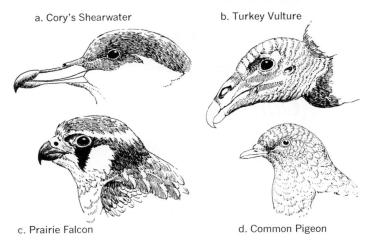

c. Prairie Falcon   d. Common Pigeon

FIG. 6.1. *External nares of birds.* a. *Cory's Shearwater, with tubular nostrils.* b. *Turkey Vulture, with perforate nostrils (no nasal septum).* c. *Prairie Falcon, with circular nostrils in a cere.* d. *Common Pigeon, with soft swollen operculum.* [Drawings by Diane Pierce.]

A few birds, such as the Turkey Vulture, lack a syrinx and the structures associated with it (Miskimen, 1957); others, such as storks and the ostrichlike birds, lack functional syringeal muscles and can only grunt, hiss, or boom. Still other birds have a relatively undifferentiated syrinx located in the bronchi (*bronchial*) or in the trachea (*tracheal*), but in most species the syrinx is *tracheobronchial* in position. This more usual type (Fig. 6.2) is composed of modified tracheal and bronchial rings, expanded to form a chamber or *tympanum,* within which sound is produced. At the bifurcation of the trachea is a bony ridge, the *pessulus,* equipped with a vibratory *semilunar membrane* that extends forward into the trachea and backward to form the membranous inner wall of the bronchi. *Extrinsic syringeal muscles* (*sternotracheal*) originate on the sternum and insert on the trachea (or on the syrinx in some birds). The *tympaniform membranes* (internal and external), forming the slitlike opening out of the bronchus, are controlled and regulated by *intrinsic syringeal muscles*, a single pair in the Common Pigeon, but at least three pairs in most songbirds and six or more pairs in versatile vocalists like the Common Crow (Fig. 6.3), Gray Catbird, and Starling (Miskimen, 1951; Chamberlain et al., 1968). Air expelled from the lungs during expiration passes between the tympaniform membranes whose tension is under delicate muscular control and is responsible for the amazing repertoire of sounds characteristic of birds.

Other structures also may assist in sound production. The semilunar membrane, once thought to be the chief means of vocalization, may

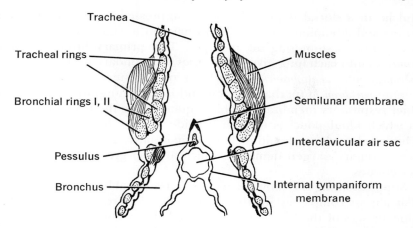

F IG. 6.2. *Section of syrinx of a magpie, showing internal structure.* [Redrawn by Diane Pierce from Walter and Sayles, *Biology of the Vertebrates,* Macmillan Publishing Co., Inc., New York, 1949.]

supplement the vibrations of the tympaniform membranes, although it can be amputated in the Starling without apparent impairment of sound (Miskimen, 1951). Male ducks have an expansive *bulla,* single or double, associated with the syrinx, which may serve as a resonance chamber, but females lack this structure. The loud trumpeting of swans and cranes is no doubt influenced by the long tracheal loop within the sternum.

The two lungs in birds are small and compact, with little elasticity,

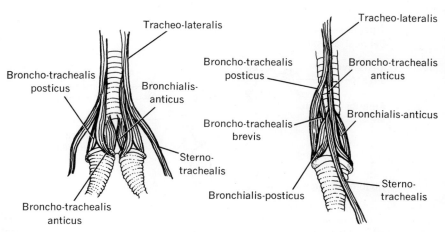

F IG. 6.3. *Syringeal muscles of Common Crow, ventral view (left) and right side.* [Redrawn from M. Miskimen, *Auk,* 68, 1951.]

and lie in a dorsal depression closely appressed to the thoracic vertebrae and adjoining ribs. The lungs have an intricate system of ramifying and anastomosing air passages. The primary bronchus (*mesobronchus*) into each lung gives off a series of secondary bronchi (the *dorsobronchi* and *ventrobronchi*), which in turn connect with numerous minute *parabronchi* (perhaps a thousand in each lung) that serve as the chief respiratory membranes. This is quite unlike the mammalian lung in which blind pouches (*alveoli*) provide the respiratory membranes. Thus a relatively small and inelastic lung in birds is able to supply their extraordinary oxygen demands, which are the greatest of any of the vertebrates.

Supplementing the lungs of birds is a remarkable system of air sacs that almost completely fill the body cavity. Figure 6.4 illustrates the nine air sacs of the Common Pigeon and shows also the lateral extension of the *interclavicular sac* supplying the humeri, and the extensions of the *cervical sacs* paralleling the trachea. There is considerable variation in the degree of development of the air sacs in different species. In many birds the air sacs permeate the hollow bones, but in the Common Loon, a diver not buoyant in flight, the air sacs are simpler and reduced: there is no cervical air sac and no penetration of any of the bones by other sacs (Gier, 1952). In the Common Grackle, and at least some other passerines, the interclavicular and the anterior thoracic sacs are fused into one sac (Kloek and Casler, 1972).

According to Hazelhoff (1951), during inspiration fresh air flows through the intricate parabronchial passageways in the lungs and out into air sacs, particularly into the larger posterior pairs (*abdominal* and *posterior thoracic*), which are more important in breathing. During ex-

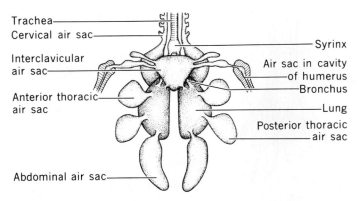

FIG. 6.4. *Diagram of the respiratory organs of Common Pigeon, showing the air sacs.* [From Stiles, Hegner, and Boolootian, *College Zoology*, 8th ed., © copyright 1969 by Macmillan Publishing Co., Inc., New York.]

piration these sacs act as bellows, forcing air back through the lungs by much the same route as that by which it entered. Salt and Zeuthen (1960) seem to agree with this interpretation. Some authors, however, maintain that much of this inspired air passes directly through the primary bronchi into the posterior air sacs and that most of the gaseous exchange takes place in the parabronchi during expiration. *Recurrent bronchi* from the posterior sacs to the parabronchi provide for the return flow. Normal activities, such as flight, probably aid in breathing by muscular contractions of the thorax forcing out used air and automatically introducing new. Unlike mammals, the avian lung is emptied with each breath, leaving little or no residual air. In a pigeon in flight the air sacs also are completely emptied with each breath, but not when the bird is at rest (Hess, 1951). There is no muscular diaphragm in birds, as occurs in mammals, to aid in respiratory movements.

Birds that sing a long uninterrupted song, like the Winter Wren, are believed to utilize reserve air in the air sacs while singing, a method used by water birds when submerged. Deep divers like the Common Loon and Oldsquaw, however, have reduced air sacs and meet their oxygen requirements in other ways (see page 106). The air sacs also function in thermoregulation.

The breathing rate in birds, as might be expected, is normally rapid, but varies so much under different conditions that figures may be misleading. The breathing rate of a hibernating Poor-will in California (Jaeger, 1949) was hardly measurable; no moisture collected on a mirror held before its nostrils. The minimum rate in torpid European Swift nestlings during their fasting periods in rainy weather (page 301) was 8 times per minute, but reached a maximum of 90 as they came out of torpidity. Baldwin and Kendeigh (1932) found rates in House Wrens varied from a low of 28 times per minute in a bird whose body temperature had been lowered to a lethal 74°F up to 340 times per minute at maximum body temperature. The rate at standard temperature was 92–112. In the more lethargic Ostrich, Schmidt-Nielson et al. (1969) determined the minimum rate at rest to be 6–12 times per minute, with a maximum of 60 after exercise. Odum (1943) found the rate in sleeping chickadees to vary with the room temperature, being lower at low temperatures (to conserve heat) and higher at high temperatures (to lose heat). (See Calder, 1968, for a long table and discussion of respiratory rates.)

Blake (1958) measured respiration rates of 30 species of birds held in the hand. Except for an unexplained low in a Starling, the rates ranged from 75 times per minute (average of three trials) in a Mourning Dove to 294 in a House Wren (263–334 in four trials). In most birds (all passerines except the dove and a Downy Woodpecker), the rates ranged between 100 and 200 times per minute. Presumably rates would be somewhat lower in a completely relaxed bird (not in the hand) but higher in a bird under stress, excitement, or in flight.

Thus it appears that respiration is very rapid in active birds, but that they can adjust remarkably to different environmental conditions, to conserve energy during fasting, sleeping, or in cold weather.

The Circulatory System

The circulatory system provides for the transportation of various products to and from all parts of the body. A circulating medium, the blood, in a complex system of arteries, veins, and connecting capillaries, transports (1) nutrients from the digestive system to tissues for growth and repair, (2) oxygen and carbon dioxide carried by hemoglobin to and from cells all over the body, (3) liquid wastes to the kidneys, and (4) hormones from the ductless endocrine glands for regulating growth and behavior. In addition to these well-known functions, Sturkie (1965) credits the circulatory system with an important role in (5) the regulation of water in avian tissues and (6) temperature regulation.

Structurally, the circulatory system of birds is quite comparable to that of other higher vertebrates, and only its unique features need to be described. The heart, relatively large and with strong ventricular walls to withstand the strain of the extremely high heart rate, is completely four-chambered, thus showing some advance over ancestral reptiles. The right aortic arch, instead of the left as in mammals, is developed as the loop leading from the heart to the dorsal aorta. Glenny (1955) discusses in detail the phylogenetic significance of the aortic arch patterns in 123 families of birds; apparently the various arrangements of the vessels, as well as serological data, are useful in taxonomic studies.

A vessel peculiar to birds is a cross vein between the two jugulars in the neck; this prevents blocking of the flow of blood from the head when one vein is momentarily shut off by twisting the neck. Figure 6.5 illustrates the principal arteries and veins in the Common Pigeon. The considerable reduction, compared to reptiles, of the *renal portal system* and the increased importance of the *brachials* and *pectorals* to and from the wings and breast muscles, respectively, may be noted.

Several authors have examined heart weights of birds and pointed out correlations with physiological needs and environmental conditions. In records on 1,340 birds of 291 species (64 families), Hartman (1955) found heart weights varying from 0.2 percent of the body

---

FIG. 6.5. *Circulatory system of Common Pigeon.* [From Stiles, Hegner, and Boolootian, *College Zoology*, 8th ed., © copyright 1969 by Macmillan Publishing Co., Inc., New York.]

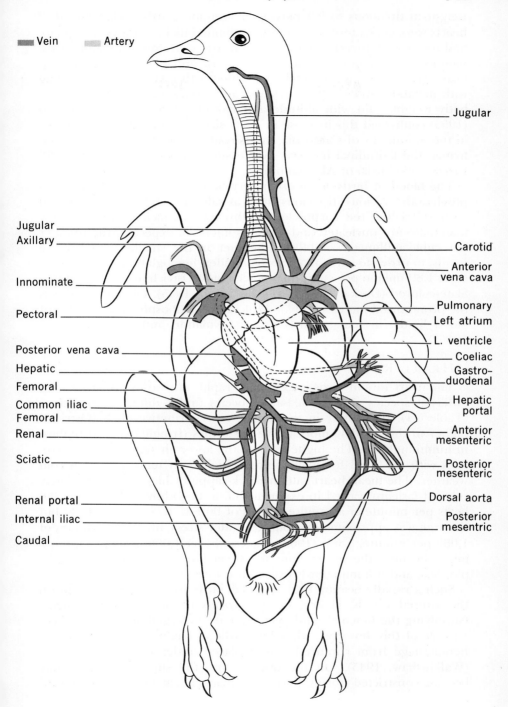

Vein   Artery

Jugular

Jugular
Axillary

Innominate

Pectoral

Posterior vena cava
Hepatic
Femoral
Common iliac
Femoral
Renal

Sciatic

Renal portal
Internal iliac
Caudal

Carotid
Anterior
vena cava
Pulmonary
Left atrium
L. ventricle
Coeliac
Gastro-
duodenal
Hepatic
portal
Anterior
mesenteric
Posterior
mesenteric
Dorsal aorta
Posterior
mesentric

weight in tinamous to 2.4 percent in hummingbirds. Relatively small hearts were characteristic of nonfliers, limited fliers (turkeys and quail), and soarers (vultures), whereas large hearts were found in shorebirds, kingfishers, swallows, warblers and vireos, and of course humming-birds. Heart size was definitely correlated with metabolism, increasing with metabolic rates. There was also an increase in the heart weight–body weight ratio with altitude and latitude. Norris and Williamson (1955) confirmed this increase in heart size with altitude in 12 species in the mountains of California and Nevada, and Johnston (1963) confirmed the latitudinal increase by measurements of a large series (77 species, 563 birds) of Alaskan birds.

The blood of birds, as in other vertebrates, consists of plasma, corpuscles, and various inclusions, such as salts and chemicals. Of special interest are the red corpuscles (erythrocytes), which in birds (unlike mammals) are nucleated and oval. As might be expected, the number per cubic millimeter of blood is higher in the smaller, more active forms, in order to carry more oxygen-laden hemoglobin; counts range from 1.89 million per cubic millimeter in the Ostrich to 7.6 million in a Slate-colored (Dark-eyed) Junco (Nice et al., 1935). Of interest also is the pronounced increase in blood calcium and phosphorus, and also of blood lipids, in female birds during ovulation and laying (Sturkie, 1965).

HEART RATE

The heart rate in birds is extremely rapid, but varies greatly among different species and in the same species under different conditions (Table 6.1). It is lower in the larger, more primitive birds (Ostrich), higher in the smaller, more active passerines (fringillids), and highest in hummingbirds. The higher rate at the lower air temperature correlates with the need for maintaining a high body temperature in cold weather. The mean heart rate in Black-capped Chickadees, completely relaxed (asleep), varied in experimental tests (Odum, 1943) from 346 beats per minute at air temperatures of 90°F to 674 beats at 43°F, but this rate was approximately doubled under stress, up to a maximum of 1,000 per minute. Yapp (1962) suggested that 1,000 beats per minute might be near the upper limit for small birds, but some fringillids, ploceids, and hummingbirds exceed this rate (Table 6.1).

Such a rapidly beating heart in small birds places a severe strain on the arterial vessels. Hence, the *innominates* leading out of the heart (supplying the brachials and pectorals) are enlarged and thick-walled. In spite of this, however, when birds "die of fright," the cause may be hemorrhage from ruptured vessels placed under too heavy a strain (Walkinshaw, 1945). On the other hand, the small capillaries may become constricted during periods of inactivity, at night or during in-

TABLE 6.1. SOME BREATHING AND HEART RATES IN BIRDS

Species	Breathing Rate		Heart Rate		Reference
	Minimum*	Maximum	Minimum*	Maximum	
Ostrich	6	40	38	176	Schmidt-Neilson et al., 1969
White Pelican	6.3		150		Calder, 1968
Mallard (Pekin)	7.6		118		Calder, 1968
Turkey Vulture	9.2		132		Calder, 1968
Mourning Dove	69†	80†	165	571	Blake, 1958; Calder, 1968
Common Nighthawk			125	330	Lasiewski et al., 1964
Blue-throated Hummingbird			480	1260	Lasiewski et al., 1966
American Robin	36.5 (av. of 6)		327.7 (av. of 6)		Lewis, 1967
Black-capped Chickadee	140†	160†	520	1000	Blake, 1958; Odum, 1941
House Wren	83	334†	445		Calder, 1968; Blake, 1958
Catbird	84		552		Lewis, 1967
House Sparrow	109†	113†	450	902	Blake, 1958; Odum, 1941
Waxbill (Estrilda troglodytes)	137.5		520	1020	Lasiewski et al., 1964
Canary	57			1000	Odum, 1941
Cardinal	45				Calder, 1968
Chipping Sparrow	143†	158†		1040	Odum, 1941
Song Sparrow	63	198†		1021	Odum, 1941

* Usually refers to basal or resting rate.
† Held in hand (Blake, 1958).

cubation; possibly stirring or stretching occasionally is required to restore circulation (Odum, 1944).

BODY TEMPERATURE

Body temperature of birds is normally high and variable. It is slightly lower in the more primitive birds, at about 100°F in kiwis, but occasionally may reach a maximum of about 112°F in passerines. There is considerable fluctuation in the temperature rhythm of most birds. In diurnal birds, body temperature is highest during the day, rising until about midday and then gradually dropping to a minimum during the night. In nocturnal birds (kiwis, owls, and goatsuckers), the daily cycle is reversed, with the highest readings obtained at night. Body temperature also varies with other conditions, rising with muscular activity and dropping during rest periods, rising with active digestion of food (full stomach) and dropping with hunger (empty stomach), rising to adjust to high environmental temperatures and dropping when air temperatures are lower.

The fluctuating range may be as much as 8 or 10°, say from 102 to 112°F in passerines, and, experimentally, can be brought much lower and a little higher without being lethal. In tests with House Wrens, one bird recovered from an experimentally induced body temperature of 74.6°F, but about 90°F was nearer the lethal minimum for others. The lethal maximum was between 115.1 and 118.2°F for four individuals (Baldwin and Kendeigh, 1932). In chickens, the lethal minimum is slightly lower than in wrens for about 16 days; thereafter, it is about the same (Sturkie, 1965). In the few birds capable of dormancy (see Torpidity, page 159) and in nestlings still incapable of temperature regulation, body temperature may be very low at times, approaching that of the environment.

In general, however, there is not much geographical, seasonal, or sexual difference in body temperatures of birds; they are about the same in tropical species as in those of high latitudes, nearly the same in winter as in summer, and show little variation in the sexes, except for the higher ventral skin temperatures (brood patch) in females during incubation and brooding.

THERMOREGULATION

Birds are such delicate entities, in such intimate association with their environment, that they need special means, physical and physiological, for precise heat or temperature regulation (*thermoregulation*). They meet the problems of adjustment to environmental conditions in many different ways.

Of course the feathers are the chief means of regulating body temperature—fluffing the feathers to conserve heat and depressing them to lose heat (page 61). Birds also have their densest plumage in winter and thinnest in summer, and arctic species are more densely feathered than tropical species. When sleeping, birds often tuck their bill in their feathers to conserve heat in breathing and the breathing rate and body metabolism are lowered. Even shivering is a temporary expedient to adapt to cold by converting muscular energy into heat. Penguins, and many other marine birds in nearly constant contact with cold water, have a thick insulating layer of subcutaneous fat for conserving body heat.

In warm weather the air sacs (page 150) are the principal means of dissipating heat. Birds have no sweat glands—such glands would be ineffective under a heavy coat of feathers—but birds can reduce body heat by expelling moist warm air during expiration. They also have unique ways of dissipating heat through the unfeathered parts of the body—the feet, tarsi, and perhaps even the ceres in raptors (Bartholomew and Cade, 1957). Incubating and nestling albatrosses on tropical islands hold their webbed feet, which are richly supplied with capillaries, up in the air to hasten heat loss. Storks and vultures defecate on their legs—in the drying process heat is released (Kahl, 1963; Hatch, 1970).

Panting and gular flutter, sometimes used synchronously, are important cooling devices. Caprimulgiform birds in particular have an extraordinarily large gular membrane richly supplied with blood vessels for dissipating heat; heat loss is effected by constantly fluttering the membrane with open mouth in hot weather. The amplitude, rate, and duration of fluttering are often increased with increasing temperatures. Gular flutter is also employed by pelicans, cormorants, boobies, frigatebirds, and many other species, especially at exposed nest sites. Few muscles are involved in gular flutter, so there is only a limited expenditure of energy. Nesting and perching birds also commonly orient themselves with their backs to the sun; they can ward off the heat of the sun more effectively that way.

Obviously birds need methods of conserving heat in cold weather as well as losing it in hot weather. In addition to control by feathers, heat loss through exposed extremities is reduced to a minimum. The feet and tarsi contain no fleshy muscles, only tough tendons with a limited nerve and vascular supply. High temperatures are not maintained in such areas in cold weather; often they drop close to the freezing point, and the blood flow is largely shut off or becomes sluggish. That is, a punctured web in the foot of an albatross on land might bleed profusely, but a severed toe of a chickadee in winter would not. Often birds' feet seem to be impervious to thermal severities. Canada (Gray) Jays have been observed to perch without apparent discomfort for 30–40 seconds on the rim of a hot pot, and for 5–8 seconds on a camp stove hot enough for a drop of water to flash into steam (Norris-Elye, 1945).

In general, the larger birds have a more favorable surface:mass ratio for withstanding cold than the smaller species. According to Bergmann's Rule — a good rule with exceptions (Kendeigh, 1969) — northern forms are usually larger than closely related southern forms and thus can withstand lower temperatures. Ptarmigans (Fig. 6.6) and Snowy Owls, for instance, are better adapted, more efficiently insulated, than the Snow Bunting, longspurs, and pipits, which nest in the Arctic but migrate south in winter. Other birds adapt in other ways. Ostriches (Schmidt-Nielson et al., 1969) can maintain a stable body temperature over a wide range of environmenal temperatures (hot deserts become cold at night). Turkey Vultures also can adapt to environmental extremes, and can lower their body temperature at night or when roosting quietly, thus saving energy (Heath, 1962). Hummingbirds at high altitudes in Ecuador and Peru roost in caves; often their nest sites are in protected situations and well insulated so that the females do not become torpid at night (see Torpidity, page 159). Burrow nesters, such as some petrels and shearwaters, can lower their body temperature when incubating (Howell and Bartholomew, 1961), although this tends to lengthen the incubation period.

FIG. 6.6. *The white plumage of the Willow Ptarmigan in winter affords both protective coloration and heat conservation, as the feathers are dense and fluffy, and well-developed aftershafts furnish additional insulation. This bird was trapped in Alberta for experimental release in the Upper Peninsula of Michigan and is not in good plumage.* [Courtesy of Ebb Warren and G. A. Ammann.]

The young of altricial species do not have the capacity for regulating their body temperature at hatching time. They are essentially *poikilothermous*—that is, their body temperature is close to and fluctuates with the temperature of the surrounding air—and they have to be brooded more or less constantly for the first few days of nest life. But the *homoiothermous* (warm-blooded) condition progresses rapidly with feather development. In nestling House Wrens it is well established by nine days of age (Kendeigh and Baldwin, 1928). Nestling Vesper Sparrows are unable to hold their body temperature more than a few degrees above environmental temperatures for the first two days, do much better at four days, and at seven to nine days of age can adjust readily to ordinary air temperatures (Dawson and Evans, 1960). Precocial birds are more advanced at hatching time, with a full covering of natal down, but their body temperature is unstable for several days. Young ducklings and gallinaceous chicks have to be brooded intermittently during cold or rainy spells. Gulls and terns, though covered with sparse down at hatching, are less efficiently insulated, and their capacity for thermoregulation fluctuates with environmental conditions; that is, it is good in warm weather but poor in cold weather (Bartholomew and Dawson, 1952).

TORPIDITY

In the 18th centry it was widely believed that swallows hibernated in the mud or spent the winter in a torpid condition in hollow trees, but such views were also widely disputed. Now it is known that hibernation in birds is possible and that torpidity is of common occurrence in caprimulgiform and apodiform birds and occurs occasionally in some other species.

So far true hibernation has been disclosed only in the Poor-will in the southwest. In 1946 Jaeger (1948, 1949) discovered a hibernating Poor-will in a rock crevice in the Chuckawalla Mountains in California; its body temperature was low (about 60°F), and its breathing rate hardly measurable. The bird was banded and it returned to the same hibernating place for three more winters. Since then much experimental work has been done with other caprimulgiform birds. Nighthawks and frogmouths are capable of entering and arousing from torpidity but have little or no real need for it, since the former are highly migratory (to South America) and the latter live in an equable climate in the East Indies (Lasiewski and Dawson, 1964; Lasiewski et al., 1970).

Hummingbirds have long been known to be capable of dormancy. Ruby-throated Hummingbirds in the northern states have been picked up, apparently dead, on frosty September mornings, but arouse quickly in a warm hand and fly away. Male hummingbirds in California become torpid on cold nights (Pearson, 1953), but females on well-insulated nests do not (Howell and Dawson, 1954). Lasiewski (1963) thinks that the males do not become dormant on all nights but do so as an "emergency measure" when necessary. At Jackson Hole in Wyoming, nights approach freezing temperatures, but the female Calliope Hummingbirds that nest there do not become torpid at night (Calder, 1971); apparently the well-constructed and sheltered nest minimizes heat loss.

Torpidity in nestling European Swifts has been well documented by Koskimies (page 301). Such dormancy has not been recorded for nestling Chimney Swifts, but White-throated Swifts in California have been found roosting "in a dazed or numb state" during a cold spell in winter (Hanna, 1917).

Many other birds—in addition to caprimulgids, swifts, and hummingbirds—are probably capable of limited dormancy on occasions. Colies or mousebirds (Coliidae) in Africa, for no apparent reason, can enter a state of torpor in which the body temperature falls 15°C (59°F) below normal, but they arouse quickly, as do Inca Doves under similar circumstances (Bartholomew and Trost, 1970). Lasiewski and Thompson (1966) reported finding three "dead" Violet-green Swallows on the ground on a cold morning in California, but the birds revived quickly after being picked up and flew away. The authors could not produce torpor in Cliff Swallows in laboratory tests. Titmice in England and Norway have been found to be capable of some dormancy in cold weather.

Excretion

The parts of the excretory and reproductive systems are closely associated in structure (Fig. 6.7) but not in function; hence, they are considered separately. The main excretory organs are the paired kidneys, which in birds are usually three-lobed, but in most passerines they appear bilobed because the median lobe is ill defined (Johnson, 1968). Some ciconiiform and charadriiform birds have four lobes, and the kiwis have five. Each lobe is divided into smaller lobules, and the whole structure lies securely imbedded in a concavity formed by the fusion of the pelvic bones with the synsacrum. The kidneys consist of a complicated array of circulatory vessels, capsules, and tubules that process liquid wastes brought in by the blood. Paired *ureters* then conduct the wastes to the cloaca where water not already removed in the renal tubules in the kidneys is resorbed; this and the absence of a urinary bladder (except in Ostriches) are further means of economizing on unnecessary weight. Nitrogenous wastes (largely uric acid) are thus reduced to a semisolid state and added to solids from the digestive tract, then voided from the cloaca as a whitish guano.

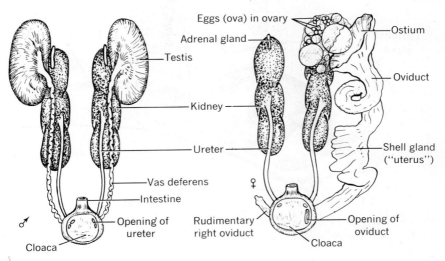

FIG. 6.7. *Urogenital system of Common Pigeon, male (left) and female (right).*
[Drawings by Vito Cangemi.]

Quay (1967) found anal glands widespread in birds (42 families). Their function is unknown, but Schleidt and Shalter (1972) noted that fecal droppings of *Coturnix* were tipped with foam from a cloacal gland and suggested that the conspicuous feces might serve as "markers" to and from the nest.

A special feature in the dorsal wall of the avian cloaca is useful to ornithologists in estimating the age of young birds: a blind pocket, the *bursa Fabricii*, develops in young birds but disappears with age. The larger game birds, such as pheasants (Gower, 1939) and waterfowl (Hochbaum, 1942), can be fairly accurately aged by probing into and measuring the depth of the bursa during their first year of life. In shorebirds this feature can be used for aging birds in the fall, but not in the spring, by which time it has atrophied (McNeil and Burton, 1972). The function of the bursa is unknown, but Ward and D'Cruz (1968) suggest that it might serve as a "cloacal thymus" to produce lymphocytes when the thymus is inactive.

SALT GLANDS

Special salt or nasal glands are found in the supraorbital region of the skull (Fig. 6.8) in birds living in marine and brackish waters. These supplement the kidneys in extracting salt (NaCl). Although birds with functional salt glands have proportionately larger kidneys than those without such glands (Hughes, 1970), the kidneys alone cannot handle high salt concentrations; the liquid oozing from the nostrils of marine

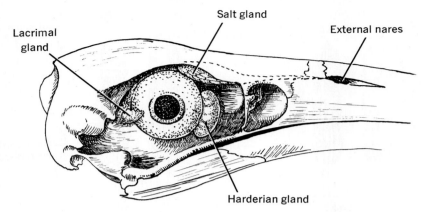

Fɪɢ. 6.8. *Salt gland of Brown Pelican.* [Redrawn by Diane Pierce from Schmidt-Nielson and Fange, *Auk,* 75, 286.]

birds has a far greater NaCl content than sea water. The original work of Schmidt-Nielson and co-workers (see Schmidt-Nielson and Fange, 1958) has been followed by studies on many other species. Not only oceanic birds—penguins, albatrosses, shearwaters, boobies, etc.—are capable of salt extraction by means of salt glands, but gulls, terns, and cormorants along the coasts, sparrows in brackish estuaries, and ducks in alkaline marshes also possess functional salt glands. In a survey of the literature Carpenter and Stafford (1970) found functional salt glands in 11 avian orders. Cade and Greenwald (1966) found functional salt glands in 24 species of falconiform birds, although the glands were much smaller than those of marine birds. The birds exuded clear fluid from the nostrils while eating; the accipitrid species disposed of it by shaking the head, the falconids by sneezing.

Ducks living in alkaline marshes have larger salt glands than those of ducks in fresh waters; in fact, the glands recrudesce and regress in response to environmental needs. Cooch (1964) believes that a functional salt gland has definite survival value for prairie Anatidae, not only by enabling them to live in highly alkaline marshes, but possibly by affording them more resistance to botulism than found in susceptible migrants from the north. Some passerines, such as Savannah, Song, and Sharp-tailed Sparrows, living in brackish water marshes have enlarged salt glands and are capable of some extrarenal salt secretion, a capacity lacking in Song and Savannah Sparrows in other habitats.

Many marine birds get their heavy salt load from feeding on squids and other marine invertebrates. Watson and Divoky (1971) found that an albatross living largely on fish instead of squids had "a remarkably small salt gland"; fish have one-half the salt load of squids.

WATER ECONOMY AND BALANCE

Birds obtain water from three sources: (1) surface water by drinking, (2) preformed water in the food they eat (e.g., insects and succulent vegetation), and (3) oxidative or metabolic water produced internally by oxidation of organic compounds containing hydrogen (Bartholomew and Cade, 1963). Probably no active bird in the wild can survive for long on metabolic water alone (a few can in captivity), but budgerygahs *(Melopsittacus undulatus)* in Australia (Cade and Dybas, 1962) and Black-throated Sparrows in the southwest (Smyth and Bartholomew, 1966) approach this capability. Nearly all young birds in the nest, and many desert species, can and do survive on preformed and metabolic water from their food, but other birds depend on drinking surface water. Carnivorous birds also can usually get along with preformed water from their prey. But rising temperatures rapidly increase evaporative water loss by respiration, necessitating increased water consumption at higher temperatures.

Water loss in urine from the kidneys is minimized by resorption in the collecting tubules in the kidney and walls of the cloaca. Johnson and Mugaas (1970) have shown that "water conservers" have many more tubules and medullary lobules and much greater urine-concentrating ability than nonconservers. Water economy is essential to birds, and this is achieved in different ways.

Birds subsisting on a dry diet of seeds usually require some intake of water, or a special ability to utilize the small amount of water in seeds, an ability most nearly perfected in Australian parakeets, which can go without surface water indefinitely, at least in captivity, by reducing their metabolism, remaining relatively quiet for periods, and conserving the limited water they get from food (Cade and Dybas, 1962). Other desert birds—and probably active parakeets in the wild—visit water holes once or twice a day, or else subsist on a diet containing more moisture than seeds.

Most American desert dwellers are not as efficient water conservers as the Australian ones, probably because the American deserts are of more recent origin and the birds have not had as much time to perfect water economy mechanisms (Bartholomew and Cade, 1963). However, most, but not all, desert birds have devised special methods of obtaining and conserving water. Doves are comparatively poor conservers, but can go long distances to water holes and drink large quantities at one time. California Quail in semidesert situations can exist without surface water when they are feeding on succulent vegetation and insects, but have to have water on a dry diet (Bartholomew and McMillen, 1961). The Black-throated Sparrow seems to be better adapted to desert life than other small North American seed-eating birds (Smyth and Bartholomew, 1966). In the laboratory they can go without water, even on a seed diet, but in the wild they eat some insects and plant material and inhabit the cooler slopes not exposed to the sun. Two races of Savannah Sparrows and two of Song Sparrows in coastal marshes are equally efficient water economizers, but have a better means of processing NaCl (page 162). Sage Sparrows in the desert survive by feeding on succulent vegetation and insects; in laboratory tests they lost weight and died in eight days on a dry diet (Moldenhauer and Wiens, 1970).

Perhaps the ultimate in utilization of water is found in sandgrouse in Africa. The males visit distant waterholes, soak up water in specially constructed water-absorbing abdominal feathers, transport it over miles of dry desert, and present it to their young who "strip" the feathers of the water content (Cade and Maclean, 1967).

Reproductive Structures

As is evident from Fig. 6.7, the chief reproductive structures in the male are the paired *testes,* two oval whitish organs that lie near the anterior end of the kidneys, and the much-coiled tubes, *vasa deferentia,* which conduct the sperm produced in the testes to the cloaca. In many birds the enlarged posterior portion of these tubes serves as a seminal vesicle or sac for the temporary storage—and possibly maturation (Middleton, 1972)—of sperm preceding breeding, sometimes causing a cloacal protuberance that makes sex determination possible. The testes of birds enlarge (*recrudesce*) several hundred times during the breeding season, then shrink (*regress*) after breeding. Often, but not invariably, the left testis is larger than the right. The small size of the gonads in fall and winter birds often makes sex determination by interanal examination difficult, which has led to many unsexed or wrongly sexed speciments in museum collections.

Transfer of the sperm to the female during breeding is accomplished by direct cloacal contact during copulation, when the male mounts the back of the female momentarily so that their cloacas are closely appressed and the transfer effected quickly. There is no intromittent or copulatory organ (*penis*) in most birds, but in waterfowl and in ostrichlike birds a special erectile structure on the cloacal wall serves as a penis. Perhaps this is an additional precautionary device in ducks, which often copulate on or under the water. Copulation is frequent during the early part of the breeding season (page 240), just prior to and during the egg-laying period, but usually wanes or ceases altogether during the later phases of the nesting cycle.

Reproductive structures in the female (Figure 6.7) are a little more elaborate, as might be expected, for the production of eggs. Though there are potentially two *ovaries*, normally only the left develops. If this is lost, however, by disease or experimental ovariectomy, the right gonad may develop, usually into an ovotestis, or more rarely into a functional testis (page 185). Raptorial birds (hawks and owls) frequently have paired ovaries, both apparently capable of producing ova (Snyder, 1948). But White (1969) found only 5 of 17 (30 percent) Peregrine Falcons he examined had paired ovaries, and there was some question as to whether or not the second ovary was functional. Like the testes, the ovaries regress after the nesting season.

Eggs (*ova*) produced by the ovary are passed into the *oviduct* where they are fertilized, if fertilization occurs (hens regularly and wild birds occasionally lay unfertilized eggs). The egg follicle (mainly yolk) then receives a thick enveloping layer of albumen from glands in the middle portion of the oviduct, and then the shell is added from glands in the lower or expanded "uterine" (*shell gland*) portion of the tract. Here also pigments are added in the many birds that lay colored or speckled eggs. The importance of calcium for shell formation is discussed on pages 123 and 458, and the importance of egg-white proteins in determining genetic relationships on page 34.

The completed egg—yolk, albumen, shell, and pigments—is then ready for deposition (Fig. 6.9). Usually an egg a day (or less frequently in the larger birds) is laid until the clutch, or set of eggs, is complete. Wild birds exercise an unexplained control over egg deposition (page 272), laying only one, two, or a dozen, as required for the clutch characteristic of the species; yet if the nest is destroyed, another clutch is produced.

The development of a chick within the egg is usually dealt with in considerable detail in embryology courses and need be summarized only briefly here. The egg, composed mainly of storage materials, contains a thin disc of protoplasm (*blastodisc* or *germinal vesicle*) on the upper surface of the yolk, which is constantly kept uppermost by its lesser weight when the egg is turned, producing the characteristic twisting in the ropelike *chalaza* (Fig. 6.9). The egg and sperm nuclei unite soon after extrusion of the egg from the ovary, and segmentation or cleavage is initiated within the warm reproductive tract before the egg is laid, but is then arrested until incubation begins, which may be several days to a week or more after oviposition. In a few cases (American Goldfinch, Song Sparrow, and Brown Thrasher) twin embryos have been reported to hatch from single eggs (Berger, 1953b).

Development proceeds rapidly with incubation. Within 24 hours in the chick, forerunners of the nervous system, digestive tract, and vascular system have appeared. Before 36 hours the heart begins beating and food is brought to the embryo by *vitelline veins* from the yolk. By 48

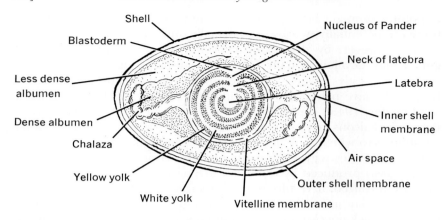

Shell
Blastoderm
Nucleus of Pander
Neck of latebra
Less dense albumen
Latebra
Dense albumen
Inner shell membrane
Chalaza
Air space
Yellow yolk
Outer shell membrane
White yolk
Vitelline membrane

FIG. 6.9. *Structure of an egg.* [Redrawn by Diane Pierce from Walter and Sayles, *Biology of the Vertebrates*, Macmillan Publishing Co., Inc., New York, 1949.]

hours many of the organs—eye, ear, aortic arches, and lobes of the brain—are in evidence. Respiration—intake of oxygen and elimination of gaseous wastes—takes place through the porous shell. Near the end of the incubation period (about 21 days in the domestic fowl) the chick almost completely fills the shell, having utilized all of the egg white and all of the yolk except that remaining in its swollen abdomen (yolk sac) for sustenance immediately after hatching. It then thrusts its beak into the air chamber of the egg for its first breath of air and gradually pips its way out of the shell by means of the egg tooth on its bill (page 282).

SELECTED REFERENCES

AMES, P. L. The morphology of the syrinx in passerine birds. Bull. 37, Peabody Mus. Nat. Hist., Yale Univ. vi + 194 pages.

BALDWIN, S. P., AND S. C. KENDEIGH. 1932. Physiology of the temperature of birds. Sci. Publ. Cleveland Mus. Nat. Hist., 3:1–196.

GLENNY, F. H. 1955. Modifications of pattern in the aortic arch system of birds and their phylogenetic significance. Proc. U. S. Nat'l. Mus., 104:525–621.

HESS, G. 1951. *The Bird: Its Life and Structure*. Translated by P. Barclay-Smith. New York: Greenberg.

ODUM, E. P. 1941. Variations in the heart rate of birds: a study in physiological ecology. Ecol. Mono., 3:299–326.

ROMANOFF, A., and A. ROMANOFF. 1949. *The Avian Egg*. New York: Wiley.

SALT, G. W., and E. ZEUTHEN. 1960. The respiratory system. Chapter X in A. J. Marshall's *Biology and Comparative Physiology of Birds*. New York: Academic.

SIMMONS, J. R. 1960. The blood vascular system. Chapter IX in A. J. Marshall's *Biology and Comparative Physiology of Birds*. New York: Academic.

STURKIE, P. D. 1965. *Avian Physiology*. Ithaca, N. Y.: Cornell Univ. Press.

VAN TYNE, J., and A. J. BERGER. 1959. *Fundamentals of Ornithology*. New York: Wiley. Chapter 2.

WETMORE, A. 1921. A study of the body temperature of birds. Smithsonian Misc. Coll., 72:1–52.

7

Coordination:
The Brain, Sense Organs,
and Hormones

T HE BRAIN, the nerves, the sense organs, and the secretions (hormones) of the endocrine glands play a fundamental role in coordinating all the activities of a bird. This is the subject matter of this chapter. Details of the nervous sytem per se are not considered here, but the brain, sense organs, and hormones in birds are described, primarily from a functional, rather than a morphological, point of view.

The Brain

The avian brain (Fig. 7.1) is similar in overall structure to that found in other vertebrates, but differs in several important aspects. The *olfactory lobes* are small, in correlation with the poorly developed sense of smell in birds; the *optic lobes* are large, in keeping with the well-known visual acuity of the avian eye; and the expansive *cerebrum,* though showing considerable advance over lower vertebrates, is relatively smooth and lacks the deep fissures characteristic of the mammalian brain. These features result in the highly stereotyped, yet selectively efficient, behavior patterns characteristic of birds. The *cerebellum,* which is responsible for precise control over movements, is well developed. The whole brain is packed into a comparatively small space in the cranium, crowded posteriorly by the extremely large orbits whose medial margins are separated only by a thin interorbital plate. Posterior crowding is carried to the extreme in some of the shorebirds; in a snipe the cerebral axis is actually vertical or at right angles to the bill, compared to the horizontal position of the brain in a smaller-eyed cormorant (Portmann and Stingelin, 1961), and in the American Woodcock (Cobb, 1959) the brain is tilted 117° so that the floor of the cerebellum faces upward!

The Grabers (1965) measured the postnatal growth of the brain in

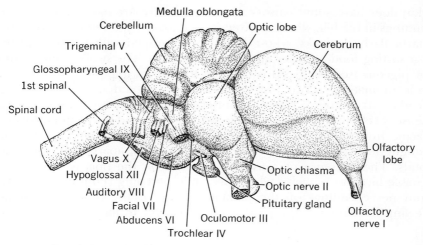

Medulla oblongata
Cerebellum
Optic lobe
Trigeminal V
Cerebrum
Glossopharyngeal IX
1st spinal
Spinal cord
Olfactory lobe
Vagus X
Optic chiasma
Hypoglossal XII
Optic nerve II
Auditory VIII
Facial VII
Pituitary gland
Olfactory nerve I
Abducens VI
Oculomotor III
Trochlear IV

FIG. 7.1. *Brain of Common Pigeon.* [Drawing by R. M. Naik.]

young birds and found larger, heavier brains in immature birds before ossification of the skull was complete; the subsequent decline was due to dehydration. Immature warblers and thrushes collected at TV towers were still adding water, lipids, and proteins to the brain, although they were hundreds of miles from their breeding grounds.

The avian brain reaches its highest development in the corvids, a group displaying unusual adaptive resourcefulness, but it is also exceptionally large in the parrots, owls, and woodpeckers. The importance of the *pituitary,* a glandular structure suspended from the floor of the brain, is discussed under endocrine glands.

The *pineal body* or *epiphysis,* a slender stalk protruding from the roof of the *diencephalon* (between-brain) but largely concealed by the large cerebrum, has long been a mystery to zoologists. However, Salyer and Wolfson (1968) found that it stimulated gonadal growth in female Japanese Quail though not in males; pinealectomy caused a delay in ovarian development. Gaston and Menaker (1968) found it essential for the circadian (24-hour daily light cycle) locomotor activity of birds exposed to light–dark cycles. Hence, it is believed to serve as a biological clock. Other experiments, however, have often produced negative results. (See Quay, 1972, for a long, but incomplete, bibliography).

The Sense Organs

TACTILE ORGANS

Tactile senses are not well developed in birds, largely because of the nearly complete investment of the body with feathers, although the skin does have some sensory receptors. There are also sensory nerve endings in the bill, which so often is used as an exploratory instrument, particularly in ducks, whose beaks are equipped with sensitive plates for sorting food particles out of bottom debris. The swollen operculum of pigeons (see Fig. 6.1d) is believed to play an important stimulatory role in mating. Rictal bristles about the bill and other protruding feather structures, while in themselves without nerve endings, have sensory cells in the feather follicle at their base, and thus can perform a tactile function when they come in contact with another object. The tongue in many birds also has tactile receptors, apart from the taste buds. The presence of sensory corpuscles on the tibia is thought to enable birds to perceive vibrations while perching (Pumphrey, 1961), and nerve endings on the plantar surface of the foot afford some feeling in perching, grasping, scratching, and in locomotor activities.

SMELL

The sense of smell is apparently of minor importance to most birds. Structural proof of this lies in the poor development of the *turbinal* bones in the nasal passages, which are usually relatively bare instead of being covered with sensory epithelium as in mammals. The external nares (see Fig. 6.1) in most birds open into a horny beak and lack the sensitive muzzle so useful for deer, for example, in testing the wind, or for a carnivore in following a trail. In the adults of some pelecaniform birds (e.g., gannets, boobies, cormorants) the external nares are practically obliterated.

Dissection of the olfactory apparatus (turbinals, olfactory tract, and lobes) of representative avian types reveals the poorest development in the more advanced terrestrial birds and somewhat better development in aquatic species. Such evidence suggests phylogenetic degeneration, perhaps correlated with increased specialization of other senses, such as vision. Some exceptions to the generalization of poor olfactory sense occur in the kiwis, procellariiform birds, and some vultures. Kiwis (Fig. 7.2), unlike most other birds, have poor vision and feed chiefly at night, groping about in the semidarkness and sniffing the ground for concealed prey. Tests conducted with earthworms buried in buckets of sand have disclosed a kiwi's uncanny ability to ferret out hidden worms, probably by smelling them, as it ignored buckets containing no worms. However, the experiments did not rule out the possibility that hearing was involved.

Bang (1966) concluded from a study of the nasal anatomy of eight representatives of procellariiform birds that olfaction functioned in the long-range identification of musk-scented food masses and short-range discrimination of specific preference items. In a later examination of

FIG. 7.2 *Kiwi* (Apteryx). *Three species of this odd creature survive in small numbers in New Zealand. Note the hairlike plumage, nearly wingless condition, and odd feet and bill. Kiwis are nocturnal, nest in burrows, and — unlike most birds — have poor vision but keen smell.* [Photo by Philip G. Coleman; courtesy of Mich. State Univ. Mus.]

the olfactory organs of 108 species of birds, Bang and Cobb (1968) found the best development in kiwis, procellariiform birds, Turkey Vultures, most water birds, marsh dwellers, and, perhaps surprisingly, in Hoatzins and swifts.

Past tests with vultures have been contradictory; some seemed to indicate that these birds were unable to detect concealed carrion by smell and used visual clues in locating it. Other tests indicated detection by smell. However, Lewis (1928) pointed out that Turkey Vultures have better olfactory equipment that Black Vultures and that this is more useful to them in warm weather than in cold.

In the early 1960's Stager (1964), in a thorough review of the whole matter, found that the Turkey Vulture had good olfactory equipment, which it used to detect food at close range, and that the Black Vulture did not. Walter (1943), after seemingly exhaustive tests with other birds, concluded that none of his avian subjects had any appreciable sense of smell, but Frings and Boyd (1952) and Hamrum (1953) concluded from feeding experiments that Bobwhites were guided in their selection of food by olfactory stimulation.

TASTE

Taste discrimination, often correlated with and sometimes inseparable from smell, is also quite variable in birds. We have seen from the scarcity of taste buds in the mouth (page 118) that the structural basis for tasting is relatively poor, and we know from general observations that most birds bolt down hurriedly the foods they are accustomed to eat, without much experimental sampling for palatability. The more adaptable birds, particularly those with somewhat omnivorous diets, do experiment at times with new or questionable foods. Birds at feeding stations are commonly observed to sample and then accept or reject unfamiliar items, although it is not certain just what part taste plays in their selections. Bené (1947) believed that hummingbirds at feeding vials could discriminate between different concentrations of sweets and ignored, after one taste, those below a certain standard. In the wild they feed at selected flowers, although factors other than taste—color, flower stucture, and the nectar supply—may influence their choice.

Davison (1962), experimenting with a wide variety of dyed and natural foods at feeding stations, concluded that selection was based on taste rather than color. Doves ate blue and green seeds readily, if their taste was not affected. They ate yellow corn, but not yellow soybeans. Of more than 150 kinds of seeds offered to the doves, he (and apparently the doves) considered 56 "choice" and 20 "fair"; the others were sampled and often refused. There were species differences also. Waxwings ate mistletoe berries (a part of their natural diet), but thrushes did not.

Insectivorous birds may have a greater need for taste than seed-eaters, as there are some distasteful insects. A young Gray-cheeked Thrush kept in captivity for a year (Wallace, 1939) became very choosey about his diet. He had decided fruit preferences, preferred steak to hamburg, and would not eat the hamburg if it was stale. When very young he learned that certain sawfly larvae (*Pristiphora geniculata*) were distasteful; he would eat them only when nothing else was offered, then shook them violently and gulped them down quickly, blinking his eyes and erecting the feathers on top of his head as the worms passed down. Nestlings, when fed these worms by their parents, commonly spit them out.

Carrion-feeders must have little or no sense of taste; at least the "ripeness" of a carcass seems not to perturb them. Many birds (several hundred neotropical species) appear to relish ants, possibly having an alkalizing digestive substance to neutralize the formic acid in the insects; yet some species will not eat them. The young thrush referred to above was apparently able to distinguish poisonous insects, for when rose chafers (*Macrodactylus*) were given him (by mistake), they were sampled, then rejected. Young pheasants in rearing pens sometimes die of rose-chafer poisoning, but adults will not eat the insects.

It has been known for nearly a hundred years that certain butterflies (monarchs and allies) are distasteful to birds. Recent studies (Brower et al., 1968; Reichstein et al., 1968) show that milkweeds in the genus *Asclepias* produce heart poisons (cardiac glycosides), which, when force-fed to Blue Jays, caused vomiting and the loss of previous meals. Some milkweeds, however, are much more emetic than others. Some species of birds seem to be immune to poisonous and distasteful insects. An Old World babbler (Timaliidae) in Africa, for instance, eats milkweed beetles and blister beetles with impunity, even seeming to prefer them. Shrikes also store up such insects.

Thus some birds seem to have a fairly discriminating sense of taste, others little or none, while most birds have a high capacity for selectivity that may be governed in part by taste.

Vision

The eye of a bird is a remarkable structure with great visual acuity, which is vital for creatures so dependent on keen sight for existence. Figure 7.3 shows the main features of the avian eye. Structurally a bird's eye is quite comparable to the human eye, but it differs in relative size, shape, and degree of development of various parts. It is notably large, for in some hawks and owls it is fully as large as the human eye, though largely concealed by lids. Rather than being strictly spherical, as in most mammals, it is flattened in most songbirds and swans, is somewhat globose with a bulging *cornea* and *lens* in the diurnal birds of prey, and is nearly tubular in the owls (Fig. 7.3c). A ring of bony *sclerotic plates* holds the eye in place, in most birds, and imparts a forward bulge to the cornea. A special muscle (Crampton's) controls the curvature of the cornea, and other muscles change the shape of the relatively soft lens, thus effecting a double means of quick accommodation, or instantaneous focus from far to near. In diving birds, Crampton's muscle is degenerate, suggesting that the shape of the cornea is not important in underwater vision, but other structures permit extreme flattening of the lens for close views under water (Pumphrey, 1961).

The *retina* of a bird, the image-forming tissue completely lining the posterior cavity of the eye, is elaborately developed. It is nearly twice as

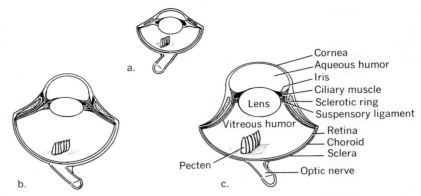

FIG. 7.3. *Structure and shape of birds' eyes.* a. *Flattened type found in swans and most songbirds.* b. *Globose, with bulging cornea, as in diurnal birds of prey.* c. *Tubular, as in owls.* [Drawings by Vito Cangemi; adapted from Walls, Cranbrook Inst. Sci. Bull. 19, 1942 (after Soemmering).]

thick as in the human eye and densely packed with minute visual cells, the *rods* and *cones.* The rods are light-sensitive cells operating chiefly at low light intensities (for night vision); the cones function at high light intensities for color discrimination and for forming sharp images. In centers of concentration these visual cells are very abundant, as many as 1,000,000 per square millimeter in some falconiform birds. All diurnal birds and most nocturnal birds have at least one sensitive spot or *fovea* on the retina for focusing sharply on an object, and many birds have two. One of these, the central fovea, is near the center of the retina for sharpened lateral (monocular) views; the other, the temporal fovea, is toward the posterior margin of the chamber for looking forward (binocularly). The former is sometimes called the "search" fovea for picking up items from the ground or foliage, the other the "pursuit" fovea for pursuing flying insects or moving prey, or, in the case of kingfishers, for underwater vision (Milne and Milne, 1950; Smith, 1945; Walls, 1942).

Most birds, obviously, depend mainly on *monocular* vision; that is, each eye is used independently for lateral views. Owls, however, have their eyes directed forward and are believed to use *binocular* vision only. There is great variation, however, in the actual position of the eyes, and most birds, other than owls, are capable of both monocular and binocular vision. Songbirds, pigeons, and waterfowl, for example, with laterally directed eyes used chiefly for scanning foliage, the ground, or the water, have a wide angle of monocular vision, nearly half a circle on each side, but only a narrow zone, ranging roughly

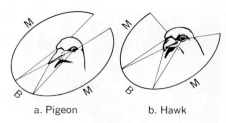

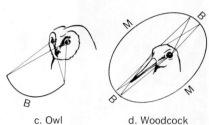

FIG. 7.4. *Range of monocular (M) and binocular (B) vision.* [Drawings by Edward Bradbury; reproduced by permission from Stuart Smith, *How to Study Birds*, copyright 1945 by Wm. Collins Sons and Company Ltd., London.]

from 10° to 25°, of binocular vision ahead (Fig. 7.4a). Hawks, however, with eyes set more nearly forward, have a smaller range of monocular vision and cannot see backward without turning the head, but have a considerably wider field of binocular vision (35° to 50°) in front (Fig. 7.4b). Owls, as already mentioned, see only forward, but over the widest binocular field found in birds (Fig. 7.4c). Their restricted field of view to the side is compensated for by swift movements of the head, which can be turned completely backward, 180° or more.

The American Woodcock is noted for its curious backward-directed eyes; apparently it is capable of binocular vision *both* forward and backward (or upward when its bill is plunged in the ground) as well as having a monocular view over nearly half a circle on each side (Fig. 7.4d). The bitterns, to cite another peculiarity, have "low-slung" eyes, and in "freezing" they can point the beak upward and look forward binocularly for danger (Fig. 7.5).

Monocular vision, however, is not without some disadvantage. It gives a poor conception of distance until the eye is brought to bear directly upon an object. Hence we see a bird cock its head, like a robin looking for a worm, to focus with one eye upon an object before snatching it. Offsetting this disadvantage, however, are the advantages of long-range vision, extraordinarily quick accommodation from telescopic to microsopic views, a wide field of vision in all directions, and, in nocturnal birds, a specialized ability to make use of dim light. According to some authors, the visual acuity of some hawks, on the basis of retinal structure of rods and cones, is judged to be eight times that of man, and an owl's capacity of seeing in dim light is about ten times

FIG. 7.5. *Three Least Bitterns "freeze" in the nest and look forward binocularly for danger, a feat made possible by their exceptionally low-slung eyes.* [Photo by Edward M. Brigham, Jr.]

that of the human eye. But Pumphrey (1961) warns against trying to compare vision in birds with that in man and, except for quickness of accommodation and the ability to survey a large field rapidly, he doubts the alleged superiority of avian vision.

Two other structures in the avian eye are the *nictitating membrane*, a semitransparent third eyelid that can be drawn posteriorly over the surface of the eye, and the *pecten*, a vascular comblike structure jutting out from the retina into the posterior cavity. The nictitating membrane serves mainly for cleansing the eyeball, but it may also function as a protective device when birds are facing or flying against the wind, and to shield the eyes of aquatic birds under water. Hill (1964) doubts the commonly alleged function of the nictitating membrane for cutting down the glare of sunlight for nocturnal birds during the day; a captive Barn Owl never protected its eye from the sun by closing the membrane. Nictitating rates are rapid in birds. Norris (1972) measured the rates in 44 species (257 individuals), mostly passerines, and obtained the highest *average* rates (40.7 times per minute) in Black-capped Chickadees and Red-winged Blackbirds and lowest in a Prothonotary Warbler (it didn't blink). Probably the rates have no phylogenetic or taxonomic significance, since a Tufted Titmouse, currently considered congeneric with chickadees, had the third lowest rate (av. 4.4 times per minute). The pecten is a highly vascular body undoubtedly associated with the nutrition of the hard-worked avian eye; some 30 other possible functions assigned to it are largely speculative, although Menner's theory that it casts a shadow on the retina when a distant object appears, thus facilitating detection of movement, has received considerable support (Pumphrey, 1961).

Various eye defects, or anomalies in development, have been frequently described in poultry and pigeons, but rarely in wild birds, presumably because the defects so often prove lethal. Wetherbee (1958), in the artificial incubation of more than 2,000 eggs of over 100 species of wild birds, found only two teratological specimens: one Common Grackle had no right eye, and a Mockingbird embryo had fused optic capsules (*synophthalmia*). Wallace (1956) reported a young American Robin successfully fledged, in which one eye had no lens and the other was only one-fourth its normal size (*microphthalmia*) with the eyelid sealed shut. Berger and Howard (1968) described a completely eyeless robin (*anophthalmia*) that fledged (normal weight) and was fed by the female at least once after nest leaving. Such anomalies would appear to make self-feeding impossible for most birds, but Lorenz (pers. comm.) says that cases of microphthalmia have appeared in his captive ducks in Germany and that the young are capable of limited self-feeding, presumably by contact with food (tactile sense) rather than vision.

Several substantial lines of evidence denote good color vision in birds. Structural proof rests in the abundance of cones in the retina for perception of colors. The colorful plumages of birds strongly suggest the role these play in species and sex recognition. Some early experiments testing color vision in poultry and pigeons by using colored grains (dyed or flooded with spectral light) showed that the birds invariably ate the grains at the red-orange-yellow end of the spectrum, but hesitated at or ignored those at the blue end, a result that was interpreted, perhaps erroneously, as "blue-blindness." The birds may have been "conditioned" to feed on red-orange-yellow grains and failed to recognize blue grains (nonexistent in nature) as something to eat. Tests by the Fish and Wildlife Service (Kalmbach and Welch, 1946) with colored poisons seemed to offer considerable promise for rodent control because color-blind rodents would eat green poison baits, whereas House Sparrows and blackbirds would not. But other species may react differently; pheasants take any color of repellents used to protect planted corn (Dambach and Leedy, 1948), and nuthatches at feeding stations provided with dyed seeds merely leave the greens and blues till other colors have been removed (see also page 172). Avoidance of the greens by some birds might be because green grains and unripe fruit are not ordinarily eaten in the wild. Kee (1964) found that "unconditioned" domestic fowl and quail (*Coturnix*) actually preferred green-colored food (cracked corn) over red and yellow.

Some ingenious experiments in England suggest that birds with yellow in their plumage may be more conscious of yellow than are brown-plumaged birds (Smith, 1945). Plasticene models of differently colored feces were placed in the nests of two closely related motacillids (the Meadow-pipit and Yellow Wagtail) and the succession in which the birds removed the models was recorded. The dull-colored Meadow-pipit usually removed the duller-colored feces first, whereas the Yellow Wagtail chose the yellow first. Bowerbirds in Australia, which decorate their bowers with petals of flowers, often, but not invariably, choose colors that occur in the plumage of the females they wish to attract. Mc-Cabe (1961) tested the selection of colored nest boxes by House Wrens over an 11-year period and found that red and green were the most preferred colors. In 98 nestings the choices were red 41, green 31, blue 16, yellow 8, and white 2.

The preference of Ruby-throated Hummingbirds for red flowers and for red and yellow feeding cups (Fig. 7.6) at feeding stations seems to be borne out by observation, but Miller and Miller (1971) thought that hummingbirds in Saskatchewan, where red flowers are not common, were influenced more by location of food sources than color, although, once the food was located, color was "an important discriminator stimulus." In regions where red is a common flower color it would facilitate recognizing food sources in migration.

Another feature of the avian eye that may influence color perception is the presence of oil droplets, primarily of red, orange, and yellow, in the retina. Though the exact function of the droplets is uncertain, it has been suggested that they act as filters, much as a camera filter, increasing a bird's perception of red, orange, and yellow, but making blues more dull or colorless. Species differ in the amount as well as in the color and distribution of these droplets; pigeons, for instance, have the yellow droplets concentrated in the lower part of the retina, sup-

FIG. 7.6. *The apparent preference of the Ruby-throated Hummingbird for red has been capitalized on by distributors of feeding vials, who sell more red cups than any other color.* [Photo by H. C. Johnson; from Nat. Audubon Soc.]

posedly for sky vision or toning down blue. Kingfishers have an abundance of red droplets (60 percent compared to about 10 percent in hawks and 20 in songbirds), which may offset the glare of water and increase visibility of objects below the surface. The percentage of red droplets is also high in early-rising passerines, and is believed to be an aid in hunting at early hours, but late-rising hawks and swallows have mainly yellow droplets.

Sensitivity to reds has led to the suggestion that birds may be able to detect infrared rays, invisible to the human eye, and that by this means nestling Wood Ducks, like turtles, can find their way to the nearest water (infrared reflection being different over water). A similar suggestion that infrared rays emanating from an animal's warm body might aid owls in detecting prey in the dark has been quite conclusively disproved by experiment (Dice, 1945). Pumphrey (1961) doubts the ability of birds to "make use of extraspectral frequencies at either end of the spectrum."

HEARING

Birds have always been credited with acute and discriminating hearing ability. The belief that owls hunt largely by ear and that some birds listen for subterranean prey, as well as the logical presumption that the great variety of vocal performances uttered by birds have real meaning to the birds themselves, are among the lines of evidence cited in support of this view.

Basically, the ear of a bird is quite comparable to the hearing organ in other vertebrates, but there are some pertinent structural differences. There is no external *pinna,* as found in mammals, for collecting sound waves. Instead, the external ear opening is covered over by a special group of feathers, the *auriculars,* which are delicately and loosely constructed, thus not interfering with sound waves. Sometimes a bird can be observed to raise its ear coverts, as if listening intently. The various ear tufts and "horns" of certain owls and Horned Larks are probably only incidental to hearing. Owls do, however, have enormous external ear openings beneath the auricular feathers; in some these are asymmetrical—that is, one cavity is larger than the other—probably for the better localization of the direction of sound.

Within the ear cavity is the vibratory *tympanum* or eardrum, which receives and transmits sound to the fluid-filled inner ear (cochlea) by means of a rodlike bone, the *columella.* Allen (1925) suggested that this direct transmission of sound waves (tympanum through columella to cochlear nerve to brain) might be responsible for the observed ability of birds in aviaries in France during World War I to detect distant sounds of battle inaudible to human ears. He also suggested that this remarkable, and now well-authenticated, sensitivity to distant explosions might be due to the quivering of perches, a fact later verified by studies of the sensory end organs in the legs, which apparently perceive earth-transmitted sound waves (Pumphrey, 1961). Beecher (1969) believes that the vanelike surfaces of the columella in the middle ear provide birds with a "motion sense" by intercepting "air shifts," which enable birds to coordinate head movements of peering and bobbing and foot movements such as scratching.

In some pioneer work on auditory responses of birds at Cornell University, Brand and Kellogg (1939a) determined the hearing ranges, in cycles per second, of Common Pigeons, Starlings, and House Sparrows and compared them with man. Subsequent work with other birds at Cornell and in Europe (Schwartzkoff, 1955) extended both the upper and lower known limits of hearing. Schwartzkoff indicated that cochlear potentials as high as 30,000 cycles per second might be reached in certain birds when greater intensities of sound than those normally encountered were used. This is considerably beyond the 20,000 cycle limit

TABLE 7.1. HEARING RANGE OF SELECTED BIRDS

Species	Cycles per Second		Author
	Lower Limit	Upper Limit	
Mallard	< 300	> 8,000	Trainer (in Schwartzkoff)
Canvasback	190	5,200	Edwards, 1943
Sparrow Hawk (Am. Kestrel)	< 300	> 10,000	Trainer (in Schwartzkoff)
Pigeon	200	7,500	Brand and Kellogg, 1939a
Ring-necked Pheasant	250	10,500	Stewart, 1955
Great Horned Owl	60	> 7,000	Edwards, 1943
Long-eared Owl (Europe)	< 100	18,000[*]	Schwartzkoff, 1955
Hairy Woodpecker	34	18,400	Ramp, 1965
Horned Lark	350	7,600	Edwards, 1943
Magpie (Europe)	< 100	21,000[*]	Schwartzkoff, 1955
Common Crow	< 300	> 8,000	Trainer (in Schwartzkoff)
Starling	700	15,000	Brand and Kellogg, 1939a
House Sparrow	675	11,500	Brand and Kellogg, 1939a
Canary	1,100	10,000	Brand and Kellogg, 1939b
Red Crossbill (Europe)		20,000[*]	Knect (in Schwartzkoff)
Snow Bunting	400	7,200	Edwards, 1943
(Man	16	20,000	Schwartzkoff, 1955)

[*] It is possible that supersensitive instruments used in the tests in Europe account for the higher values at upper levels (Frings and Slocum, 1958).

he set for man. Table 7.1 gives the range of hearing for some representative birds and man.

Tests of the songs of passerine birds (Brand, 1938) show that they range from about 1,100 to 10,000 cycles per second, indicating, as might be expected, that birds sing songs well within the the range of their own hearing as well as within the acoustic capacity of most other birds. It seems probable, however, that birds with a low range of hearing cannot hear the high-frequency notes of warblers and kinglets, or the highest notes of the Starling, and that the low-pitched voices of other birds may not be audible to birds whose hearing is defective at the low end of the scale. Ramp (1965) concluded from tests on the hearing range of a Hairy Woodpecker (Table 7.1) that it could probably hear the stridulations and chewing noises of the insects on which it feeds.

Some authors, however, believe that the larger predatory birds cannot hear the high-pitched alarm notes of small songbirds; hence the latter can sound an alarm without exposing themselves to danger. Hailman (1959) suggests that the high notes may not actually be inaudible to the predator but merely difficult to localize as to source, a significant safety factor for the prey. It is also thought that young chicks can hear the low clucking notes of their mother but *not* the high peeps of their siblings, which would be a needless distraction.

Owls have long been credited with a sensitive ear that aids them in hunting, and recent information indicates that this ability may be even more fantastic than formerly supposed. Dice (1945) tested the ability of four species of owls in finding prey (under laboratory conditions) in low light intensities and concluded that light in natural habitats "must often fall below the minimum at which the birds (owls) can see their prey." Later, Payne and Drury (1958) demonstrated that Barn Owls could locate prey almost unfailingly in *total darkness* by hearing alone (Fig. 7.7). Mice released in a darkened room were pounced upon — as soon as the rustling of the mouse ceased (not during its movements), the owl dropped and was successful in 13 out of 17 strikes. The possibility that odor or heat (infrared) was involved was ruled out by dragging a paper "mouse" along the floor; it was attacked like the living mice. When one ear of the owl was plugged, it missed its prey by 18 inches, probably indicating the value of asymmetrical ears in the Barn Owl for localizing sounds exactly.

Another acoustic ability evident in some birds is echolocation. It has long been known that bats can detect supersonic sounds reflected back

FIG. 7.7. *Barn Owls can locate moving prey almost unerringly in total darkness by hearing alone. Vision is useful to them in dim light, but mice rustling in the leaves or grass are detected by sound.* [Photo by A. D. Moore; courtesy of *Jack-Pine Warbler*.]

to them from obstacles in their path, enabling them to dodge through a maze of wires in the dark. Studies on the Oilbird (*Steatornis*), which nests in caves in northern South America and on Trinidad, show that these birds use a type of echolocation to find their way to and from their nests and diurnal roosting places; their calls (not supersonic) are reflected back from walls and other obstacles and picked up in the sensitive ears of the birds (Griffin, 1953). Swifts nesting in subterranean caves in the East Indies use this same device, but Medway and Wells (1969) report that not all swifts in that region are capable of echolocation, possibly having lost the ability in favor of increased vision for nocturnal hunting.

Endocrine Glands

Though the endocrine glands are commonly regarded as constituting a separate system, their products, the hormones, are definitely associated and interrelated with all the other systems. These hormone-bearing secretions of the ductless glands are distributed by the blood to all parts of the body where they act as excitors or inhibitors in regulating and coordinating various life processes such as growth, development, and behavior. Minute amounts produce profound effects; too much or too little of the secretions results in abnormalities. Extracts of many of the endocrines are now widely used in experimental work, and some hormones have been prepared synthetically. This brief account of avian hormones points out some of their influences on bird life.

The more important endocrine glands in birds, together with the hormones they produce and some of their functions, are as follows.

Pituitary ("master" gland)—controls and regulates secretions of other glands.
  Anterior lobe—secretes following hormones:
    Thyrotropin (TSH)—stimulates and influences secretions of the thyroid.
    Adrenocorticotropin (ACTH)—acts on the cortical tissue of the adrenal gland.
    Gonadotropin—controls growth of gonads and their secretions.
    Prolactin—stimulates production of "pigeon's milk," parental behavior.
  Posterior lobe—produces several hormones related to physiological activities (blood pressure, urine flow, egg expulsion).
Thyroid—produces thyroxin, for feather development, pigmentation, molting.
Parathyroid—produce parathormone, for calcium and phosphorus regulation.

Adrenals—produce (1) adrenalin from medulla, for blood pressure, digestive processes; (2) cortin from cortex, for carbohydrate metabolism.

Gonads—produce (1) androgens (testosterone) from testes, secondary sexual characters; (2) estrogens (estradiol) from ovaries, female functions.

From the list of the glands and their secretions, the importance of the *anterior pituitary* is immediately apparent. It stimulates and regulates the activities of the other endocrine organs, which are discussed below. In addition it produces a lactogenic hormone (*prolactin*) that stimulates the production of "pigeon's milk" in pigeons by a sloughing off (desquamation) of cells in the walls of the crop. Prolactin also regulates broodiness, influences various aspects of parental behavior, and, in the presence of estrogen, is responsible for the full development of the brood patch (see pages 272–74) in incubating birds. The absence of broodiness and the failure of brood-patch development in such parasitic birds as certain cuckoos and cowbirds are thought to be due to hereditary failure in prolactin secretion (Hohn, 1961). Selander and Kuich (1963) and Robinson and Warner (1964) found that the abdominal skin of Brown-headed Cowbirds was insensitive to prolactin injections. Injected cowbirds could be induced to initiate but not complete the nesting cycle (Selander and Yang, 1966).

The *posterior pituitary,* less studied in birds, apparently influences a variety of physiological functions. It regulates blood pressure, may cause premature expulsion of eggs, even of soft-shelled eggs, and reduces urine volume, probably by increasing resorption in the kidney (Höhn, 1961).

Another fundamental endocrine gland in birds is the *thyroid,* which is responsible for the proper growth and development of many structures. Via the pituitary, its secretions (chiefly *thyroxin*) control the growth and development of the gonads; deficiencies (*hypothyroidism*) retard and inhibit development of the gonads and males remain immature. An excess (*hyperthyroidism*) accelerates molting, increases barbule formation in feathers, and causes heavy pigmentation, especially deposition of melanins, whereas a deficiency has the opposite effect (inhibits molting, produces barbless, fluffy feathers, and causes loss of pigmentation). The thyroid is also important in the maintenance of proper metabolism and in temperature regulation.

The *parathyroids* in birds are closely associated physically with the thyroid. Their secretion (*parathormone*) regulates the calcium and phosphorus level in the blood and aids in bone formation. As we shall see (page 458), calcium deficiencies caused by DDT have a profound and often disastrous effect on egg-shell formation. Removal of the parathyroids produces convulsions and death.

The avian *adrenals,* small yellow or orange-colored glands situated on the anterior-ventral surface of the kidneys, consist of poorly defined and intermingled cortical and medullary tissue, quite unlike the well-defined outer *cortex* and inner *medulla* in mammals. Their secretions and functions seem somewhat debatable, too, but several cortisonelike secretions from the cortical tissue influence various metabolic processes, such as carbohydrate metabolism and the functions of some internal organs. *Adrenalin,* or an adrenalinlike substance, from the medullary portion also helps regulate metabolism and can raise blood pressure in chickens by constricting the walls of blood vessels. It also raises the blood sugar level and increases heart rate. Emotions like anger and excitability, known to be related to adrenalin flow in mammals, probably do occur in birds, but their relation to a specific hormone is uncertain.

The avian hormones most studied have been those associated with sex. The gonads, in addition to producing sex cells, also produce hormones, primarily an androgen (*testosterone*) and an estrogen (*estradiol*) that, in coordination with other hormones, control the development of the sex organs and the secondary sexual characteristics (male plumage, spurs, and comb), and profoundly influence sexual behavior. In domestic fowl and many other birds, male hormones are also present in the females, for if deprived of ovaries by disease or castration, females develop male characteristics; in rare cases sex reversal goes all the way so that in the common fowl, at least, egg-producing hens have been known to change completely into functional cocks. Sexual plumage reversals are also known in many other birds. Yet in the House Sparrow (and presumably other species), the plumage of the sexes is genetic and neither gonadectomy nor administration of hormones alters it. Apparently in species subject to plumage reversal, a female hormone normally suppresses the influence of the male hormone, and when the effect of the former is removed the latter asserts itself. Castration in the male, however, usually produces only minor effects, if any, on the male plumage, though the development of some secondary sexual characters is arrested and male behavior largely nullified. Thus administration of testosterone to females, in some cases at least, produces profound effects—hens become cocks and female Canaries sing—but attempts to feminize males with estrogen are only partially successful.

Bill color changes in the breeding season are usually due to hormones; the horn-colored bill of the male House Sparrow becomes blue-black because of increased testosterone, but that of the female remains the same because of the suppressive effect of estrogen. Mundinger (1972) has shown experimentally that the dark bill color in young American Goldfinches and in winter adults is caused by testosterone withdrawal and an increase in melanin deposition, and that the bright yellow bill color attained in spring is caused by both melanin withdrawal and deposition of carotenoid pigments. Similarly the dusky brown or blackish bill of the Starling in winter becomes a bright yellow in spring.

Sex reversal in phalaropes is well known, the females having the brighter colors. Apparently the male hormone testosterone is at a high level in females prior to the breeding season and low in males, and prolactin, responsible for brood-patch formation, is high in incubating males and low in females (Johns, 1964). Injection of testosterone into the duller colored males produced bright feathers.

SELECTED REFERENCES

ALLEN, G. M. 1925 *Birds and Their Attributes*. Francestown, N.H.: Marshall Jones. Chapter X.

HOHN, E. O. 1961. Endocrine glands, thymus, and pineal body. Chapter XVI in A. J. Marshall's *Biology and Comparative Physiology of Birds*. New York: Academic.

PETTINGILL, O. S., JR. 1970. *Ornithology in Laboratory and Field*. Minneapolis, Minn.: Burgess. Pages 91–96, 122–33.

PORTMANN, A. 1961. Sensory organs: skin, taste, and olfaction. Chapter XIV in A. J. Marshall's *Biology and Comparative Physiology of Birds*. New York: Academic.

———— and W. STINGELIN. 1961. The central nervous system. Chapter XIII in A. J. Marshall's *Biology and Comparative Physiology of Birds*. New York: Academic.

PUMPHREY, R. J. 1961. Sensory organs: vision and hearing. Chapter XV in A. J. Marshall's *Biology and Comparative Physiology of Birds*. New York: Academic.

SMITH, S. 1945. *How to Study Birds*. London: Collins. Chapters 8 and 9.

STAGER, K. E. 1964. The role of olfaction in food location by the Turkey Vulture, *Cathartes aura* (Linnaeus). Los Angeles Co. Mus., Contr. in Sci., No. 81. 63 pages.

WALLS, G. L. 1942. The vertebrate eye and its adaptive radiation. Cranbrook Inst. Sci. Bull. 19. Bloomfield Hills, Mich.

WALTER, W. G. 1943. Some experiments on the sense of smell in birds. Arch. Nierland. Physiol. 27:1–72.

WELTY, J. C. 1962. *The Life of Birds*. Philadelphia and London: Saunders. Chapter 5.

8

Behavior

REVISED BY HAROLD D. MAHAN

T HE PAST three decades have witnessed a major emphasis on detailing the behavior of birds in both field and laboratory. This work been of great importance not only to ornithologists but also to those who study human behavior. Two of the 1973 Nobel Prizes in medicine/physiology, in fact, were awarded to animal behaviorists (Niko Tinbergen and Konrad Lorenz) for work in which birds were used as principal subjects. In ornithology the study of bird behavior has great ramifications with regard to a bird's anatomy, physiology, ecology, life style, and evolution. During the past 20 years, in fact, many taxonomic changes have been made as a result of a critical examination of behavior (Table 8.1).

Definitions and Concepts

As with any new science, the rapid development of new concepts in animal behavior in general, and bird behavior in particular, has led to a vast and confusing vocabulary. The best summary of these concepts and terms as applied to birds has been given most recently by Thorpe (1963), Hinde (1966), and Ficken and Ficken (1966). Also of interest is Crook's 1970 summary of social organization and its relationship to individual behavior and environment.

The earliest studies in animal behavior simply attempted to explain behavior using *stimulus-response* terminology. The stimulus (internal or external, environmental or social) was responded to by a set pattern of behavior with a certain degree of variability. Later, Scott (1958) suggested that animal behavior studies should portray the adaptive significance of behavior, and should be "concerned with the activity of the whole organism and of groups of organisms." He felt that such studies should involve anatomical, physiological, ecological, genetic, and even embryological and taxonomic approaches. His categories of behavior that have adaptive significance include ingestive, shelter-seeking, agonistic, sexual, care-giving (*epimeletic*), care-soliciting (*et-epimeletic*), eliminative, and contagious (*allelomimetic*) behavior.

TABLE 8.1. EXAMPLES OF THE USE OF BEHAVIORAL
CHARACTERS IN TAXONOMY*

Activity	Level Used	Taxonomic Discrimination
Method of head scratching	Familial	*Recurvirostra, Himantopus,* and *Haematopus* related to charadriids rather than scolopacids
	Subfamilial	Suggest Psittacinae may be polyphyletic
Holding food with the foot	Familial	Separates *Icteria virens* from Parulidae
Dust bathing	Subfamilial	Supports relationship of *Passer* to ploceids
Water bathing and method of oiling feathers	Familial	Separates timaliids from most other passerines
Visual agonistic displays	Generic	Splits *Hylocichla mustelina* from *Catharus* spp.
Flight call notes	Subfamilial	Suggest relationship of *Fringilla* and carduelines
Form of courtship display	Subfamilial	Shows affinity between *Fringilla* and carduelines
	Interfamilial	Presence of tail quivering suggests possible relationship among corvids, estrildines, and ploceids
Nest construction	Ordinal	Evidence for relationship between hummingbirds and swifts
Participation of sexes in parental care	Subfamilial	Separates *Anseranas semipalmata* from other Anatidae
Feeding of nestlings	Subfamilial	Separates estrildines from ploceids
Nest sanitation	Familial	Separates *Peucedramus* from Parulidae

* Condensed from Ficken and Ficken, 1966, which should be consulted for the references.

More recent behavior studies, however, made mostly by European behaviorists (who named this new science *ethology*), have attempted to express behavior more precisely from a functional (what it does for the animal) as well as from a causal (the forces that cause it to occur) standpoint. For a history of the study of animal behavior and ethology the student should consult Armstrong (1947), Lorenz (1950), Tinbergen (1951), Thorpe (1963), and Etkin (1964).

Ethologists, who concentrate mostly on *instinctive* (as opposed to *learned*) behavior, attempt to categorize each single act that the animal performs as well as to explain the evolutionary importance of each act. Thorpe (1951) and Tinbergen (1951) redefined the concept (first proposed by Craig in 1918) of a two-phase behavioral sequence that applies to many such acts. The first phase, *appetitive behavior,* stems from an inner *drive* (which may be analogous to *stimuli* or *tendency to perform*) that leads to a *consummatory act.* For instance, Tinbergen describes the hunting behavior of a Peregrine Falcon as consisting of hunger (inner drive), random hunting, location of prey, pursuit, capture, kill, plucking, and eating, as a "chain of consummatory acts."

Lorenz (1950) and Thorpe (1951) developed the concept of *fixed action patterns,* or *stereotype* behavior of an inherited type. Hinde (1959) used this concept to define *instinctive* behavior as those stereotyped actions that are inherited and hence are species specific. Lorenz also theorized that there is an internal neural pathway (*innate releasing mechanism* or IRM) that, according to Tinbergen and Perceck (1951), is "selectively responsive to a specific stimulus situation, even though that particular stimulus has never been encountered before." A related concept is that within the environments of birds and other animals there are certain kinds of *releasers* that evoke *sign stimuli* for the release of instinctive behavior. The red spot on the lower mandible of an adult Herring Gull almost invariably serves as a sign stimulus to release the pecking response in chicks of this species. Tinbergen (1953) showed that the red spot alone was most important to release such behavior, and that the body, head, and/or bill were not as significant as the red spot from the standpoint of eliciting this particular response. Sign stimuli may have not only a visual nature but also an auditory configuration in birds. There are even cases where configuration of the sign stimuli involves size or numerical quantities. Tinbergen's (1951) work with oystercatchers, for example, indicates that this species prefers a clutch of five eggs to its normal three and prefers to incubate larger eggs than it normally lays. Such sign stimuli have been termed *supernormal.* Occasionally, also, two conflicting sign stimuli are simultaneously presented and may result in *ambivalent* behavior. For example, incubating Herring Gulls have a tendency to sit on egg-shaped objects and to reject red objects. A red-colored, egg-shaped object will be removed from the nest, sat on outside the nest, and then moved again. Also, there are some sign stimuli situations that are subnormal from a quantitative standpoint and lead only to *intentional movements* rather than to the complete normal response.

LEARNING VS. INNATE BEHAVIOR

Unlike instinctive behavior, *learning* in birds and in other animals depends on what they experience in their environment. Although there is still some controversy concerning the differences between instinctive (innate) and learned (acquired) behavior, most workers agree that only a fine line separates the two concepts in many cases. For example, some controversy exists as to whether or not young birds instinctively sing the song of their species or acquire it by imitation of adults, but now many studies indicate that both inheritance and learning are involved. Certainly a crow will caw and a duck quack whether or not it has ever heard its kind. But it is likewise true that elaborate songs require practice and imitation before reaching perfection. Wild songbirds reared in captivity often sing initial songs that are only partially suggestive of their species. A hand-reared Gray-cheeked Thrush, which had probably never heard the songs of the adults, sang a thrushlike, but decidely off-tune, song all winter. However, when taken back to its native haunts in the spring, it learned to sing the song characteristic of the species (Wallace, 1939). Lanyon (1957) reports that the primary (territorial) song of meadowlarks (both Eastern and Western) is learned rather than inherited, since juveniles acquire the songs of other species with which they are associated. Nice (1943) divided the song-learning process of Song Sparrows into five stages before the full adult song was perfected. Many other experiments with hand-reared birds indicate that calls and simple songs are instinctive but that more complex songs require some practice and learning before perfection is attained.

Learning apparently is involved, too, in the case of a young bird recognizing its parent's voice. Beer (1970) found that Laughing Gull (*Larus atricilla*) chicks gradually learned the individual characteristics of their parent's calls during the first six days after hatching. Belcher and Thompson (1969), however, found that young Indigo Buntings apparently do not learn to recognize their species song until after they have left the nest.

Some learning, of course, takes place prior to hatching. Welty (1962) refers to the cessation of peeping in grouse chicks before hatching after the incubating female gives a distress call. Oppenheim's study (1970) of embryonic behavior in Pekin ducks indicated a rapid increase in coordination of behavior after 20 days of embryonic life in that species.

At hatching a not uncommon learned behavior is *imprinting*. This is a learning process in which young birds, often within a few hours of hatching in precocial birds, get impressions that may remain with them throughout life. Thus young birds learn to recognize their parents or foster parents, or become attached to one of the first objects they notice after birth. Incubator chicks often "adopt" the person who feeds or takes care of them, even to the extent of attempting to "mate" with their adopted parent and refusing to accept mates of their own species. Hess (1959), in a study of imprinting in 5,700 birds, observed distinct species differences and classified his subjects from excellent to poor. Canada Geese and Mallards were examples of "excellent" imprinters and Wood Ducks "poor," although some other investigators have obtained better imprinting responses from Wood Ducks. Gottlieb (1965) concluded that Wood Duck and Mallard young do not leave the nesting box or nest until after the proper auditory cue ("following response") is given by adults. Later, Gottlieb (1966) found that chicks and ducklings "can discriminate the maternal call of their species in the absence of specific previous experience" with it. Apparently imprinting has to come early, the capacity for it being lost within a few days of hatching.

Imprint learning differs from *trial-and-error* learning, which starts later in life and teaches the young bird to distinguish edible from inedible substances, how to fly, or how to sing. For although feeding, flying, and singing are to some extent inborn characteristics, they can be modified or perfected by trial-and-error learning.

Habituation is a term sometimes applied to the process of learning what *not* to do. This often involves, among other things, enemy recognition. A bird startled by a harmless butterfly or a dangerous predator soon learns that one is harmless and that the other is not. Starlings become "habituated" to artificial sounds (page 411) designed to frighten them from a roost and cease to react to such disturbances (Fig. 8.1). Habituation has survival value as it teaches the bird what to fear or avoid and what to ignore or accept—it is "uneconomical" for a bird to be alarmed by a harmless object.

F IG. 8.1. *The Starling adjusts quickly to new situations. Sometimes it can be dispersed from roosts by playing sound recordings of Starling alarm notes, but it soon learns that such sounds are harmless and ceases to react to them. The white flecks indicate winter plumage.* [Photo by Allan D. Cruickshank.]

Habituation, like other behavioral acts, responds to selection pressure. For instance, Hailman (1968) mentions that cliff-dwelling kittiwake chicks avoid falling from cliffs by crouching, whereas ground-inhabiting gull chicks run and hide. Crouching in ground-level gulls presumably would not have adaptive value since it would increase the chance of being captured by predators. There can be no trial and error in the case of predators, if the bird is to survive.

In experimental work, animal subjects are often "conditioned" to respond in particular ways (*conditioned reflexes*). Pavlov's (1927) classic experiments with dogs, which salivated at the sound of a buzzer whether food was provided or not, are oft-cited examples of this, but since then many experimental animals have been conditioned to respond, or not to respond, to various visual or auditory signals.

Communication and Displays

Although much of a bird's behavior is a result of inner forces that determine its capacity to respond to external stimuli, the outcome of its acts affects the behavior of other birds (and other animals). Such communications between birds are usually sterotyped forms of behavior termed *displays*.

Displays and their apparent significance have been described for many species. Some recent examples are Storer (1969) for grebes, Johnsgard (1965) for waterfowl, Hamilton (1965) and D. Snow (1968) for hummingbirds, B. Snow (1970, 1972) for cotingids, Slud (1957) and D. Snow (1971a) for manakins (*Pipridae*), Hardy (1961) for jays, and Nero (1956, 1963) for Red-winged Blackbirds. (See Chapter 9 for courtship displays.) Ficken and Ficken (1966) have provided the best current summary of the significance of displays, especially in regard to vocal communication.

Songs and Calls

Songs and calls are one of the most obvious forms of bird communication. Not only do they express a variety of emotions but also they serve a variety of purposes; even the somewhat anthropomorphic idea that birds sing for pleasure is perhaps not entirely out of place. In listening to the amazing repertoire of the Mockingbird one gets the impression that he is really enjoying himself. But Lack (1943) appropriately comments that if birds sing because they are happy, then female English Robins are happy only in the fall, males are more happy before acquiring a mate than after they get one, and are happiest of all when fighting.

Though it is well known that singing is characteristic mainly of male birds, there is ample evidence that the females in many species sing. To be sure, some records of apparent singing females are rightly open to suspicion on the grounds that the singers might have been immature males (e.g., Purple Finch, American Redstart), but there are many cases not open to doubt. Female Cardinals sing quite regularly (the senior author has observed one singing immediately after laying an egg), but they have a shorter season of song than the male. Female Gray-cheeked Thrushes have been observed singing on their nests during incubation, hatching of the eggs, and while brooding young (Wallace, 1939). In the titmice, both sexes sing; the softly whistled "fee-be" of the Black-capped Chickadee is employed by both sexes, but more frequently by the male. Female winter territory holders (page 228), such as English Robins, Mockingbirds, and Loggerhead Shrikes, sing quite regularly. Antiphonal singing, or the answering songs of a male and female stationed some distance apart, is common to some birds, particularly in the dense tropics where location songs are needed to inform a bird of the whereabouts of its mate.

An interesting relationship appears to exist between testicular development and singing in male Budgerigars (*Melopsittacus undulatus*). Brockway's data (1967) indicate that males of this species may "self-stimulate" gonadal endocrine activities by hearing their own songs. She adds that "since vocalizations are ethologically regarded as displays, the current thinking about displays may need expansion."

Obviously birds have different songs and calls for different purposes. Laskey (1944), in her study of the Cardinal in Tennessee, distinguished 28 different Cardinal songs. Aretas Saunders (1951), who has spent nearly a lifetime studying bird songs, has recorded 884 variations in Song Sparrow songs. Odum (1942) lists and describes, with attempted explanation of meaning, 16 types of songs or calls used by Black-capped Chickadees, 8 related to breeding behavior, 8 primarily social. The American Goldfinch, according to Stokes (1950), has an "off-territory" courtship song, a territorial flight song, and a canarylike warbling song used in flock formation. Nice (1937) describes a "signal" song used by male Song Sparrows to call their mates off the nest for feeding purposes. Hann (1937) describes an "all's well" song used by male Ovenbirds to assure the females, which sit in ovenlike nests with a view in one direction only, that a certain danger has passed. House Wrens have a "stimulation" song, used to stimulate the young to gape for food when it is brought.

Davis (1959) considers the familiar "chebec" of the male Least Fly-catcher primarily a "position" call advising the female of his location. R. Smith (1959) describes three vocalizations in the Grasshopper Sparrow: (1) a "grasshopper" song used early in the season for territorial defense and threat (female doesn't respond to it); (2) a "sustained" song used to attract the female and to maintain the pair bond (female responds to it); and (3) a "trill" used by both sexes to notify each other and the young of their presence. The male does not respond to the trill of another male but challenges another "grass-hopper" song.

Song Analysis

During the past decade a vast literature has accumulated detailing individual song patterns for a number of species. Most of this work is the result of technological advances in the field of electronics that have led to vastly improved portable recorders and song-analyzing instruments. Most used are machines called *spectrographs,* which transform the sound into electrical energy and plot sound frequency against time on pressure-sensitive paper (producing a two-dimensional sound picture called a *spectrogram;* Fig. 8.2). In this way each element of the song can be measured in a linear fashion. Such measurements allow more precise analysis of an individual bird's song as well as comparison with the song elements of other species. Recent examples may be seen in Thompson (1965) for a brief analysis of bunting songs, Belcher and Thompson (1969) for Indigo Buntings, Mengel and Jenkinson (1971) for Chuck-will's-widow, and Marler et al. (1972) for song development in Red-winged Blackbirds; for a brief review of electronic acoustic analysis using oscilloscope tracings see Boudreau (1968). In addition, numerous commercial tapes and disc records of bird songs have been available for more than a decade (Fig. 8.3).

Daily Cycle of Song

It is common knowledge that birds sing most in the early morning. Birds have definite awakening hours, apparently determined by light intensities. Some birds, like robins, are notoriously early risers; others, like crows, jays, chickadees, and various birds that roost in cavities, may sleep a little later. Robins break into song as soon as they awaken, then cease singing for a period of feeding; other birds may forage before singing, or, like vireos, do both simultaneously. Singing wanes through the later morning hours, and reaches a minimum in early afternoon when birds, especially in northern regions where summer days are long, take rest periods. Then they may renew song in late afternoon or evening, though usually this does not equal the morning climax. Some species, like vireos and wrens, or species proclaiming territory, may sing persistently throughout the day. A vireo's leisurely method of gleaning insects from leaves and branches does not interfere with singing, for it sings between bites; a wren, unlike most birds, sings while carrying food to its young.

Many birds have special times for singing, often using a distinctive song for a particular time of day. The Eastern Kingbird has a special

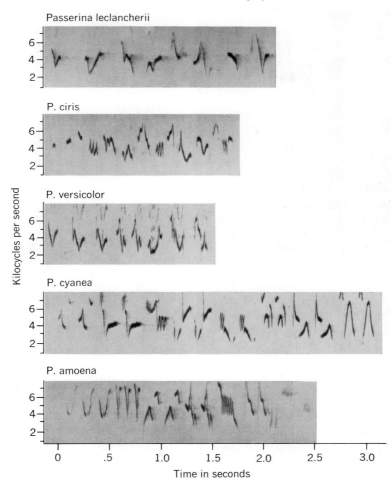

F IG. 8.2. *Spectrograms of buntings. These electrical recordings indicate differences in sequence, duration, and amplitude of the songs of five species of buntings.* [From W. L. Thompson, *Behaviour*, 31, 267, 1968.]

"matin" song used for 10 to 15 minutes before sunrise (Mayer, 1952). The Willow (Traill's) Flycatcher has a peculiar flight song, different from the daytime songs, used after sunset (McCabe, 1951). The Ovenbird and Common Yellowthroat also have special evening flight songs and the Tree Swallow performs long before dawn. Thrushes, the most renowned songsters among American birds, are justly famous for evening performances. Gray-cheeked Thrushes in the New England mountains have a 10- to 15-minute period of flight songs, just after sunset, but seldom resort to flight singing at other times (Wallace,

F IG. 8.3. *Recording bird songs in Mexico. The parabolic mirror with inset microphone is directed to pick up songs, which are recorded on tape by the man at the left.* [Photo by Dale A. Zimmerman.]

1939). The Whip-poor-will and American Woodcock (Fig. 8.4) are noted for the punctuality of their evening performances in relation to sundown.

Persistency and frequency of song have been recorded by many observers. Perhaps the record goes to a Red-eyed Vireo in Ontario that sang 22,197 songs during a 10-hour period (de Kiriline, 1954). D. Davis (1959) counted 700 "chebecs" of a Least Flycatcher in a 15-minute period in the early morning, but song output declined after 5:30 A.M. J. Davis (1958) counted 3,390 songs of an unmated (mated birds sang less) Rufous-sided Towhee during an 8-hour observation period (8:15 A.M. to 6:25 P.M.); it sang 95 percent of the time between 8:15 A.M. and 9:15 A.M. but only 64 percent between 5:25 P.M. and 6:25 P.M.

SEASONAL CYCLE OF SONG

Birds likewise have a seasonal cycle of song, usually closely correlated with nesting activities. Song is at its peak early in the season, during territory establishment and pair formation, then wanes abruptly, or may cease almost entirely in some species. In spring a White-throated Sparrow sings when he sees or hears *any* species nearby, but he will attack only members of his own species; late in the season he responds *only* to his own species, visually or vocally (Milligan, 1966).

Some birds sing during the period of incubation, some do not; a few birds sing through the whole nesting cycle but most do not. Curiously, Common Yellowthroats sing, at least by spurts, through the whole nesting cycle, *except* during courtship when song ceases (Stewart, 1953).

Any threat to a male's territory, or the loss of his mate or nest, causes him to renew or increase singing, perhaps to assert his continued claim to the territory, perhaps to stimulate the female to nest again. Feeding a brood of young has a telling effect on the males, and song practically ceases at this stage. Of course, the reason may be primarily physiological, associated with regression of the gonads, but also there is little environmental incentive to sing, for the territories are now being abandoned and pairs separated. If all birds were in the same stage of the nesting cycle at the same time, there would probably be little singing in summer, but interrupted nests as well as early and late nesters produce

F IG. 8.4. *The American Woodcock or "Timberdoodle," as it is known to many sportsmen, is noted for its punctual evening performance in relation to sundown.* [Photo by Lawrence A. Ryel.]

great variations. That is, while one male may be silent, feeding young, another may be singing for the beginning of his second or third brood. Some birds are all through nesting in June, but American Goldfinches may be still nesting, and singing, in September. Perhaps the record for year-round persistency in singing belongs to the Andean or Rufous-collared Sparrow (*Zonotrichia capensis*). Wallace (unpublished) observed an almost constant output of song (6–10 songs per minute) in Colombia from 6 A.M. to 6 P.M., with occasional outbursts during the night, from September 10 to July 18, though the total output of marked individuals was not determined.

Fall singing is fairly frequent among some birds. Often the songs are mere snatches, quite unlike the full songs of spring; often, too, they are the efforts of young birds trying out their inexperienced voices. There is also a slight but temporary recrudescence of the gonads in some species in the fall, which may account for autumn songs. Some birds, especially icterids, have abortive, or sometimes successful, attempts to breed in the fall. English Robins, both male and female, establish sharply defended territories in the fall, and sing vigorously.

Breeding Behavior

Many interesting traits of behavior develop, or are most readily observed, during the breeding season. Several of these merit more explanation than is accorded them in the chapters on the annual cycle (Chapters 9 to 11) and are described below.

AGONISTIC BEHAVIOR

A conspicuous feature in the early phases of the breeding cycle is the development of a hostile attitude between certain individuals, particularly rival males (Fig. 8.5). Birds that feed or associate together more or less peacefully during fall and winter become antagonistic during the breeding season. Males in particular tend to isolate themselves and to defend plots of ground known as territories (page 223). Increase in a male hormone at the onset of the breeding season is believed to be largely responsible for this aggressiveness. Apparently the potential for this behavior exists even at an earlier age in some species, since recent studies with Herring and Black-backed Gulls (Delius, 1973) indicate the presence of "agonistic sites" in the brains of juveniles of these species.

During the breeding season, males resort to a great variety of displays, sing almost constantly, and, when necessary, attack and expel intruders from their small domain. Both displays and song help to

FIG. 8.5. *Herring Gull in agonistic or threatening display.* [Photo by Allan D. Cruickshank.]

minimize the need for fighting, however, and thus conserve the bird's energy for other purposes. Shrikes, for example, can inflict serious injury on birds their own size, sometimes with a single peck. Several inhibiting displays have been evolved, however, that prevent this "ability to easily injure" from fully expressing itself in shrike behavior (S. Smith, 1973).

Birds vary greatly in aggressiveness, and perhaps in the need for it. English Robins studied by Lack (1943) were particularly aggressive; both males and females attacked dummies set up in their territories. The red breast seemed to be the provocative stimulus, as juvenile robins or adult robins whose breasts were painted brown usually were not attacked, whereas a tuft of red feathers mounted on a perch was attacked. One particularly aggressive female continued to fight the empty space after the feathers were removed, and three birds deserted their nests because they could not drive dummies out of their territory. On the other hand, some species do not defend territories even while nesting, and still others seem to lack the aggressiveness characteristic of so many birds.

One apparently irrelevant outcome of agonistic behavior is *image-fighting* in which birds attack their own reflection in a window or other reflecting surface. Sometimes such an aimless pursuit is abandoned after a few trials, or engaged in only at infrequent intervals, but in some cases the attackers literally exhaust themselves in the futile attempt to drive a potential rival from their territory.

Most aggressive actions are intraspecific, but many interspecific hostilities develop, not only among closely related species that might be competitors for food or other requirements but also against potential or actual enemies. Potential nest robbers and predators are often surrounded and threatened by their would-be victims, whether the latter are being molested or not (Fig. 8.6). This is known as *mobbing* behavior, and is well illustrated by the long-standing enmity between crows (or jays) and owls. Altmann (1956) experimented with stuffed owls set up in various situations and studied the reactions of 39 species that might have reacted. Nineteen species paid no attention to the dummies; 13 attacked only after seeing others do so; the others mobbed the owls fairly regularly, although not all members of a given species did so. Crows and Red-shouldered Hawks, without any expression of hostility between them, join other birds when they mob owls. Apparently crows can distinguish between Red-tailed and Red-shouldered Hawks, however, because they attack only the former (Kilham, 1964).

An interesting aside to mobbing behavior was suggested by J. Emlen (1969) in an experiment designed to test the response of birds to the squeaking sound used by ornithologists to lure birds. He found that spaced repetition of such sounds resulted in a rapid decline of bird

FIG. 8.6. *Young owls at their daytime roosts are prime targets for mobbing by other birds, especially by crows and jays. This photograph of a young Great Horned Owl seems also to catch the spirit of the April woods.* [Photo by Leon Alger, Jr.; courtesy of *Jack-Pine Warbler*.]

responses and suggested that this could be "attributed to habituation in the absence of reinforcement normally obtained by participation in the mobbing of a predator."

Nesting Behavior

The various nesting activities of birds (Chapters 9 to 11) provide many interesting illustrations of both innate and acquired behavior. Nest-site selection, choice of nesting materials, and nest-building techniques are fundamentally fixed species characteristics, but many birds adapt to new or changing situations. Many birds now take advantage of man-made facilities, nesting in chimneys or mine shafts (Chimney Swifts), in or on buildings (Starlings, robins, swallows, phoebes), or in artificial nest boxes (Fig. 8.7). Sometimes substitute nest sites run into unforeseen difficulties, as in the frequent cases where robins construct nests on moving or movable vehicles. Birds likewise often accept artificial nesting materials—string, paper, rags, tinsel—though such selections appear to have no real value other than availability. Oriole nests of colored yarn or a nest of bright tinsel might be disadvantageous because of conspicuousness.

Fig. 8.7. *Cliff Swallows still utilize natural nesting situations, such as cliffs and caves, but more often they adopt man-made structures, plastering their globular nests under the eaves of barns. They are colonial nesters and highly gregarious at all times. Note the stages of nest building.* [Photo by John T. Emlen, Jr.; courtesy of the *Auk*.]

Recent studies (Collias and Collias, 1964, 1973) indicate that learning plays a significant role in the nest-building abilities of some species. Male Village Weaverbirds (*Ploceus cucullatus*) that had been deprived of nesting materials during their first few years of life were not able to build nests when later given the opportunity.

Filching nesting material from unguarded neighboring nests is a fairly common trait among some birds. "Pebble-stealing" is traditional in Adelie Penguin colonies. The adults spend a week or more preparing nest sites, methodically collecting stones to mark the nest location, and even showing marked preferences for size and color of pebbles. In general, the females arrange the stones that the males transport, but many of the stones are stolen from under the hens when their backs are turned. Alert females viciously attack intruders, but unaggressive birds have most of their stones pilfered. Goldfinches commonly dismantle their own or other nests, active or inactive, to get suitable nesting material quickly.

Birds also have some capacity to retrieve eggs that have rolled out of or been experimentally moved from ground nests. Gulls and terns will roll eggs over short distances back to their nests. Red-tailed Tropic-birds (*Phaëthon rubricauda*) almost always retrieve their own eggs when displaced (Howell and Bartholomew, 1969). Common Murres, which build no nest and lay a single egg, can recognize their own egg when it is transferred to another location and will roll it back to its original position from a distance of one to five yards (Johnson, 1941).

In a study of 42 adult Royal Terns in which eggs had been switched to adjacent nests, Buckley and Buckley (1972) found that incubating adults in the colony consistently recognized (but did not try to move) their own eggs after they had been switched to adjacent nests, but failed to recognize three of four young that had been switched. Although many popular accounts (including those by Audubon) have been given concerning the transportation of eggs and young by the Chuck-will's widow, Ganier (1964) could find no evidence to support this, either by his own observations or through a search of the scientific literature. Apparently, Audubon based his evidence on hearsay.

Some birds, of course, cover their eggs when they leave the nest. The Pied-billed Grebe constructs a "blanket" of plant materials that it pulls over the eggs when it departs. More unusual behavior is found in the Kittlitz's Sandplover (*Charadrius pecurarius*) and the White-fronted Plover (*Charadrius marginatus*), both of which flick sand over their eggs when flushed from their nests. The Egyptian Plover (*Pluvianus aegyptius*), however, normally covers its eggs during the day. Although the foregoing examples have led some to believe that such behavior is for concealment or protection from predators, Conway and Bell (1968) believe that this behavior may have been selected as protection from the hot sun and may have evolved from "intention-settling" and "intention-leaving" movements that were later incorporated into behavior that effectively provided protection of the eggs in the parent's absence.

Even young Black Skimmers excavate shallow burrows and kick sand over their backs, partially concealing themselves (Hays and Donaldson, 1970); the adaptive significance of this might be related as much to protection from the sun as to concealment from predators.

Brood parasitism is an intriguing nesting specialization that has developed in a few groups of birds. Cuckoos, on all continents except North America, honeyguides and some weaverbirds in the Old World, and several species of cowbirds in the Americas regularly lay their eggs in the nests of other birds. Such eggs are almost invariably accepted and hatched by the foster parent, often at a considerable sacrifice, since their own young may perish. In the case of the Brown-headed Cowbird only a few species, chiefly those with conspicuously different eggs, ever throw out the impostor's eggs. A few others, notably Yellow Warblers, sometimes ingeniously bury a cowbird's eggs by building a second story on the nest, although it would be much simpler to throw out the eggs. Five-story nests of the Yellow Warbler, with a cowbird egg or eggs (and sometimes warbler eggs) in each story, have been found (Fig. 8.8).

To assure successful parasitism, too, the adult Brown-headed Cowbird female occasionally shows a "proprietary interest" in the host nest (Mayfield, 1961). Balda and Carothers (1968) related the case of a female cowbird that attacked a feral cat at a Lazuli Bunting's nest where possibly another cowbird had laid an egg.

Hardy (1970) described a Hooded Oriole's nest that consisted of two chambers, one of which contained a cowbird egg; the other appeared to have been used for the oriole's young. Friedmann (1963) lists 29 species that avoid cowbird parasitism by burying eggs in false lining, but none with displaced construction.

Of 12 species of glossy cuckoos in Africa, Australia, and New Zealand, 9 are known to be parasitic. The parasitic Black-headed Duck (*Heteronetta atricapilla*) of South America possibly owes its success to laying its eggs in the nests of marsh birds and coots rather than in other duck nests. Although some duck embryos die because coots often cover the foreign eggs with new nesting material, survival of young Black-headed Ducks reared in coot nests is good, since the young ducks can take care of themselves soon after leaving the nest (Weller, 1968). Attempts to "artificially parasitize" American Coot nests with chicken eggs, however, failed because the coots removed the eggs (Weller, 1971).

An Old World cuckoo (*Cuculus canorus*) has developed brood parasitism to such a high degree that it lays eggs of a color that matches those of the host species, even to the extent of laying a reddish egg that simu-

F IG. 8.8. *Three-story nest of Yellow Warbler. The bottom compartment contained one cowbird egg, the second story had two cowbird eggs and a warbler egg.* [Photo by Walter P. Nickell.]

lates the reddish egg of a host warbler (Sylviidae) in Japan. In addition, the young cuckoos have a saddle-shaped depression on their backs by means of which they shoulder their nest mates and heave them out of the nest. Honeyguides lay their eggs in nests of woodpeckers and barbets in cavities from which the host species cannot eject the young honeyguides; moreover, the latter have a murderous hook (which disappears soon after hatching) on the bill that they use to butcher their nest mates.

Injury-feigning or *distraction display*, a reaction in which a flushed parent bird appears to try to divert the attention of a visitor away from its nest by fluttering half helplessly just out of reach (Fig. 8.9), is a trait of nesting behavior whose interpretation has long been a controversial issue. The earlier naturalists credited the bird with a deliberate and conscious attempt to protect its nest by simulating injury to lead a predator astray, but later ornithologists suggested that it might be purely automatic, a partial paralysis of the locomotor apparatus brought about by two conflicting emotions, a fear for itself and an impelling desire to protect its nest (Friedmann, 1934). Perhaps the truth is somewhere between the two extremes. Some call it a "displacement activity," that is, substituting a new behavior pattern for brooding when prevented from doing the latter. Other distraction movements include "protean" displays or erratic movement made by prey when under attack to confuse and disorient predators (Humphries and Driver, 1967) and "pretense" displays that simulate the act of carrying some object. Although woodcock, for instance, occasionally carry their young when flushed, they will often simulate this behavior even when no young are present. Possibly this serves as a means to distract predators (Lowe, 1972).

Although there has been some controversy concerning whether or not adult birds can carry young, Johnsgard and Kear (1968) feel that too many observations have been reported to discount this phenomenon entirely. It has, of course, long been known in loons and grebes, but many Anseriformes (three species of swans, two sheldgeese, and seven duck species) also carry young on their backs while swimming. Carrying young in flight, also, has been reported for 16 waterfowl species, most of which involved young being carried by the bill of the adult.

Another habit that may be a distraction technique is that seen in certain species of parrots that hang upside down. Although Green-rumped Parrotlet (*Forpus passerinus*) young hang upside down for periods of 10 minutes (Buckley and Buckley, 1968), all species of the hanging parrots (*Loriculus* spp.) sleep while hanging upside down and even assume a sleeping position when threatened by a predator (Buckley, 1968).

Injury-feigning is more common, as might be expected, among ground-nesting birds of open spaces, particularly shorebirds, and is rare among treetop nesters and colonial birds, where it would serve little practical purpose. Most flightless birds would jeopardize their own lives as well as maintenance of the species by injury-feigning, but in

FIG. 8.9. *Piping Plover employing distraction display. Note the scarcity of nesting material and the protectively colored eggs.* [Photo by Bertha Daubendiek; courtesy of *Jack-Pine Warbler.*]

others it would have survival value by elimination of those not doing it well.

Some predatory birds with staggered hatching (page 284) sometimes resort to *cannibalism*, the older nestlings in a nest attacking and eating one or more of their weaker nest sibs; or, if the latter die, they may be fed to the survivors by their parents. Juvenile cannibalism is common in Marsh Hawks (see Fig. 10.20), Barn Owls, Short-eared Owls, perhaps other birds of prey, and some sea birds. Ingram (1959) believes that this practice has survival value in species with large clutches of eggs; sacrificing the younger members of the family for the benefit of the older ensures maximum fledging in relation to the available food supply.

Behavior and Feeding Habits

The feeding habits of birds afford some striking examples of behavioral adaptability. The feeding sequence of nestlings, for example, is determined *by the nestlings* rather than by the adults. The hungriest young lifts its head higher than its nest mates (see Fig. 11.2); hence it is fed first by the parent with food. This phenomenon is referred to as automatic apportionment (see page 303). The lack of behavioral adaptability is illustrated by specialized feeders accustomed to live on a restricted diet, and apparently unable to change. Probably the few surviving Everglade Kites in Florida will perish rather than turn to a new source of food, if the snail supply is exhausted by draining the Everglades. Griscom (1945) relates that Brant wintering on the Atlantic Coast were threatened with extinction (90 percent perished) when a blight nearly wiped out the eel grass on which they subsisted. Partial recovery of the eel grass and the adaptiveness of a few Brant in learning to eat other foods by watching other anatids apparently saved them.

By contrast, some other birds are quite resourceful at exploiting new food supplies. Many birds now utilize highway kills or other food supplies formerly not available. Mockingbirds and Gray Catbirds will glean insects, still warm, off radiators of parked cars. An American Kestrel, perhaps in an unusual display of learning, has been reported to watch pigeons feeding on bread crusts in a park, and then, after some trial and error, descend to the ground and feed with the pigeons (Warburton, 1952). Sawyer (1959) describes another case of kestrel resourcefulness; the hawk perched on the antlers of a bull elk (the only perch available) and caught mice flushed by the elk as it foraged in haystacks in Yellowstone National Park. Baird and Meyerriecks (1965) relate a sequence of feeding activities during an ant mating swarm in which Starlings, the first predators, caught the ants in the air, duplicating flycatcher or swallow feeding behavior. They were replaced temporarily by Common Nighthawks. Later, other feeders included Cedar Waxwings, Blue Jays, nuthatches, catbirds, grackles, and Red-winged Blackbirds, all of which exhibited irregular aerial feeding behavior, possibly stimulated by the Starlings' unusual aerial predation.

Many birds now take advantage of farming practices: swallows follow mowers for flushed insects, blackbirds and Franklin's Gulls follow the plow, and cowbirds and anis follow cattle. Rand (1953a) has shown that Groove-billed Anis are more successful in making catches when using cattle as beaters for insects than when feeding alone. Dawn (1959) observed Cattle Egrets (Fig. 8.10) feeding ahead of grazing cattle but becoming inactive when the cattle rested. Sometimes the egrets got restless and appeared to prod resting cattle into doing more grazing to stir up more food. Similarly ant-tangers in the tropics follow army ants, not to feed on the moving ant hordes, but to capture other insects flushed by the ants (Willis, 1960). Gulls and other birds that follow boats are quick to snatch up food thrown overboard by passengers and

FIG. 8.10. *In the United States, Cattle Egrets commonly follow cattle, using them as beaters for insects, but in Africa they associate with native ruminants, as shown here by this egret with a rhinoceros.* [Photo by Dale A. Zimmerman.]

crew. Several species of Galapagos finches brace their heads against stones and apply force with their feet to uncover new food sources (DeBenedictis, 1966).

Crows are remarkably resourceful. They catch fish at fishing holes in winter, dig up newly planted corn in spring, and haul young mice out of their nests in season; there are even well-authenticated cases of cooperative hunting, perhaps planned out in advance. Perhaps the most ingenious of these exploits is described by Homberg (1957) for Hooded Crows and a raven in Scandinavia. The corvids learned to pull up fishermen's lines for the bait by running back over the ice with the line in the beak, then walking forward on the line, repeating the process until the bait was out on the ice. Others have reported that some crows learn to recoganize a red flag as a signal when a fish is on the line.

A similar display of resourcefulness has been described by Smith (1945) for titmice in England. Several individuals learned to reef up a dangling string with a peanut tied on the end by standing on the loops of string as it was hauled up with the bill. More recently Dickinson (1969) observed this in the wild when a Tufted Titmouse obtained a caterpillar from the silken web of a spider by pulling up the web-strand. Titmice in England also learned to steal cream by pecking holes in the caps of milk bottles on porches in the early morning, until milk companies replaced the cardboard caps with metal.

Equally "imaginative" is an example of Green Herons in Florida fishing with bread (Lovell, 1958). The herons retrieved bread crusts intended for gulls on shore, placed them in the water, and caught fishes that came to nibble on the bread. In one case a heron actually moved the bread from an unsuccessful fishing site to a new location where several fish could be seen breaking the surface.

Fry (1969) reports that bee-eaters treat venomous worker bees and nonvenomous drones differently, removing the venom of the former by rubbing their abdomens against a branch. Gulls and Common Crows sometimes carry hard-shelled molluscs into the air and drop them on rocks or pavement to crack the hard shells, and buzzards in Europe are said to use the same strategy with turtles. In fact, Pliny relates that a Greek poet (Aeschylus) lost his life when a buzzard (the Lammergeier) mistook the poet's bald head for a rock and dropped a turtle on it. Egyptian Vultures (*Neophron percnopterus*) in East Africa break open Ostrich eggs by throwing stones at them with their bills (Lawick-Goodall, 1967); smaller, less resistant eggs are picked up and thrown on the ground.

Tinbergen (1960) theorized that songbirds in pine forests will not accept an item as food until they have developed a "specific search image" for it (i.e., learn to recognize it as appropriate food) and that they cannot acquire this until after they have many chance exposures to it.

Because of unforeseen circumstances, tapping new food resources does not always pay dividends. Hawks long ago discovered that a farmer's chickens were a new source of food, but from this knowledge have brought disaster upon themselves, as have fish-eating birds attempting to utilize the abundant supply of food at hatcheries. Even greater ingenuity was shown by the Kea in New Zealand when it learned to pick through the hide and into the groins of living sheep for kidney fat, a discovery that nearly proved its undoing, since sheep herders have almost exterminated the giant parrots. In the days of plume hunting, egrets soon learned to flee in panic from their nesting rookeries at the first blast of a gun, but resourceful crows merely waited nearby to plunder the nests while the egrets were gone.

Variability also occurs in the feeding techniques of birds. For instance, Meyerriecks (1962) describes the feeding behavior of the Reddish Egret as consisting of underwing feeding, open water method, wing-flicking, canopy (wings-extended) feeding, standing and waiting, stalking and foot-stirring. Often the feeding behavior of one species benefits another, as in the case of Belted Kingfishers that catch fish disturbed by fishing egrets (Meyerriecks and Nellis, 1967). Siegfried and Batt (1972) described a "feeding association" of Wilson's Phalaropes with shovelers. Each of the 25 shovelers present was attended by approximately 5 phalaropes that took advantage of the stirring of the water by the ducks.

Storage of food materials by birds has never reached the degree of development that it has among certain mammals (squirrels, pikas, beavers), but some birds show similar specializations. Woodpeckers, jays, nutcrackers, titmice, nuthatches, and shrikes are among those that practice the habit quite regularly. Presumably many of the stored items are never relocated, or are stolen by other animals, but some birds show a phenomenal place memory for recovering cached items. Chettleburgh (1952) watched 30 to 40 jays in England bury 200,000 acorns, carrying some of them three-fourths of a mile; apparently they remembered the burial sites. Clark's Nutcrackers in the Rockies will dig through snow up to 8 inches deep to recover items buried several months previously (Cahalane, 1944). Some of the recovery ability shown by birds may be due to the habit of searching typical hiding places rather than remembering where a particular item was stored.

Sometimes woodpeckers employ special techniques for storing nuts. Red-headed Woodpeckers seal in their winter stores (Kilham, 1958), not only wedging them so tightly into the bark of trees that they are difficult to dislodge, but actually sealing them in with damp splinters of wood so that they are well concealed. Further protection is accorded their supplies by scattering them rather widely. By contrast the stores of the Acorn Woodpecker are more exposed; the acorns are tucked into a little cavity drilled for that purpose and are often concentrated in a single tree. Dawson (Dennis, 1957) described a case of California (Acorn) Woodpeckers storing 50,000 acorns in a single pine; recovery of acorns stored in trees with rings dating back to 802 A.D. indicates that the habit is very old. Red-bellied Woodpeckers store items in deep crevices that other woodpeckers, with shorter tongues, cannot reach. Like Acorn Woodpeckers, they protect such storage areas from other birds (Kilham, 1963).

At winter feeding stations, carrying away and storing seeds becomes a game of hide-and-seek. Dennis (1957) credits chickadees, but not nuthatches, with great craftiness in hiding their stores; they "pretend" to hide a seed in a particular place, then move it to another less conspicuous site. Brown Creepers are not hoarders (they are chiefly insectivorous), but in their methodical search over the bark of trees they often locate items stored by other birds.

Joseph Grinnell long ago pointed out an indirect useful result of the burying of nuts by birds in the western mountains. He noted that birds often carried nuts and seeds *up* the mountain for burial and that unrecovered items sprouted into seedling trees; otherwise there would be no nut trees at high altitudes, since subsequent generations would always be *below* the parent tree. He called such birds "up-hill planters."

Shrikes (Fig. 8.11) have the well-known habit of impaling surplus prey (grasshoppers, mice, small birds) on thorns of trees and barbed-wire fences. Hawks and owls frequently accumulate surplus items at their nests, a practice perhaps carried to the extreme in Barn Owls, which have been recorded with 80 mice stockpiled in the nest with six eggs, and 73 mice in a nest with four young (Wallace, 1948). A neotropical thrush (*Catharus*) kept by the senior author for several months stored up food items in various corners of the room in which he had his freedom; when we cleaned house we found his little stockpiles hidden behind furniture. Such a habit could prove useful in the wild.

Finally there is recent evidence that breeding in some species might be determined by ability to feed effectively. Recher and Recher (1969) suggest that immature Little Blue Herons do not breed as early as they

FIG. 8.11. *Young Loggerhead Shrikes. Being predatory on grasshoppers and small vertebrates, the Loggerhead Shrike has a strong, hooked beak for dissecting prey. Surplus prey is often impaled on thorns, either for anchorage or for storage.* [Photo by Allan D. Cruickshank.]

otherwise could because they do not develop efficient feeding habits until a later period.

As with feeding behavior, the drinking habits of birds, in many cases, indicate stereotype actions that are characteristic of certain phylogenetic levels. Passeriform birds, for example, tilt their heads up after ingesting water, but Blue-crowned Hanging Parrots (*Loriculus galgulus*), and possibly several other species, sweep their bills through water to draw it up (Buckley, 1968). Pigeons and doves also draw up water, like mammals; sandgrouse sometimes, but not invariably, use this method.

Maintenance Activities

Several other traits of behavior, not necessarily related to breeding or feeding, have been described for birds. One of these is a habit popularly called "*anting*." Many birds, particularly passerines, have a curious habit of picking up ants (or sometimes such ant substitutes as black walnuts, moth balls, soap suds, and burning cigarette stubs) and rubbing them over their feathers, especially the underside of the primaries, sometimes falling over and rolling on the ground in an effort to get at less accessible parts of their anatomy. Early reports of this inexplicable behavior were regarded as fabrications, but it has now been verified by so many observers in so many species that its actual occurrence is unquestionable.

Whitaker (1957) reviewed the earlier literature and did some additional experimental work on an Orchard Oriole. Her 11-page bibliography and list of 148 species of birds known to ant indicate the extent of the literature on the subject. She divided anting birds into *active* and *passive* types, the former those that select ants and apply them to their body, the latter those that merely allow ants to invade their plumage. Apparently only ants that have a repugnatorial fluid (formic acid), which can be sprayed or exuded from the body, are acceptable to active anters. Heated, dead, or frozen ants, which had lost their taste and odor, were not used. This lends support to one of the supposed functions of anting, namely that the ants have a thermogenic property that gives a pleasurable sensation to the bird through contact with the skin. It is thought that sun-bathing and the habit some birds have of exposing themselves to heated pavement or other warm surfaces may have a similar function. However, some early, little-known experiments in Russia indicate that secretions from ants are lethal to feather mites; the dying mites dropped onto the substratum (paper) when ants invaded the birds' plumage. Potter (1970) gives one of the most recent summaries of anting behavior.

Goodman (1960) describes the possible connection between anting and accidental fires. Firemen attribute some fires in old buildings to Starlings and House Sparrows carrying smouldering cigarettes or matches to their trashy nests. Rooks and other corvids in Europe have been accused, since Roman times, of starting fires with matches and burning embers, sometimes even fanning the flames with their wings. It is conceivable that some forest fires are started by birds toying with unextinguished cigarettes along roadsides and transporting them to more distant and more flammable situations.

Wing flashing in Mockingbirds and its possible function form another subject of recent debate. The most obvious function, suggested by the earlier observers, is that the white wing patches exposed by flashing wings of foraging Mockingbirds startled insect prey into revealing themselves. Several recent authors, however, have pointed out that some birds without wing patches also resort to wing flashing and that securing prey might not be the explanation of the habit. Selander and Hunter (1960) thought that it might be an agonistic or threat display (it was used against dummy owls), but Hailman (1960) favors the food-gathering theory as the primary function, since in his observations 256 of 258 observed wing flashes were followed by foraging. Horwick (1965, 1969) presents the most recent complete summary for this phenomenon.

Foot quivering is a similarly perplexing phenomenon. Dilger (1956) interpreted it as a form of hostile behavior in thrushes, perhaps a low motivation of attack and escape drives in balance, for he had not observed it in foraging birds. Brackbill (1960), however, noted that eight out of nine Hermit Thrushes engaged in foot quivering while foraging on his lawn. Wallace (1939) thought that the similar "dancing" of a captive Gray-cheeked Thrush on rustling surfaces, such as paper, was related to auditory or musical effects of the sound produced (the thrush cocked its head and appeared to listen while tapping its feet on paper).

Foot stirring and *foot paddling* in herons and some shorebirds, however, seem definitely related to feeding, for such actions are invariably followed by captures of prey. Meyerriecks (1959) notes that only the yellow-toed species of egrets (the Snowy Egret in North America and Little Egret of Europe) use foot-stirring motions in searching for prey. The Snowy Egret drags its feet over water, but often catches prey with its bill without foot-dragging (Kushlan, 1972) or sometimes by flicking its tongue in the water (Buckley and Buckley, 1968).

Head scratching, whose obvious function is relief from or reaction to ectoparasites or other irritation, is of interest because of the methods employed and their possible phylogenetic significance. Two methods are used: *indirect,* with the foot placed over the wing while scratching, and *direct,* with the foot under the wing. Brown (1959) points out that Wrentits (Chamaeidae) in California use the less common direct method, a fact that tends to support their probable relationship to Old World babblers (Timaliidae), which also use this method. Nice and Schantz (1959), however, point out that the habit is less stereotyped and rigid than formerly supposed and that many species, especially warblers, use both methods.

Other maintenance activities include various grooming techniques such as preening (allopreening and heteropreening; Sparks, 1965; Goodwin, 1965), bill sweeping (Clark, 1970; Kilham, 1971), yawning (Sauer and Sauer, 1967), bathing (Slessers, 1970), and activities associated with defecation (Brackbill, 1966; Recher, 1972).

USE OF TOOLS

Although the use of thorns or spines, presumably to drive insects from holes bored in trees, has been well publicized for Galapagos finches, a similar more recent record is that of a Double-crested Cormorant using a shed secondary feather to poke its preen gland. After obtaining oil on the feather the bird used it to brush oil onto other feathers (Meyerriecks, 1972). Morse (1968) recorded the use of bark scales from long-leaf pine by Brown-headed Nuthatches to remove other bark so as to expose insects underneath. Orenstein (1972) observed a New Caledonia Crow (*Corvus moneduloides*) use a twig or leaf petiole to probe under bark and into the end of a hollow twig.

Social Organization

A prominent feature of social behavior is *gregariousness* or *flocking* behavior. Late summer flocking is a striking feature of the postnesting season. Species that were segregated on inviolate territories during the breeding season often become gregarious again, indicating that sociability is of a fundamental nature. Some flocks are built up primarily of a single species; others are mixed aggregations of species brought together by similar food or habitat requirements. Game birds assemble into coveys (groups) usually composed of a single species, but waterfowl collect in suitable feeding areas with several to a dozen or more species somewhat intermingled (Fig. 8.12). After the breeding season, Starlings congregate in large flocks, scattering to forage by day but reassembling at night. Such roosts may be entirely of Starlings, or have components of Common Grackles and Red-winged Blackbirds.

Large flocks of swallows, usually of several species, congregate on telephone wires, but associated with them may be flycatchers, bluebirds, various sparrows, and sometimes warblers. The social nature of swallow

FIG. 8.12. *Ducks assemble in large flocks during migration, usually with several to a dozen species somewhat intermingled, yet group associations of species within the larger flock are often maintained.* [Photo by Allan D. Cruickshank.]

flocks is indicated by their perching habits; there may be uninhabited miles of wires, but newcomers alight in the center of the flock, if space permits, rather than on the periphery (Emlen, 1952). Flocking gives birds certain advantages both in foraging for food and in escaping predators. Individuals within the flock may detect new sources of food or sound an alarm when an enemy appears. Avian predators, for instance, commonly pick off stragglers that have strayed from the group, but ignore the flock as a whole. (For a good current review of flocking see Wiley, 1971.)

Close social bonds, particularly observable in late summer and fall, continue through winter in gregarious species (page 320). Winter roosts of Common Crows, Starlings, blackbirds, and robins, in some cases numbering millions of birds at a single roost, are perhaps too well known to need further comment. Often social ties extend into the breeding season in colonial birds (Fig. 8.13), but in spite of severe space limitations, small defended territories prevail. That the close social organization is fundamental in colonial nesters is indicated by the fact

FIG. 8.13. *Weaverbirds (Ploceidae) are often highly gregarious, even in the nesting season. Here a nesting colony of weavers* (Ploceus cucullatus), *with 1,200 nests estimated in one tree, is found in Kruger National Park in South Africa. The leaves were plucked from the trees by the birds.* [Photo by John T. Emlen, Jr.]

that unaggressive birds crowded to peripheral nesting sites are often unsuccessful at breeding; sites near the center are sought by the dominant birds.

Dominance and peck-right play a fundamental role. In social organizations, dominance is sometimes distinguished from peck-right, since the former may be incomplete; the dominant bird merely may win the majority of conflicts, whereas peck-right imples more complete subjugation of subordinate birds (Goforth and Baskett, 1971). A third type, *supersedence,* is merely a replacement at a feeding table, such as a satiated goldfinch being replaced by another on a thistle head. In a simple linear peck order, as in a flock of 10 birds, No. 1 may peck the remaining nine birds, while No. 2 pecks only the eight below him—and so on down the line to No. 10, a completely subordinate bird pecked by all and pecking back at none. The peck order is determined by fights which may be short and quickly settled without future conflicts; or two birds, close together in the hierarchy, may be at one another constantly, frequently changing rank. Usually birds low in the hierarchy lack aggressiveness and do little fighting, yet sometimes such a bird may suddenly assert himself, probably owing to a hormonal change, and fight his way up through the ranks or even displace the head bird in a

single conflict, thus finding himself in a new and glorified position. Injection of testosterone into a weakling may cause him to become dominant. Further complications arise when a bird low in the peck order
toadies to one high in the series and thus receives some immunity from
other birds by basking in the shadow of its protector. Eventually some
stability within the flock is attained; this decreases the need for fighting
and permits more time for other activities, which is probably one of the
chief functions of dominance.

Not all peck orders are of the linear type. Some are triangles in
which A pecks B, and B pecks C, but C pecks A and not B. In geese the
family is the unit—one family, usually the largest, dominating other
families by sheer superiority of numbers (Hansen, 1953).

Formerly most observations on dominance were made on domesticated birds (poultry, pigeons, and canaries), but such studies have
now been extended to native birds in confinement, semiconfinement,
or in the wild. Useful studies of this type have been published by
Collias and Taber (1951) on pheasants, Tordoff (1954) on crossbills,
Sabine (1959) on juncos, Dilger (1960) on redpolls, Thompson (1960)
on House Finches, and Brown (1963, 1970, 1972) on jays. A linear
peck order, with some triangles developing within certain flocks, seems
to prevail among these birds. In some species males are dominant over
females (Thompson gives a list), in others the females dominate their
mates. Changes in the peck order may take place at the onset of the
breeding season owing to male aggressiveness.

Dominance of one species over another is characteristic among
closely associated species. Sharp (1957) found pheasants dominant over
Prairie Chickens, but not over Sharp-tailed Grouse. Fighting techniques seemed to determine the relationship. Pheasants fought with
spurs and had an advantage over the shorter-legged, shorter-spurred
Prairie Chickens, but Sharp-tails ducked under the pheasants when the
latter jumped, and plucked at their tail and under tail coverts. A pheasant thus treated beat a hasty retreat and never renewed the attack.
Sharp-tails were also dominant over Prairie Chickens but quite tolerant
of them.

On the other hand, a rather loose social hierarchy, without definite
dominance, seems to prevail in some aggregations of birds. Norris
(1960) found five races of Savannah Sparrows intermingled on their
wintering grounds in South Carolina without apparent social or ecological segregation; certain individuals were dominant over others, but
many showed no signs of hostility at all.

Although it has long been recognized that synchronization in flocks
depends largely upon visual contact, recent evidence (Evans and Patterson, 1971) indicates that with several species, at least, it depends on
vocal as well as visual releasers.

SELECTED REFERENCES

ARMSTRONG, E. A. 1947. *Bird Display and Behavior.* New York: Oxford Univ. Press.

COLLIAS, N. E. 1952. The development of social behavior in birds. Auk, 69:127–159.

EIBL-EIBESFELDT, I. 1970. *Ethology.* New York: Holt, Rinehart and Winston.

FICKEN, R. W., and M. S. FICKEN. 1966. A review of some aspects of avian field ethology. Auk, 83:637–661.

HINDE, R. A. 1970. *Animal Behavior.* New York: McGraw-Hill.

JOHNSGARD, P. A. 1965. *Handbook of Waterfowl Behavior.* Ithaca, N.Y.: Cornell Univ. Press.

LACK, D. 1943. *The Life of the Robin.* London: Witherby.

LORENZ, K. Z. 1952. *King Solomon's Ring,* New York: Crowell.

SCOTT, J. P. 1958. *Animal Behavior.* Chicago: Univ. of Chicago Press.

SMITH, S. 1945. *How to Study Birds.* London: Collins. Chapter 5.

THORPE, W. H. 1963. *Learning and Instinct in Animals.* London: Meuthen.

TINBERGEN, N. 1948. Social releasers and the experimental method required for their study. Wilson Bull., 60:6–51.

———. 1951. *The Study of Instinct.* Oxford: Oxford Press.

———. 1953. *The Herring Gull's World.* London: Collins.

WELTY, J. C. 1962. *The Life of Birds.* Philadelphia: W. B. Saunders Co.

WHITAKER, L. M. 1957a. A resume of anting, with particular reference to a captive Orchard Oriole. Wilson Bull., 69:195–262.

9

The Annual Cycle:
Arrival, Territory, Courtship, and
Mating

THIS series of three chapters on the annual cycle describes the various activities of birds through the four seasons: their arrival in spring, nesting activities, late summer behavior, the fall journey, and a brief account of their winter habits. This first phase in the series covers the various preliminaries that lead to nesting.

Spring Arrival

The coming of the birds in the spring has long been regarded as one of the foremost harbingers of the season. Most migratory birds in the northern hemisphere respond to the lengthening days of spring by advancing northward and nonmigratory species exhibit preliminary indications of breeding. Some of the initial responses may appear premature. A resident Cardinal may burst into full song on a deceptively sunny day in midwinter, and invariably a few transient robins will put in a premature appearance, only to be forced into temporary seclusion by a return of winter conditions.

The factors that initiate migration and subsequent breeding are poorly known. In brief, an internal rhythm, under hormonal control but often modified to keep in tune with prevailing environmental conditions, governs the breeding and migrational cycles of birds. The arrival of birds on their nesting grounds, described below, is only one phase of this cycle. Day length appears to be the chief external factor stimulating hormone flow and recrudescence of the gonads, but favorable or unfavorable weather conditions as well as the availability of the right kind of food and suitable nest sites can apparently modify the inherent internal rhythm. Chapter 12 on migration analyzes these factors more fully.

The span of spring migration over the central and northern states covers the period from February to June, usually with each species appearing at a given time, so that its coming can often be quite accurately predicted. The early comers are less punctual in their time of return, for they may need to adjust their arrival schedule to local weather conditions. The first comers in the northern states may arrive during the first mild spell of early spring, whether it be in late February or in March. But April migrants, which in general wintered farther south, are less influenced by the local weather picture and are more punctual in their time of return. May migrants, many of which wintered in South America, have even more precise arrival dates, often appearing year after year on practically the same day. When one keeps careful arrival records over the years, however, many variables are found to enter into and confuse this picture of precision; even the much-publicized Cliff Swallows at Capistrano in California and the less-publicized Turkey Vultures at Hinckley, Ohio, are probably not as punctual as reports make out. The latter are reported to have arrived on March 15 for the last 150 years; public officials claim to have checked this precise arrival date for 40 years (Thomas, 1969).

Thus, though the urge to migrate is probably initiated internally, there is considerable correlation between the northward movement of birds and vernal conditions. Waterfowl often linger along the southern fringe of frozen waters, then surge northward as rapidly as thawing lakes and streams permit. Sometimes, birds advance too rapidly and then may have to "stall," reverse their migration, or suffer disaster. Many birds are victims of storms during their early spring migration, and others get blown off course by unfavorable winds. In the spring of 1954, Indigo Buntings were apparently airborne on a storm track from Yucatan to Maine and arrived nearly a month ahead of schedule (Bagg, 1955).

A feature of spring migrations of which bird watchers are well aware is that birds often appear in waves. That is, they may make a fairly long flight under favorable weather conditions (southerly winds, warm temperatures, low barometric pressure), then, meeting a cold front to the north, may be stalled for days before a change in weather favors another flight. Birds also accumulate fat reserves before a long flight, then when these are depleted they stop to replenish the supply before undertaking another stage of the journey. The net result of these various factors is that migrants appear in waves which are only partly predictable. (See Saunders, 1959, for precision and irregularities in arrival schedules over a long period of time.)

Though some birds make long flights, as across the Gulf of Mexico or over long stretches of the Atlantic or Pacific, the daily rate of progress for most land birds is slow. Lincoln (1950) gives the average speed of all species travelling up the Mississippi Valley as only 23 miles per day, but of course this includes rest stops. Many species, and many individuals, would achieve much faster rates. Some birds proceed rather leisurely on the first laps of their journey, then speed up on the final stages. The Blackpoll Warbler (Fig. 9.1), which winters in South America and nests in the sub-Arctic conifer belt, averages about 30 miles per day through the states, but speeds up to about 200 miles a day over the more northern stretches (Lincoln, 1950). Actually there are few records of measured flights of banded birds from one station to

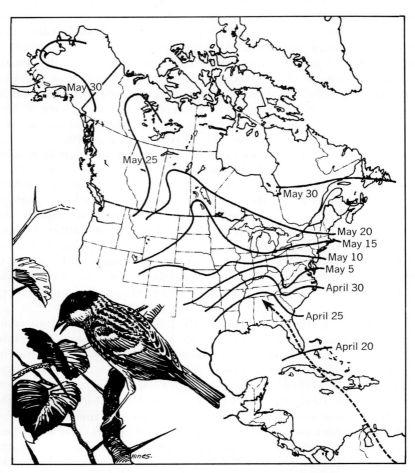

F ɪ g. 9.1. *Isochronal lines show the rate of advance of the Blackpoll Warbler during spring migration. Starting out at a leisurely pace of about 30 miles per day, it speeds up to more than 200 miles per day before the trip is finished.* [Drawing by Robert W. Hines, Bureau of Sport Fisheries and Wildlife.]

another (but see page 342) and the few records available are not necessarily typical.

A frequently overlooked feature of spring migration is that all the individuals of a species do not arrive at the same time. In a study of the Red-winged Blackbird at Ithaca, New York, Allen (1914) found the different status groups arriving at different times from late February to early June, thus covering practically the whole span of spring migration. The vagrants arrived first, in late February and early March, followed by the migrant and resident adult males throughout March, migrant and resident females through most of April, and immature birds throughout May and into early June. Some forty years later the arrival schedule of Red-wings was restudied in Wisconsin (Beer and Tibbitts, 1950) and found to be quite similar to that ascertained in New York. Immature birds, in particular, may be irregular and late in their time of appearance, since their gonads may not be fully developed early in the season and the stimulus to migrate and breed may be largely lacking.

It should be pointed out that this description of spring arrivals applies mainly to land birds in the northern hemisphere. In the southern hemisphere, where migration is less pronounced because of smaller land masses, the seasons are reversed, and tropical and oceanic birds are influenced by different external factors (see pages 328, 329–30, and 338).

Territories

The first objective of a male bird arriving on its breeding grounds is to find a suitable plot of land, *the territory,* that will provide a home for himself and prospective family during the nesting season. In a general way a territory may be defined as "any defended area," to which Tinbergen (1936) adds the concept of "sexual fighting" on a "restricted area." Lack and Lack (1933) define it as an "isolated area defended by one individual of a species or by a breeding pair against intruders of the same species and in which the owner of the territory makes itself conspicuous."

The concept of territories among birds is by no means new. The food territories of the eagle and the raven were described by Aristotle (about 350 B.C.). One of the earliest descriptions of a defended nest site was by François Legaut, who, in 1691, wrote an extensive, though somewhat fanciful, account of territorial defense in the Solitaire on the island of Rodriquez, near Madagascar (Hann, 1945). Audubon's description of territorialism in the Eastern Meadowlark in 1835 is perhaps the earliest American reference of this sort (Allen, 1951). But the importance of territories to birds was largely overlooked until the publication of Howard's classic work, *Territory in Bird Life,* in 1920, gave new impetus to the topic. Since then, the concept of the territory has been subjected to much criticism, redefinition, and continued study, and is now an important part of most life-history studies.

There are many variations in territorial arrangements, but certain well-defined patterns can be described. Perhaps the most common is a territory that serves as a mating, nesting, and feeding area. In most migratory species, the males arrive from several days to several weeks before the females, and have established territories by the time their prospective mates appear. The first male to arrive may be a bird seeking to reclaim his previous year's holdings. If he finds his territory already occupied by a newcomer, which is not likely, as older birds are apt to be the first to return, he may evict the intruder. Then he proceeds to advertise his holding by almost constant song, by some other type of display, or both. These serve as a warning that the territory is occupied. Other males, or even females at this stage of the cycle, that venture to trespass are expelled. Aggressive species, particularly under conditions of crowding, do much fighting or at least chasing off of intruders, but in most species the song or display is adequate advertisement and minimizes the need for fighting. Sometimes a particularly pugnacious invader can drive out the original owner, but ordinarily the first bird definitely established on a plot is invincible. Figure 9.2 illustrates the partitioning of an old field into Song Sparrow territories as shown by Mrs. Nice.

Songbirds usually adopt one or many singing posts in or around the periphery of their territory and make themselves conspicuous. Nonsinging birds resort to a variety of performances (see page 234) that serve as a substitute for song. Boundaries are sometimes sharply fixed, but perhaps more often are rather fluid lines wandered over first by one male and then another. "Buffer" zones between territories may separate breeding pairs more effectively than when boundary lines are in contact.

Ordinarily, territorial fights occur only among members of the same species; other species are not strictly competitive, and interspecific controversies are infrequent. That is, a pair of Eastern Meadowlarks, Bobolinks, and Field Sparrows may occupy the same plot of ground for nesting, but not two pairs of Eastern Meadowlarks, Bobolinks, or Field Sparrows. Species vary greatly in this respect, however; the Song Sparrow, for instance, is particularly aggressive toward other species and tries to keep them off his territory (Nice, 1937). Coots are pugnacious to the extent of attacking other vertebrates as well as birds. Ryder (1959) found Coots in Utah attacking 11 species of ducks, 18 other species of birds, a fish, a turtle, a weasel, and a muskrat—a list

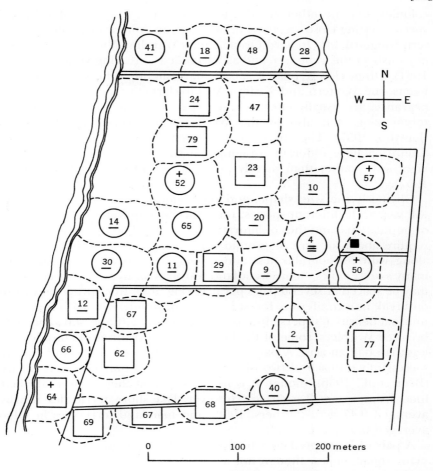

FIG. 9.2. *Territories of Song Sparrows. Number indicates individual bird; line under number indicates bird of the present year; each added line under number indicates subsequent years; circle around number indicates resident; square around number indicates summer resident; plus sign indicates first-year bird.* [Redrawn by Diane Pierce from M. M. Nice, *Trans. Linn. Soc.,* 4, 71, 1937.]

that could be extended by including Californian and European records.

Some closely related species have interspecific problems. In the mountains of Tennessee the Black-capped and Carolina Chickadees are competitive in the breeding season; the Black-caps keep the Carolinas from occupying the higher altitudes (Tanner, 1952). Ring-billed and Herring Gulls usually maintain a strict segregation in their nesting colonies. On the other hand, there are many instances of different species nesting close together; in one case, an American Robin, an Eastern Kingbird, an Orchard Oriole, and a Warbling Vireo nested simultaneously in the same tree and all succeeded in rearing young (Stamm, 1951). Often closely related species have slightly different habitats or habits; the Western and Eastern Meadowlarks, nearly identical in appearance, are usually effectively segregated not only by distinctly different songs but also by the Western's preference for drier sites (Lanyon, 1957). The Brown and Spotted (Rufous-sided) Towhees in California have different foraging habits, the latter scratching more vigorously, rather than merely pecking, in more open spaces (Davis, 1957).

Obviously the size, shape, and distributional patterns of territories will vary with different species, their specific requirements, the density of populations, and the configuration of the land occupied. Many birds require about an acre of ground for nesting purposes, but some get along with considerably less and others range over several acres or many square miles. Some woodland warblers in areas of concentration may be spaced at several pairs per acre (Fig. 9.3). In a spruce budworm outbreak in a coniferous forest in Ontario, Kendeigh (1947) found an unusual density ("supersaturation") of Tennessee, Cape May, and Bay-breasted Warblers, some of the latter having territories of considerably less than half an acre. A pair of Song Sparrows has been found nesting successfully on each of two islands in Minnesota only 0.04 acre in size (Beer et al., 1956), though it is not certain that they secured all their food from so small a space. However, lake-shore territories nearby averaged 0.47 acre (Suthers, 1960) and mainland territories in Ohio averaged 0.67 (Nice, 1943).

A pair of large predatory birds, on the other hand, may range over many square miles, though not all of the occupied territory (hunting space) is defended. Dixon (1937) mapped Golden Eagle territories in California varying from 19 to 59 square miles; McGahan (1968) recorded an average density of one pair per 74.2 square miles in Montana in 1964, possibly an indication of the decline of the species in the intervening years. Other predatory birds, like the Red-shouldered Hawks in flood plains and along rivers (Stewart, 1949) and Marsh Hawks in certain marshes (Hecht, 1951), may congregate on small territories where favorable nest sites are available but range more widely for food.

Size of territories may change markedly during the season. The Mourning Dove may stake out ambitious claims at the beginning of the season, but defend only· the immediate area of the nest site after nesting is underway. Stephanski (1967) found Black-capped Chickadees in Utah utilizing only 22 percent of the original territory while

FIG. 9.3. *Where a variety of insect food is readily available, woodland warblers may be concentrated in small territories. This female Black-and-White Warbler forages mainly over the trunks and branches of trees and is not strictly competitive with other species of warblers in the same woods.* [Photo by Allan D. Cruickshank.]

feeding young. Yarrow (1970) found redstarts gathering food for nestlings on only part of the original territory, but they increased the territory again while feeding fledglings. Presumably such arrangements function in shortening trips and saving energy while feeding nestlings. Ricklefs (1971) reported that two pairs of Mangrove Swallows (*Iridoprocne albilinea*) on Barro Colorado Island, Panama, had two feeding areas, one near the nest for feeding young, another further afield for themselves. Such an arrangement also appears to apply to northern Tree Swallows.

Territories may conform to the lay of the land. Typically they are somewhat oval-shaped if the habitat permits, but a kingfisher may control fishing rights to a linear mile or so of a stream and have little interest in land away from the bank. The Dipper (see Fig. 4.12) in western mountains and the Louisiana Waterthrush in the east may have similar interests in linear strips of water courses. A pair of Common Loons may control an entire small lake, regardless of contours, but larger lakes may accommodate more than one pair.

Territories are ordinarily breeding season phenomena, established in the spring and deserted at the close of the nesting season, but some nonmigratory species remain on or near their holdings over winter. Male Song Sparrows in Ohio remain on or near, or at least periodically visit, their nesting territory at all seasons, but do not defend it as vigorously in winter, and often seek more sheltered swamps during cold spells (Nice, 1937, 1943). Many nonmigratory birds—woodpeckers, Mockingbirds, Cactus Wrens, Wrentits, Loggerhead Shrikes in the southwest, and English Robins—set up defended winter territories which may or may not coincide or overlap with the nesting territory. Often the sexes are segregated in such winter arrangements, but in spring the barriers break down and erstwhile rivals become mates on a single territory.

Studies in both the Old World and the New World tropics now show that many migratory birds have winter territories to which they return year after year; Snow and Snow (1960) recorded a banded Northern Waterthrush returning to the same winter quarters in Trinidad for four successive winters. Schwartz (1964) points out that the winter territory of the Northern Waterthrush in Venezuela, to which it usually returns in September and from which it leaves in May, is only one-fifth the size of the breeding territory, but of course it serves only one bird.

MODIFIED TERRITORIES

A common modification of the standardized territory outlined above is one that is used primarily for mating and nesting but not exclusively for feeding. Many birds, even of the noncolonial types (see next paragraph), do at least part of their feeding outside the defended area. This, as already noted, is particularly characteristic of predatory birds. Other birds with separate, usually undefended, feeding grounds, include Black-capped Chickadees, Cedar Waxwings, Tree Swallows, and Barn Swallows.

A more obvious and striking modification of the territory is found in the often severely restricted nest sites of colonial birds. Cliff-dwelling sea birds, for example, may be so crowded on a nesting ledge that each bird has only a square foot or two for a nest (Fig. 9.4). Often, indeed, the space occupied is determined by the length of reach of the owner's

FIG. 9.4. *Available ledges on Bonaventure, historic bird rock in the Gulf of St. Lawrence, are densely packed with nesting Gannets—a $100,000 boost to the tourist trade annually.* [Photo by Allan D. Cruickshank.]

sharp bill. A colony of Common Murres in Oregon is said to have a density of 22,000 pairs per acre. Table 14.1 (page 394) gives some other notable nesting densities of alcids. Each bird in a sea-bird colony has a defended nest site—a true territory, albeit a small one—but the whole colony gathers its food out at sea usually on neutral feeding grounds. Gulls and terns often, but not invariably, have closely packed nest sites, but allow enough room for change-over of mates during incubation and feeding. Gulls often engage in sparring matches across a boundary line, and may even attempt to pull their opponent over the line for further punishment. Bank Swallows in a bank, Cliff Swallows on a ledge or under the eaves of a barn, or herons in a heronry have their respective defended nest sites, but perform many of their activities elsewhere. Cliff Swallows from different nesting colonies may intermix in foraging, mud-gathering, and perching on wires, but return to their respective colonies after such social affairs (Emlen, 1952).

Some marsh dwellers, like rails and Red-winged Blackbirds, have a semicolonial arrangement of territories. Nero (1956) found Red-wing territories near Madison, Wisconsin, averaged about 1/12 of an acre, with great variability in different areas. Berger (1951a) found five nests of the Virginia Rail and four of the Sora on less than a half acre of marsh in Michigan. In the Red-winged Blackbird the male's territory, usually including several nesting females, is zealously defended from trespassing males and unwanted females, but the birds do most of their feeding in surrounding fields and do considerable singing from undefended perches scattered about the marsh. Common Grackles, either in a marsh or in a grove of conifers, and Mourning Doves in orchards sometimes reach high nesting concentrations.

Ducks, with a few exceptions, have a strikingly different arrangement. Though there are many variations, the drake commonly selects and defends a "loafing bar" where he and his mate can find isolation and freedom from other drakes, but it is mainly a trysting place. The hen usually nests elsewhere (Fig. 9.5), perhaps a mile or more away at a site probably unknown to the male; but daily or more often during the egg-laying period and early stages of incubation she visits the male at his loafing bar. In hummingbirds the male and female maintain separate areas; courtship and mating may take place on the male feeding territory or elsewhere, but the female has a separate defended nest site, seldom visited or at least not used as a feeding station by the male (Pitelka, 1942).

Goldfinches exhibit further modifications (Stokes, 1950). They delay nesting until late summer, when thistledown for their nests and new seeds for their young are available, but they have paired much earlier. Long after the pairs have formed, the *female* selects a territory which the male, usually but not invariably, proceeds to defend vigorously, although he apparently cannot dominate for his sole use such a priceless commodity as thistledown. Cedar Waxwings at least initiate mating activities while still in flock formation, then seek out a territory together where further courtship takes place. House Sparrows, Black-capped Chickadees, and some other hole-nesting species often mate before the territory is selected, then go house hunting after the pairs are formed.

Some gallinaceous birds (Sage Grouse, prairie chickens, Sharp-tailed Grouse) and the Ruff in Europe have developed elaborate display grounds or "arenas" (pages 234 and 240) where they mate promiscuously with whatever hens visit them. The hens then repair to nest sites of their own selection which are not defended or even known to the males. The territory in this case serves as a social gathering place for

FIG. 9.5. *This female Redhead constructed the nest, will incubate and hatch the eggs, and will rear the young without help from the drake. He frequents a "loafing bar" some distance away, but will soon abandon it to go into an early eclipse molt.* [Photo by Allan D. Cruickshank.]

the males in winter and early spring before the females visit them, and as a place of union for the sexes later.

Apparently some birds are essentially nonterritorial. Sharp-tailed Sparrows studied by Woolfenden (1966) were semicolonial, and the polygamous males did not defend a nest site or territory. Brewer's Blackbirds, also polygamous when surplus females are available, attend and protect their females, but not the nest or a territory (Williams, 1952).

To summarize, then, territories appear to be divisible into about five categories: (1) territories which provide for all the requirements of breeding and nesting—courtship, pair formation, nesting and feeding the young; (2) territories, usually more restricted, which provide for all breeding requirements except food (predatory birds, chickadees, Cedar Waxwings, noncolonial swallows); (3) more severely restricted territories of colonial and semicolonial birds which usually defend only the immediate nest site (most sea birds, gulls and terns, rails, many icterids and ploceids); (4) territories that serve mainly or exclusively for pair formation and associated events, not nesting or feeding (some grouse, some shorebirds, some bowerbirds and birds of paradise); and (5) winter territories.

The Function of Territory

The functions of territories are probably apparent from the preceding account, but these may be summed up under the following headings:

1. Pair isolation, or providing a place for courtship and mating without interference.
2. Assuring an even distribution of birds over available habitats, so that their carrying capacity is not exceeded.
3. Facilitating food gathering for the young.
4. Protection from predators.

Isolation for mating purposes is essential for many birds. Among ducks, where there is often an unbalanced sex ratio with excess males threatening to break up mated pairs, it appears necessary for the drake to isolate a female from other ducks onto a private territory if subsequent breeding is to be successful. In other species the territory plays an important role in pair formation, courtship performances, and later for undisturbed nesting. The song of the male informs a female of the whereabouts of an available mate and territory; this is thought to be of particular importance in dense cover, as in dark tropical forests where the sharp calls and sometimes elaborate ceremonies of tropical birds aid the females in finding the males.

Territorial arrangements serve a second important function in assuring a more even distribution over available habitats, so that breeding pairs will not be overcrowded in one region and absent in another. That is, if about an acre of meadow is required for a pair of field birds, this would encourage a distribution over all the available acres within the range of the species rather than in congested sections. Populations are regulated to some extent by such a distribution of territories. In spite of this concept that territories even out distribution, birds sometimes tend to "bunch up" in selecting breeding sites, leaving some apparently suitable spaces unoccupied. Many birds are somewhat sociable even in the nesting season. Proximity of neighbors in colonies in fact stimulates breeding; the most sought after nesting sites are toward the center of the colony, and periperal nesters are often unsuccessful at breeding.

A third and perhaps more obvious function of the territory is to facilitate food gathering from a restricted area without intraspecific competition. Species which feed their young several hundred times a day need to gather the food hurriedly in places close at hand, not gleaned over by competitors. However, even this important requirement is not always fulfilled by the territory, as many birds go outside the defended area for food, and many male birds relax their vigilance during this phase of the nesting cycle and allow other birds to trespass. Perhaps the explanation for this lack of vigilance at a time when it would appear to be most urgent is due to the waning of sexual aggressiveness, or because the males are too busy feeding young to defend their territories.

A fourth suggested function of the territory is to provide some protection from predators; the occupant soon becomes thoroughly familiar with the territory and may know of quick escape routes and places of concealment.

Courtship and Mating

COURTSHIP RITUALS

The preliminary ceremonies in which the sexes engage to attract and stimulate each other are commonly referred to as *courtship,* as opposed to *mating,* which implies pair formation but not necessarily sexual union (copulation) which may come later. Courtship is not necessarily restricted to the period preceding or accompanying pair formation, though in some species it is. In the Blue-winged and Golden-winged Warblers the period from pair formation to copulation may last only one day (Ficken and Ficken, 1968). In other species courtship is continued throughout the breeding season or even, sporadically at least, throughout the year and seems to be the bond that keeps the pair intact. In ducks, penguins, and some other sea birds, courtship begins early, often in the winter preceding the nesting season.

Some courtship antics may appear ludicrous to us, as in the feather fluffing and wheezy gurgling of the Brown-headed Cowbirds and Common Grackles, but some birds engage in spectacular exhibits. Usually such performances are by the male, the female appearing to be only mildly or not at all interested. But in some birds, such as the grebes and penguins, the sexes play a nearly equal role in courtship ceremonies. In still other species, such as the thrushes, courtship is so simple it is hardly noticeable, and nesting proceeds without apparent preliminaries.

The great variety of courtship rituals performed by birds precludes the possibility of describing many of them; a few selected examples will serve to illustrate the scope of such activities. Among the songbirds, of course, the song is the principal means of sex attraction. Many songbirds, however, especially those with inferior voices, supplement vocal performances with displays and posturing. American Redstarts show their orange-red wing and tail patches both to intimidate other males and to attract females. Kinglets, the Eastern Kingbird, and the Oven-bird are examples of those that display bright brown patches. Brilliantly plumaged males such as tanagers and orioles parade before the females as well as rival males. Many field birds (larks, Bobolinks, long-spurs, pipits, and some sandpipers) have evolved elaborate flight songs which make them conspicuous over their territory. In general, with many exceptions, males with weak or simple songs (most fringillids) sing from conspicuous perches, but males with elaborate songs (thrushes) may sing from concealment. Grassquits (*Volatina*), in tropical fields and pastures, perch on a tuft of grass or weed, vault into the air and deliver a weak ecstatic song or call, then flutter down to their perch again; a whole field of grassquits has been likened to popcorn in a popper. Alderton (1963) once recorded 105 hops by one male in 24 minutes, each hop accompanied by calls, then he added 148 calls from his perch.

Birds that do not sing substitute other performances. Woodpeckers drum on dead limbs, and have strengthened neck muscles and thickened skull bones to withstand the pounding. The male Ruffed Grouse mounts a log or stump (Fig. 9.6) and fans the air with his wings, slowly at first, then with increased tempo, producing a reverberating "thunder" which carries long distances. Among the shorebirds the Common Snipe and the American Woodcock in particular have developed spectacular aerial displays. The latter performs every evening during the breeding season, rising on whistling wings some 300 feet above the ground, circling, maneuvering, and "chippering" loudly before plunging into a precipitous descent. The related European Ruff stages its dance in groups on an arena or "hill"; the ruffs encircling the necks have even evolved into different colors on different individuals. Prairie chickens congregate on hereditary dancing grounds, where the males strut and dance, inflating their colorful neck pouches to produce their characteristic "booming" (Fig. 9.7). Sham battles are staged, but seldom with any physical injury to the participants, suggesting that the ceremonies at the outset at least are primarily social or to determine dominance. The similar ceremonies of the Sage Grouse in the west are even more highly evolved, with a complex hierarchy of master cocks,

---

Fig. 9.6 [opposite, top]. *A male Ruffed Grouse mounts a log and "thunders" a proclamation of territorial ownership by fanning the air with his wings.* [Photo by Mich. Dept. Nat. Res.]

Fig. 9.7 [opposite, bottom]. *A Greater Prairie Chicken issues a challenge by "booming."* [Photo by Mich. Dept. Nat. Res.]

subcocks, and guard cocks each with his special role to play (Scott, 1942).

Among the most elaborate courtship ceremonies are those of birds of paradise in New Guinea and bowerbirds in Australia. The latter (Fig. 9.8) prepare "bowers" of twigs or leaves which they decorate with bright petals of flowers, fruits, shells, coins, silverware, jewelry and other trinkets. One species "paints" its bower with plant secretions and saliva applied with a "brush" of bark or a wad of leaves. In these artistic pavilions they dance and display throughout the breeding season. Birds of paradise (Fig. 9.9) possess incredibly fantastic and often brilliantly colored plumage decorations — ruffs, capes, false wings, and wiry plumes protruding from various parts of the body (Ripley, 1950). They display, turn somersaults, and dangle upside down in their finery. But like some of the bowerbirds, they are so preoccupied with these performances that the males of most species take little or no part in subsequent family affairs.

Equally fantastic, but less publicized, are the dances of the manakins (Pipridae) in the American tropics. Since there are 59 species, the performances vary considerably, but usually the males assemble on traditional perches, hop from branch to branch, flutter in the air like butterflies, flash bright-colored crowns or thighs, and pop up and down like balls in a juggling act. Some performances are solo, but more often several males participate together. The females, often not in evidence at first, finally appear and seem to be stimulated by the acts of the males (Slud, 1957; Snow, 1971a). Somewhat similar performances are staged by some hummingbirds which gather on display grounds called "leks," form "singing assemblies" and perform aerial dances (Snow, 1968; Barash, 1972). Some cotingids (there are 90 species) have equally remarkable displays and calls, utilizing their wattles, beards, and plumages to good advantage (Snow, 1970, 1972).

Another courtship phenomenon, notable particularly in gallinaceous birds, is "tidbiting" or courtship feeding, in which the male attracts the female by proffering choice morsels of food (Stokes, 1971). Such acts appear to overcome the fear or reluctance of the female and are usually followed by pair formation and copulation. Stokes and Williams (1971)

FIG. 9.8 [OPPOSITE, TOP]. *This Fawn-breasted Bowerbird* (Chlamydera cerviniventris) *seems to have assembled largely natural objects at its bower, but often other items — silverware, jewelry, bottle tops, shells, coins, etc. — are confiscated to decorate the site.* [Courtesy of Amer. Mus. Nat. Hist.]

FIG. 9.9 [OPPOSITE, BOTTOM]. *Male Lesser Birds of Paradise* (Paradisaea minor), *like most other paradiseids, dance and display their fantastic plumes throughout the breeding season, but take no part in other nesting activities.* [N. Y. Zool. Soc. photo.]

recorded courtship feeding in 60 species of gallinaceous birds. Krebs (1970) thought that courtship feeding in titmice functioned to provide the female with extra nutrition needed for egg production. It increased her intake two and a half times.

PAIR FORMATION

The processes involved in selection of mates are not well understood, but courtship ceremonies generally culminate in the establishment of mated pairs on territories. Of course the territory is usually the place of union; that is one of its functions. Characteristically the male attacks or threatens any bird, male or female, that enters his domain. Attacked females usually retreat, transients not returning, but would-be residents may remain in or near the area until the male modifies his tactics from aggressive threat to invitatory displays, Sexual flights, with the male in pursuit of the female, are of common occurrence. This is particularly noticeable in ducks, where excess males in many species pursue and harass the females; in Mourning Doves where "three bird chases" usually involving two males and one female are common (Goforth, 1971); and in the "leader flights" of Common Grackles where two or more males follow a female and display their deeply cupped "V-tails" to good advantage (Ficken, 1963). The female is not territory conscious at this stage and may wander from one area to another, but the song of the male may call her back. Some territory-holding males appear to distinguish transient males from potential rivals (residents) and attack only the latter, and also appear to distinguish transient (nonbreeding) females from the more receptive resident females, showing interest only in the latter.

Pair formation takes place early — often in the winter flock — among ducks (Weller, 1965; Smith, 1968), in the American Goldfinch, and in certain sea birds. Richdale (1951) states that in most penguins new pairs are formed and old pairs reunited in the winter before they return to the nesting colonies, but that some pairs may be formed at any time of the year. In the Yellow-eyed Penguin some individuals show definite affinities for a particular mate; one female waited six years for the "male of her choice," and "divorces" and "flirtations" played an important part in the determination of pairs.

SEX RECOGNITION

Sex recognition often poses problems for birds. For though it is a simple matter among birds in which the sexes have strikingly different plumages, in many birds there are only slight visual differences or none at all. Experiment shows that birds use visual clues, at least in part, where such clues are possible. The male and female Yellow-shafted form of the Common Flicker, for example, differ chiefly in the black malar stripe or "mustache" possessed by the male. If a black mustache is experimentally painted onto a female, the male mistakes her for another male and attacks her. Such errors in recognition are usually soon rectified by characteristic reaction, however, for an altered female still reacts as a female and is thus identified by the male. In the Common Yellowthroat the black mask of the male apparently is not necessary for sex recognition; black masks painted on females did not fool the males who seemed to recognize their mates by calls (Lewis, 1972).

When plumages are identical in the two sexes, birds apparently depend on behavioral characteristics to identify each other. Once mated, most males soon learn to recognize their mates by vocalizations, actions, or other behavior patterns.

COPULATION

Sexual union or copulation usually follows pair formation, that is, though there are many variations, the usual sequence of events is a preliminary period of courtship in which selection of mates is determined, often followed by more rituals after the pair is formed before actual coition takes place. Usually the female initiates copulation by an invitatory call, song, display, or posture. She may have repulsed the earlier aggressive advances of the male, and the male in turn does not always respond to the invitation of the female. "False" or incomplete mountings and "reverse mountings" may be frequent before effective copulation takes place.

Copulation is usually repeated at frequent intervals during the breeding season, mainly before and during the egg-laying period in order that the eggs may be fertile. Ducks, American Kestrels, Cactus Wrens, and House Sparrows may start such activities in winter, long before the eggs are laid, and continue throughout the breeding season, but Swallowtanagers (Tersinidae) studied by Schaefer (1953) in Venezuela delayed copulation until after the nest was built. Female Snow Buntings observed by Tinbergen (1939) in Greenland were not receptive during the egg-laying period.

Frequency, and regularity, of copulation vary greatly. Haverschmidt (1946) states that a pair of Little Owls (*Athene noctua*) in his garden came and perched on a branch near the nest tree at about six o'clock each evening and that the male mounted the female from one to four times (perhaps not always effectively) each evening during March, April, and early May. The eggs were laid in late April. American Kestrels, House Sparrows, and Chipping Sparrows are notorious for frequency of coition. Hartman (1959) counted 20 copulations in a pair of captive American Kestrels before noon in one day, and Berger

(1957) observed 11 mountings in rapid succession in a pair of House Sparrows.

Prolonged sexual activities are believed to function more in strengthening pair bonds than to insure fertility to eggs (Willoughby and Cade, 1964). In some cases, however, prolonging such activities may be necessary to assure fertility of later clutches. Unlike domestic hens which have been known to lay fertile eggs three weeks after separation from the rooster, and domestic turkeys which have laid fertile eggs 73 days after separation from the tom, Elder and Weller (1954) found some Mallard eggs infertile after 6 days. They suggested that the function of excess unmated ducks might be to insure fertility of second clutches, if the first clutch failed after the original drake deserted and was molting (see Fig. 9.5 and page 243).

Among "arena" or "lek" birds (some grouse, the Ruff, bowerbirds, and many others) there is no regular pair formation, and copulation is usually promiscuous. Hen prairie chickens visit the dancing grounds of the males, pair temporarily with a dominant male, then withdraw to their respective nesting sites unaccompanied by the males. Similarly in the Sage Grouse, the females visit the ceremonial grounds and are served mainly by the more virile master cocks; one master cock has been known to serve 21 hens in a single morning, but the subcocks and guard cocks are relegated to subordinate roles (Scott, 1942).

Perhaps needless to say, even monogamous birds are not always faithful to their mates. Meanly (1955) reported that male Little Blue Herons in a colony in Arkansas frequently visited neighboring females on their nests and were accepted by them. Ficken (1962) reported that female redstarts accepted, or even solicited, neighboring males when their own mates were absent. Sick or injured birds sometimes seem to stimulate other birds to attack or mount them and "dummy" females set up for experiments in a male's territory usually elicit attempts at copulation.

Types of Pairing Bonds

It is probably apparent by now that monogamy for one breeding season is the prevailing relationship between the sexes, but that other types of pair bonds exist. The various kinds known to occur in birds and examples of each are outlined below, although any attempted classification of pair bonds is quite arbitrary, since individual birds do not necessarily conform to the role assigned to them.

Monogamy (paired to a single mate, may be of short or long duration):
1. Pair for life—swans, geese, eagles (perhaps most raptors), cranes, probably parrots, Wrentits.

2. Retain, or reunite with, same mate for more than one season — many sea birds, Paridae, Corvidae, probably some woodpeckers, nuthatches, and creepers.
3. Pair for one whole breeding season — most passerines and the smaller nonpasserines (the majority of species).
4. Pair for single brood (change mates between broods) — House Wrens (commonly), bluebirds and sometimes other hole-nesters, Dunlin.
5. Shorter term than one brood — most ducks, some hummingbirds (e.g., Hermit), some scolopacids.

Polygamy (paired to two or more mates at the same time, either sex):
1. Conventional polygyny (male with two or more mates) — Ostrich, rheas, pheasants and many other phasianids, many icterids (Redwinged, Brewer's, and Yellow-headed Blackbirds, sometimes Bobolinks and meadowlarks), Marsh Hawks, Bittern (of Europe), Winter Wren, Long-billed Marsh Wren, Dickcissel.
2. Successive (serial) polygyny — Some tinamous (e.g., Boucard's), some oropendolas, some ploceids.
3. Polyandry (female with two or more mates) — rheas, some tinamous, jaçanas, button-quails, some plovers, Spotted Sandpiper, sometimes other scolopacids, sometimes cuckoos and cowbirds.

Promiscuity — most "lek" birds (Prairie Chickens, Sharp-tailed Grouse, Sage Grouse, Ruff), solitary grouse (Ruffed, Spruce, Blue), bowerbirds, some birds of paradise, Boat-tailed Grackles, Sharp-tailed Sparrows, others irregularly (Little Blue Herons, cuckoos and cowbirds, American Redstart).

MONOGAMY AND ITS VARIATIONS

PAIRING FOR LIFE. It has long been supposed that many of the larger birds — sea birds, cranes, falconiform and strigiform birds — mate for life, but conclusive evidence in most cases is lacking. Mating for life usually means until one of the pair dies, when the other may take a new mate, but there are known cases, particularly among captive swans and geese (Fig. 9.10), where the surviving bird never remated. Perhaps the idea of such birds never remating has been overemphasized, however, as there are known examples, particularly among captive geese, of voluntary changes during the life of the individuals as well as remating after the death of one partner.

Parrots, at least in captivity, remain paired for long periods, a condition also believed to apply to the wild species. Among the smaller birds the Wrentits are known to form permanent pairs and some corvids (Jackdaw, magpies, Carrion Crow, ravens) as well as Paridae either

Fɪɢ. 9.10. *Canada Geese are noted for the permanence of their pairing bonds. The gander guards the female and nest, and the whole family stays together until the following spring, when the young are driven away.* [Photo by Miles D. Pirnie.]

form long lasting bonds or else reunite in successive seasons. A Carrion Crow has been known to have the same mate for ten years, and mate changes in magpies appear to be due chiefly to replacements.

RETENTION OF MATE FOR MORE THAN ONE SEASON. Perhaps a more common pairing bond than actually mating for life is the frequent retention of or reunion with the same mate in successive seasons. Richdale (1951), in a 10-year study of Yellow-eyed Penguins, found that they retained the same mates in successive seasons in 82 percent of the possible cases, as compared to 18 percent of separation or "divorces." Known pairs were intact for 7, 9, and 11 years. He believes that other penguins and most albatrosses and petrels follow a similar pattern. Associations between some gulls and terns are often of long duration. Perhaps the nonmigratory Paridae as well as nuthatches and creepers are of this nature; usually it is not known whether they remain paired over winter or merely so closely associated that remating in the spring is facilitated. Downy and Hairy Woodpeckers maintain separate winter sleeping quarters but get together again in early spring. Other winter territory-holders — Mockingbirds, Loggerhead Shrikes, and English Robins — also frequently reunite for the breeding season.

The occasional reunion of migratory birds, however, is probably purely fortuitous, aided by the return of both sexes to their former nesting area. Brackbill (1952), for instance, found only 1 (12 percent) of 8 pairs of returning robins remating, compared to 8 out of 30 (27 percent) in Mrs. Nice's Song Sparrows, and 11 out of 26 (42 percent) in Kendeigh's House Wrens. Brackbill attributed reunion in the robin to the faithfulness of both members of the pair to their previous year's territory and lack of reunion to unfaithfulness to territory.

PAIRING FOR ONE SEASON. In general the way of life among birds favors seasonal monogamy as the most convenient and practical plan. Territory-holding birds pair with a single female and remain together during the nesting season, then separate and go their respective ways. When they return in the spring, males attempt to reclaim their former territory, but the females are more inclined to scatter. Mrs. Nice found that female Song Sparrows in Ohio often returned to their former territories, but usually found the male already mated to a female that did not migrate. Most ducks belong in this category of monogamy for one season, for though the males of most species desert the nesting hens early in the season, they do not remate.

CHANGE OF MATES BETWEEN BROODS. Some birds, such as the House Wren and sometimes Eastern Bluebirds and other hole-nesting species, may change mates between broods, perhaps as a matter of convenience, as one parent, often the male, may take charge of the first brood while the other remates with another partner for a second brood. Some long-term studies of the Eastern Bluebird, however, (Peakall, 1970a; Pinkowski, 1971), did not reveal any cases of separation of mates between broods. But Soikkeli (1967), in a four-year study of Dunlins in Finland, had three females that raised two broods by remating with another male while her first mate took care of the first brood. In the Cedar Waxwing, Putnam (1949) found that overlapping of broods favors continuity of the pairing bond. The male renews courtship before the first nest cycle is finished, and the female abandons her young for the male to finish rearing while she renests. This, incidentally, gives the male an unusually rigorous schedule; he is involved in courtship, building a second nest, and rearing young all at the same time.

SHORTER TERM THAN ONE BROOD. In a restricted sense, only birds that desert their mates early in the nesting season but do not remate — most ducks, some hummingbirds, and various sandpipers — belong in this category, but often it is difficult to distinguish between short-term monogamy of this type and successive or serial polygamy and promiscuity (pages 245 and 247). Among the ducks, drakes may go into an early eclipse and not remate, but a woodcock (Fig. 9.11) may move to a new singing ground and a new mate

FIG. 9.11. *These woodcock will never know their father. He "sang" to and consorted with their mother for a brief period in early spring and then (presumably) moved on to other singing grounds.* [Photo by L. H. Walkinshaw.]

(Sheldon, 1967). The much-studied Ruffed Grouse has been called monogamous, polygynous and promiscuous by different authorities. In Arctic-nesting scolopacids it is not unusual for the "unemployed" sex (the one not incubating) to desert the territory and start its long migration, leaving the food resources in the territory for its mate and young.

POLYGAMY

CONVENTIONAL POLYGYNY. Among the regularly polygynous birds the pheasants are perhaps the best known example. In captivity cock pheasants are commonly mated with five or more hens, but in nature the male's crowing grounds may not be spacious enough, or sufficient hens available, to permit so large a harem. Many other Old World phasianids are regularly polygynous, but our native grouse, though often referred to as polygynous, are perhaps more properly termed promiscuous, and our native quail are essentially monogamous. Ostriches are polygynous, each male having several hens contributing to a family nest over which he takes supervision; in rheas the males are polygynous and the females polyandrous.

Some icterids, notably Red-winged and Yellow-headed Blackbirds nesting in semicolonial situations, are usually, but not invariably, polygynous. In the Red-wing each male controls an area including several nests which he zealously guards, but both sexes frequently leave the territory for feeding and other purposes; the females, indeed, may consort with other males while "off duty," so that the males may not be the fathers of the young they so carefully protect (Beer and Tibbitts, 1950). In Williams' (1952) study of the Brewer's Blackbird some males were monogamous, some polygynous. Cases of polygyny increased when the number of females increased. Kluijver et al. (1940) found polygyny common in Winter Wrens in Europe. Hecht (1951) reported it for Marsh Hawks in semicolonial nest situations in Manitoba, although more commonly a single pair occupies a given marsh. Most monogamous species may on occasions take an extra mate; sometimes when a male disappears, his neighbor takes over the vacated territory and its female occupant in addition to his own.

SUCCESSIVE OR SERIAL POLYGYNY. This type of pair bond is especially well developed in ploceids in Africa, in neotropical oropendolas, and some tinamous. Collias and Collias and co-workers, in a series of papers, and Emlen (1957) have shown that in many African ploceids the male builds a series of nests and by vigorous displays entices a different female to occupy each nest. Indeed, a female may eject the male from the nest he built, if she occupies it, but he merely builds another nest and gets another female, almost ad infinitum. Similarly, some male oropendolas, neotropical relatives of our orioles, build a series of nests (see Fig. 10.7), each one taken over by a different female. This permits maximum productivity where there is a surplus of females. Lancaster (1964) disclosed an even more intriguing situation in the Boucard's Tinamou in British Honduras. The males have calling or whistling territories to which they attract several females who deposit eggs in a nest cared for by the male. But the incubation period is short (16 days), the young are precocial, and the breeding season is long, so that the males are able to repeat this cycle one or more times with the same or other females who, successively, visit other males. Such an arrangement permits maximum reproduction.

POLYANDRY. As in polygyny, polyandry may be of the harem type, as in jaçanas and button-quails (Fig. 9.12), or serial, as in some tin-

FIG. 9.12. *Barred Button-quail* (Turnix suscitator) *are polyandrous. The larger, more brightly colored female lays several clutches of eggs, leaving a male in charge of each.* [From Delacour, *Birds of Malaysia,* copyright 1947 by Jean Delacour, Macmillan Publishing Co., Inc., New York.]

amous, rheas, and sandpipers, although overlap in the two types is inevitable since it takes time for the female to produce more than one clutch. It has long been known that jaçanas, at least in some species, are polyandrous, but recently Jenni and Collier (1972) have disclosed that in the American Jaçana (*Jacana spinosa*) in Costa Rica the larger female sets up and defends territories occupied by different males, but does not feed on the nesting territory, thus leaving all the food for the small male (males averaged 91 grams, females 161 grams) and his young. Most females had two males, some three or four, some only one. Button-quails in the East Indies are also regularly polyandrous, each female, in most cases, having two or more clutches of eggs being incubated more or less simultaneously by males.

In Lancaster's tinamous referred to above, the polyandrous females visited several whistling male territories in succession and deposited eggs in each nest. Polyandry is not universal among tinamous, however. Although William Beebe originally reported that males outnumbered females four to one in British Guiana, Pearson and Pearson (1955) found sex ratios about even in Peru, and Schaefer (1954) found some cases of polygyny among the older males in Venezuela. In rheas the males are polygynous, each having a temporary harem of hens, but the females are polyandrous and contribute eggs to several different nests attended by different males. In Spotted Sandpipers, and some other shorebirds in which the male incubates the eggs, the female may lay two clutches of eggs for two different males (Hays, 1972; Oring and Knudson, 1972), although cases are known where the female takes charge of one of the clutches. Raner (1972) disclosed one case of polyandry in the Northern Phalarope, but Höhn (1967b, 1971) has shown that all three species of phalaropes are usually monogamous in spite of their well-known sex reversal.

As in the case of occasionally polygynous males, females may become polyandrous when extra males are available. Wynne-Edward (1952) found an excess of males among longspurs on Baffin Island; in one case three males were attending one nest-building female. Gross (1952) reported a female Hicks' Seed-eater (*Sporophila aurita*) in Panama with two mates during the nesting season; both were seen to copulate with the nesting female and both had singing posts and fed the young. Hann (1940) reported a similar polyandrous relationship in Ovenbirds. In most species with an excess of males, however, the extra males remain unmated, but are available for replacements in case a female loses her first mate.

PROMISCUITY

In the three currently recognized genera of our native grouse (prairie chickens, Sharp-tailed Grouse, and Sage Grouse) courting, mating, and copulation are communal and promiscous affairs and there is no regular pair formation. In other grouse (Ruffed, Spruce, and Blue Grouse) the males have individual (rather than communal) performing areas, but are promiscuous in mating. Birds of paradise on display perches, bowerbirds in their bowers, and honeyguides on "stud" trees are usually, but not invariably, promiscuous. McIlhenny (1937) states that in Boat-tailed Grackles in Louisiana the females visit the male flocks, which are not stationary, and mate at random with them. Woolfenden (1956) reports that Sharp-tailed Sparrows are promiscuous but that Seaside Sparrows are not.

Replacement of Lost Mates

Another factor that often prevents long-lasting bonds between birds is high mortality during nesting, which in many cases necessitates quick replacements if nesting is to be successful. If a female is lost, the male renews singing and displaying and soon secures another mate. If a male is lost, the vacated territory is promptly taken over by another male.

An experiment in collecting birds on their territories in a spruce-fir forest community in northern Maine showed that the collected males were usually replaced overnight; 455 birds, mostly males, were taken where only 148 males held territories at the beginning of the study. The following year 154 pairs established territories and 518 birds were collected (Stewart and Aldrich, 1951; Hensley and Cope, 1951). Both Allen (1961) and Odum (1942) tell of a female Black-capped Chickadee that had three successive mates in one season, the males presumably having perished. Griscom (1945) describes an investigation in which male Indigo Buntings were collected to see how quickly the lost birds would be replaced. Nine males in succession were taken from a single territory; the tenth was left to help the oft-widowed female raise her young. We have noted similarly quick replacements of lost mates in American Robins following "die-offs" in local spraying programs.

Helpers at the Nest

In many cases where more than two adults are found attending a nest, it is not polygamy, promiscuity or a broken pair bond, but merely an unmated bird joining in voluntarily to help out in family affairs. In the tropics these "helpers at the nest" are of common occurrence; often it is a case of one or more of the young of the previous nesting remaining with their parents and helping in the rearing of another brood. In a survey of his own voluminous Central American records and of the literature (Skutch 1961) listed more than 130 species known to serve as "helpers." Extra-parental cooperation is common among Chimney Swifts (Dexter, 1952), the helpers actually sharing in incubation and brooding as well as feeding, and sometimes renewing such cooperative arrangements in successive years. A study of this peculiar relationship in Ohio disclosed 22 "threesomes" and 6 "foursomes." Tree Swallows also often form "threesomes."

Trios in the Yellow-eyed Penguin are fairly common, usually due to a bird which has lost its mate trying to join with an already mated pair (Richdale, 1951). Bleitz (1951) reported two pairs of Pygmy Nuthatches attending a nine-egg nest (normal maximum for this species) and all four feeding the young. Lawrence (1960) cited three cases of trilateral relationships at nests of the Gray Jay in Ontario, but suggested that in one case where two females were sitting on the same nest that the male in the triumvirate may have been lost and that the two females tried to carry on as a pair (the eggs never hatched).

Selected references follow Chapter 11.

10

The Annual Cycle:
Nests, Eggs, Incubation
and Hatching

The Nest

THE SELECTION of a nest site and construction of a nest, if any, are the first major tasks of a pair of birds in the breeding season. Typically the nest site is selected by the female on the territory defended by the male, but there are numerous exceptions. In some cases the male chooses, or at least designates, suitable locations. In hole-nesting species the male may lead the female to available nesting cavities within his territory, or, as in some of the wrens, the female may take over and finish a nest already largely constructed by the male. In Prothonotary Warblers the males select the nest cavity and start carrying moss before the females arrive (Walkinshaw, 1953). In the Cedar Waxwing, nest-site selection appears to be a cooperative project, the pair seeking a suitable place together, and sometimes forming the territory around the chosen location (Putnam, 1949). In the American Redstart (Ficken, 1964) the female explores the whole territory defended by the male, then examines more specific sites, rejects the "suggestions" of the male, and tries out various crotches in trees before gathering material for the nest.

Nests occur in many different locations, from burrows in the ground to the tops of large trees. They may be on any type of ground surface, in herbaceous or shrubby plants, in a great variety of arboreal situations, in banks and caves, floating on or suspended over water, and in or on many man-made structures. Some of these many nest situations are indicated in the following account of nest structures.

NEST STRUCTURES

The nests of birds exhibit great variation and levels of workmanship, from no prepared structure to the amazingly fantastic creations of some tropical birds. The obvious and often indispensable function of the nest is to house the eggs, but the facts that many ground-nesting and cliff-nesting birds build none and that poorly constructed nests often seem to serve as successfully as the most elaborate types, indicate that the art of nest building has, in many cases, gone far beyond purely utilitarian requirements.

Various authors (e.g., Herrick, 1911; Collias, 1965) have attempted to classify nests, but such classifications are artificial and have little evolutionary significance. Many tree nesters, including the American Robin and Mourning Dove, sometimes nest on the ground, and ground nesters sometimes build elevated nests, often modifying the structure to fit the new location. Nickell (1958), who collected 20,000 nests of 169 different species over a period of 35 years, outlines some of the engineering features involved in the various types of nests. The following paragraphs describe some of the many variations in structure and placement of nests.

NO PREPARED NEST. Among the birds not constructing a nest are some shorebirds that lay their eggs on the bare ground; they may or may not place a few stalks of vegetation, pebbles, or shells about the eggs. Likewise Whip-poor-wills deposit their two eggs on the forest floor (Fig. 10.1), on whatever litter happens to be available; and their close relatives, the nighthawks, make similar use of gravel beds, pavements, or the flat-topped roofs of buildings, sometimes experiencing difficulties in keeping the eggs from rolling away. The Turkey Vulture builds no nest as such, but usually seeks the protection of a hollow log or stump. Some cliff-dwelling sea birds lay on rocky ledges with no nest materials to mark the location. Primary hole-nesters like woodpeckers use no nest materials except incidental chips or sawdust, but secondary users such as titmice and wrens build substantial nests.

Several unusual situations may be cited. The Fairy (Elegant) Tern (Fig. 10.1) on islands in the Pacific precariously balances its single egg on the horizontal branch of a tree with no nest material to hold it in place. A jaçana in Africa, a "lily-trotter" with an 8″ × 6″ expanse of toes as long as the bird itself, places its four eggs on a broad lily pad or cluster of floating water weeds; later, it may pull a few supporting stems about the eggs. When the nest site submerges during incubation, the eggs are taken up under the wings (Miller, 1952). Emperor Penguins,

---

FIG. 10.1. *No prepared nest.* TOP *Whip-poor-wills lay their eggs on the bare forest floor, usually with no semblance of nest materials.* [Photo by Edward M. Brigham, Jr.] BOTTOM *Fairy Terns* (Gygis alba) *sometimes place their egg on a bare branch, where it is incubated and the young chick brooded without apparent loss. The young are well equipped with sharp claws for gymnastics on the branches.* [Courtesy of Amer. Mus. Nat. Hist.]

after a long inland trek from the sea, nest on the Antarctic ice during the long *winter night* in temperatures ranging from 40 to 70°F *below* zero (Rivolier, 1959). With no nesting materials available, the male stands or shuffles around on the ice for two months while holding the single egg on the web of its foot with a fold of skin from the abdomen enveloping the egg to keep it warm.

GRASS AND STICK NESTS. The next step in the evolution of nest structures is a simple affair of grass or twigs. Ground-nesting birds like gulls and terns have little need for nest materials and often use only a few grasses or bits of vegetation surrounding or underlying the eggs. Some ground-nesting thrushes, warblers, and sparrows build more substantial though not necessarily elaborate nests. A further step in the development of ground structures is the arched-over grassy nest of the meadowlarks and the ovenlike grass and leaf nest of the Ovenbird, which of course have considerable protective value. Roseberry and Klimstra (1970) have shown that roofed-over nests of the Eastern Meadowlark in Illinois were more successful than meadowlark nests without a protective canopy.

For tree-nesters some material is usually a necessity, yet birds like the Mourning Dove get along with a minimum. Often a dove's nest is so skimpy that the eggs can be seen through the interstices from below; if it is on a prostrate log or large horizontal branch there may be only a few twigs about the eggs (Fig. 10.2). North American cuckoos, the Gray Catbird, and Cardinal do a better job, yet often the nest is shallow and insecurely fastened to the supporting branches so that the nest may get dislodged. A crude platform of sticks seems to serve the needs of tree-nesting egrets, herons, and other rookery birds, but they may reoccupy the nest in successive seasons and add more materials.

Predatory birds (Fig. 10.2) commonly build even more bulky structures of sticks and leaves and may add to them from year to year, though a particular nest may serve for a crow one season, and a hawk or a Great Horned Owl the next. Bald Eagles, if unmolested, usually reoccupy and enlarge their eyrie annually until the huge structure may collapse. A nest at Vermilion, Ohio, was occupied for 35 years and was estimated to weigh about two tons when it finally crashed (Herrick, 1934). A Bald Eagle nest in Florida, perhaps the largest known, measured 20 feet deep and 9½ feet wide (Broley, 1947).

MUD NESTS. The use of mud as a construction material is well ex-

---

FIG. 10.2. *Simple twig nests.* TOP *Mourning Dove nest, usually a skimpy affair in trees, was here placed on a log.* [Photo by Edward M. Brigham, Jr.] BOTTOM *The Osprey builds a bulky nest of sticks, usually in dead trees, but often on telephone poles and power lines in the East and on rocky crags in the Western mountains.* [Photo by L. H. Walkinshaw.]

emplified in the familiar mud and grass nest of the American Robin. It is a fairly solid and substantial structure, which rain or wind only occasionally dissolves or dislodges. When mud is not immediately available, robins have been known to manufacture mud by dipping dry earth into water or to convey water on their feathers to the dirt. The Barn Swallow (Fig. 10.3) and the Eastern Phoebe are likewise mud-daubers and take the additional precaution of nesting on buildings and bridges for protection. The nest of either species may be *statant* (supported from below) or *adherent* (plastered to a vertical surface). Cliff Swallows (Fig. 10.3) build a globular adherent structure of clayey materials that eventually harden to a mortarlike consistency. Flamingos use mud even more elaborately. They heap up pyramidal cones one or more feet high, hundreds or even thousands of nests in a colony, and place their usually single egg in the concave top of the nest.

NESTS OF PLANT DOWN.   Among the many birds that use plant down in their nests the Yellow Warbler (see Fig. 8.8) and American Goldfinch (Fig. 10.4) are particularly noted for well-made exquisite structures composed of the down of willows, poplars, and thistles or other Compositae. So durable are these nests, and so securely anchored to supporting twigs, that they survive more or less intact over winter; often they are remodeled for winter nests and storage places of deer mice (Nickell, 1951). The deeply cupped nest of the American Goldfinch is relatively waterproof; if filled with rain it holds water for 24 hours, causing abandonment of unattended nests, or drowning of the young if they are neglected by their parents during storms (Lewis, 1952).

LICHEN-COVERED NESTS.   The nest of the Ruby-throated Hummingbird (Fig. 10.5), neatly constructed of bud scales and spider webbing, and attractively camouflaged with gray-green lichens, is a well-known example of a lichen-covered nest. The Blue-gray Gnatcatcher and Eastern Wood Pewee build similarly attractive and substantial structures that blend well with the branches on which they are placed. Nests of the Parula Warbler in the northern states are commonly tucked into a dangling tuft of lichens (*Usnea*); in the southern states similar use is made of Spanish moss (*Tillandsia*—a flowering plant). Where the "moss" does not occur, the Parula Warbler is said to be absent.

FEATHER NESTS.   Many birds use feathers in their nests, either plucked from their own bodies or collected from other sources. Examples of the former are the down-feather nests of ducks and geese. The females of these species grow a special nuptial down that is plucked for lining their nests. Among Arctic-nesting eider ducks, the production and use of this down are so profuse that it completely surrounds and covers the eggs when the female is absent; it can be har-

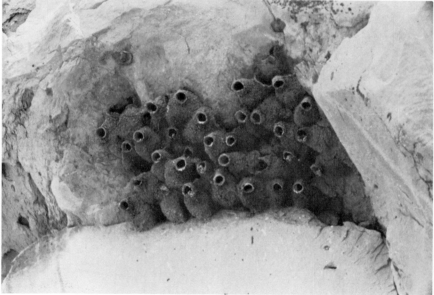

Fɪɢ. 10.3. *Mud nests.* ᴛᴏᴘ *The Barn Swallow builds a solid substantial nest largely of mud and almost invariably adopts man-made structures for additional protection.* [Photo by Dale A. Zimmerman.] ʙᴏᴛᴛᴏᴍ *The Cliff Swallow, though often nesting under the eaves of buildings (see Fig. 8.7), sometimes utilizes natural nesting sites on cliffs or ledges.* [Photo by John T. Emlen, Jr.; courtesy of the *Auk*.]

F I G. 10.4 [OPPOSITE, TOP]. *Felted nest. The American Goldfinch builds a compact, substantial nest of felted materials that weathers the elements well.* [Photo by Walter P. Nickell.]

F I G. 10.5 [OPPOSITE, BOTTOM]. *Lichen-covered nest. The Ruby-throated Hummingbird builds a decorative yet well-camouflaged nest of bud scales, lichens, and spider webbing.* [Photo by Edward M. Brigham, Jr.]

vested commercially (page 402) without apparent loss of hatchability of the eggs. Phoebes, wrens, and swallows, among other birds, pick up stray feathers to incorporate into their nests; the frequent use of chicken feathers often results in infestation of the nest and its occupants with lice and mites.

PENSILE AND PENDULOUS NESTS. Vireos build a pensile nest (Fig. 10.6), a compact bark-fiber affair suspended from a terminal fork of a limb. Like the nests of the American Goldfinch and Yellow Warbler, vireo nests weather the elements well. Pensile nests are also characteristic of kinglets, seven species of blackbirds, the Parula Warbler, and Acadian Flycatcher (Nickell, 1958). Chamberlain (1954)

F I G. 10.6. *Pensile nest of Warbling Vireo.* [Photo by Allan D. Cruickshank.]

tested the strength or "safety factor" of Yellow-throated Vireo nests by placing weights in them. Minor tearing (rearrangement but not breakage of fibers) took place at 45 ounces and total failure at 76 ounces (4.75 pounds).

Pendulous nests are characteristic of some orioles and other icterids, many tropical flycatchers, broadbills, some sylviids, and some weaverbirds. Northern Orioles construct a deeply cupped cradle, which swings in the breeze. It is woven of long strands of grasses, bark, hair, and yarn or string sometimes provided by homeowners for the birds. Oropendolas, colonial neotropical relatives of our orioles, build a drooping basket, 3–6 feet deep, perhaps 50 nests to a tree, with the entrance near the top and the eggs nestled in the toe of the structure (Fig. 10.7). Nests of the Common Tody-flycatcher (*Todirostrum cinereum*) in Colombia are similar, but only a few inches deep with a neat roofed-over entrance in the upper third of the otherwise completely enclosed structure.

Broadbills (see Fig. 13.7), a passerine family found in Africa and southern Asia, build a globular grass and fiber nest suspended from a branch by a single thread, often over water; the side entrance is frequently protected by an overhanging portico. Some weaverbirds also build a suspended nest, but have the entrance at the *bottom*, with an ingenious platform within provided with a guard rail to prevent the eggs from falling out. Pensile and pendulous nests, and other ingenious devices, are more common among tropical birds, perhaps as safety measures from the many predatory hazards they face.

HOLE- AND BURROW-NESTERS. Many species of birds throughout the world utilize holes in trees, crevices, or burrows in the ground for nesting. These provide relatively safe sites compared to nests in the open. Woodpeckers, nuthatches, and some titmice excavate holes in dead trees, stumps, or branches, and lay their eggs in the cavity. The original builders usually abandon these holes after one nesting, so that they are available for bluebirds, Tree Swallows, House Wrens, and other hole-nesters not equipped to excavate their own. Bank-dwellers, such as kingfishers and Bank Swallows, construct deep tunnels in sand or gravel banks and deposit their eggs at the rear of the cavity. Tunnel or burrow nesting is common among tropical birds, perhaps as a protection against snakes, and among oceanic birds like petrels and shearwaters, perhaps as a protection against nest-marauders. Kiwis in New Zealand nest in underground burrows that they dig with their strong claws (see Fig. 7.2).

FIG. 10.7. *Pendulous nest of an oropendola* (Psarocolinus angustifrons). *The entrance, not shown here, is at the top of the expanded portion.* [Photo by Bruce Frumker; courtesy Clevelend Mus. Nat. Hist.]

Some oddities among hole-nesters include the Burrowing Owl, which often adopts prairie dog or ground squirrel holes for underground nesting; Gila Woodpeckers and Elf Owls, which nest in cactus stubs; and many tropical species (49 species in 7 different families according to Hindwood, 1959), which tunnel into large termite nests, presumably acquiring some immunity from predators by the presence of the biting insects. Wrens get into all sorts of curious nesting situations; not the least of these was a human skull hung up at a doctor's home in Kentucky and said to have been occupied by wrens from 1888 to 1945 when the skull was moved to Indiana and reoccupied there (Hobson, 1952).

AQUATIC NEST SITES.   Marsh and water birds resort to a great variety of nest sites. They may utilize muskrat houses or beaver lodges, placing their nests on top or in the walls. Canada Geese, normally ground-nesters, sometimes make use of tree nests of Ospreys and herons or adopt artificial platforms—even washtubs or baskets—placed in trees (Yocum, 1952). Least Bitterns (see Fig. 7.5), American Coots, rails, gallinules, and Black Terns construct simple platforms composed of aquatic plants and usually placed over shallow water. Problems of flooding are often partially circumvented by anchoring the nest loosely among tail reeds so that it will adjust itself to changing water levels. Nests of the Red-winged Blackbird, however, are more firmly attached so that unequal growth of the supporting stems sometimes tilts the nest and spills the eggs or young. Long-billed Marsh Wrens forestall such mishaps by building an enclosed nest. The nest of the Pied-billed Grebe, sometimes not anchored to vegetation at all, is a large, floating, water-soaked mass of material (Fig. 10.8); when the incubating bird departs, the eggs are covered over with wet plants without apparent impairment of their hatchability. The nest of the Horned Coot (*Fulica cornuta*), in cold water lakes in the high Andes, at 13,000 or more feet, is a huge mound of stones, perhaps a ton or more, with its base in the water and the top of the rockpile lined with aquatic plants.

CAVE AND CHIMNEY DWELLERS.   Many swifts, some swallows, and the Oilbird (Steatornithidae) nest in caves, or suitable substitutes, such as mineshafts and chimneys. The Chimney Swift builds a platform of twigs glued together with a secretion from the salivary glands (see Fig. 11.1). The shallow structure is then glued to some surface, formerly in a hollow tree but now almost exclusively inside chimneys or similar shafts. More remarkable are the edible nests of swifts in subterranean caves in the East Indies. The first nests of these birds are almost entirely of salivary secretions, and are collected by the natives and by visiting Chinese for making a highly coveted but not very nutritive birds'-nest soup. Replacement nests, which contain more foreign materials, make a lower-grade soup and are usually left for the birds.

FIG. 10.8. *Floating, water-soaked nest of Pied-billed Grebe. The conspicuous eggs are covered with damp vegetation when the incubating bird departs.* [Photo by Allan D. Cruickshank.]

Some Javanese people build platforms for the swifts in their homes, thus assuring themselves of some income from the nests.

OTHER NESTS. The Dipper builds an extraordinary nest (Fig. 10.9). It is a spherical mass of green moss with an inner bowl of dry grasses, usually tucked into a crevice beside a swift-flowing stream or waterfall so that the outer walls of the nest are often kept moist and green from the spray. Several species of swifts build their nests behind waterfalls, sometimes flying through the water to reach their nests. The nests of the Black-billed Magpie, in trees or thorny bushes, are huge affairs several feet deep and completely domed over, often with a crown of protective thorns. Abandoned magpie nests, which may distract predators from active nests (Erpino, 1968), are often used by other birds for shelter in inclement weather and for breeding purposes in spring. Social weaverbirds in South Africa are gregarious even in the nesting season (see Fig. 8.13); some combine their efforts in the construction of a common domicile. It is a huge, umbrella-shaped, roofed-over apartment house honey-combed with separate noncommunicating, warmly feather-lined cavities that serve for both nesting and shelter.

FIG. 10.9. *Mossy nest of Dipper beside a Colorado stream.* See page 104. [Photo by H. W. Hann; courtesy of *Condor.*]

Some birds build supplementary, auxiliary, secondary, or dummy nests (terms used more or less interchangeably) for a variety of purposes. Many tropical birds construct special nests or dormitories for sleeping (page 319). Black Terns build auxiliary nests to which the young may move, voluntarily or when disturbed, when three to five days old, and continue to be cared for (brooded and fed) by the adults (Cuthbert, 1954). Coots and grebes may build special copulatory platforms, not necessarily used for nesting. Wrens are particularly noted for building supplementary nests. Cactus Wrens (Anderson and Anderson, 1960) work jointly in the construction of a breeding nest; then the male builds one or more secondary nests, which the female uses for second broods and the male uses for sleeping. Male House Wrens commonly fill up all available crevices in their territories with sticks; then the female completes one of them for her nest. Polygynous Long-billed Marsh Wrens build many dummy nests, as many as 20 in a breeding season, display in front of the nests to prospective females who may or may not adopt one of them; those not used for nesting serve for sleeping quarters for adults, secondary shelters for nestlings, and for winter roosts for wrens that do not migrate (Verner, 1965).

Nest-Building

Nest-building techniques and construction materials vary greatly. Materials used are commonly, but not invariably, those close at hand, but the innate requirements of the species may also govern selection; a dove must have its twigs, a robin its mud and grass, a duck its special down—materials all readily available to the builder. On the other hand, a Chipping Sparrow may be hard pressed to find horsehair or a suitable substitute for lining its nest, a Tree or Barn Swallow may risk its life for the chicken feathers it uses, and the Great Crested Flycatcher may possibly have to go far afield to find a cast-off snakeskin for decorating its doorway. Snow Buntings often range widely for coveted items; in one case all the buntings in the neighborhood visited a dead gull behind an Eskimo tent for feathers for their nests (Tinbergen, 1939). McCabe (1965) identified feathers from 25 species of birds in House Wren nests. Thistledown is so highly prized for nest construction by American Goldfinches that it cannot be controlled for exclusive use by a pair on its own territory; in fact, the down is often acquired by the birds dismantling their old nests and those of other species. Red-breasted Nuthatches and Red-cockaded Woodpeckers smear pitch around the entrance to their nests, perhaps for protection from nest robbers though the precaution is often ineffective (Dennis, 1971).

Thus innate requirements play an important role in the nest-building habits of birds. A list of the natural materials used by birds would be long and include items from nearly every species of higher plants, as well as the hair, feathers, skin, shells, or skeletal fragments of many animals. Barn Owls utilize both fur and bones from regurgitated and dissected pellets of their prey for nest materials. Kingfishers make similar use of fishbones and scales. Goertz (1962) records a Tufted Titmouse plucking hairs (at least 1,500) from two live opossums; the animals protested at first but eventually gave up. More than 30 species of birds are known to use cast-off snakeskins in their nests; 8 of them do it so regularly as to suggest that the incorporation of the skins in the nest is deliberate (a species habit) and not accidental (Rand, 1953b).

Many birds have learned to supplement natural items with materials made available by man. Pieces of cleaning tissues and other roadside and picnic litter, unfortunately, are readily available. Gross (1958) describes a nest of a bananaquit (Coerebidae) on Tobago Island (West Indies) that was built chiefly of tinsel in a Christmas tree in a hotel. In spite of some inconveniences (the birds also nested on a chandelier and left droppings on the carpets), the management permitted the birds to stay because guests were amused. Broley (1947) found a strange assortment of curios in Bald Eagle nests in Florida. These included golf balls, electric light bulbs, a bottle, a fish plug with 70 feet of line, a child's dress, a shoe, and a white rubber ball that an eagle "incubated" for six weeks after her real eggs hatched.

Birds use their bills and feet for collecting, carrying, and weaving together the materials used in nest construction. Collias and Collias (1964, 1973) describe the intricate weaving process of nest-building and the development of nest-building behavior in an African weaverbird, which builds an enclosed pensile nest with a brood chamber and entrance tube. In birds with more open nests (e.g., thrushes) rotary motions of the body make the usually cup-shaped cavity conform to the size of the sitting bird, thus forming a more efficient watershed during incubation and brooding. Birds that drill holes in decaying trees chisel out chips and carry them away or drop them at the nest site. Kingfishers and other tunnel-nesters dig with their bills, but kick the dirt out behind them. An Old World tailor-bird (Sylviidae) is said to puncture 16 holes in each edge of the single leaf that encloses its nest, then to sew the leaf up with plant thread, even to tying the knots. Sometimes two or more leaves are joined in this fashion.

The time required for nest construction of course varies with the elaborateness of the nest and the diligence of the workers. Song Sparrows (Nice, 1937) were seen to take from 3 to 13 days to build their fairly simple nest, but 4–5 days was usual. Ovenbirds (Hann, 1937) required about 5 days in all to build their first nests; the main body was constructed rather rapidly, in about 2 days, while the lining was added more leisurely. Robins have been observed to take from 6 to 20 days, but intervening spells of rainy weather accounted for the longer periods. One nest was built in a single day, with both sexes working, perhaps of necessity, for an egg was laid the next day. Stokes (1950) found that American Goldfinches averaged 13 days for July nests, but only 5–6 days for August nests.

More elaborate nests require more time. Black-billed Magpies studied by Erpino (1968) in Wyoming took about 40 days to build their huge stick-basket nest. Hammerheads (Scopidae) in Africa may spend 3–4 months constructing their gigantic, clay-lined grass and stick domicile, but once built it is used year after year. Cayenne Swifts (*Panyptila cayennensis*) may spend 6 months or more constructing their amazing nest—a felted sleevelike tube 2 feet long made of the feathery tufts of plant seeds glued together with saliva and fastened to the side of a building or tree trunk (Haverschmidt, 1958). Male megapodes (Megapodiidae) in Australia may spend most of the year building, repairing, or at least attending to the mound in which the females lay their eggs.

In general, northern birds, often working under pressure of a short breeding season, take less time for building than southern species. Derby Flycatchers in Mexico took 24 days for nest construction (Pettingill, 1942), whereas their northern relatives require less than half that time. A seed-eater (Fringillidae) in Panama took 18–19 days to build one nest (Alderton, 1961), about three times the time required for the more northern fringillids. Arctic-nesting passerines commonly save time by reoccupying old nests; one Wheatear had an eight-layered nest, which suggested occupancy over an eight-year period (Wynne-Edwards, 1952).

Nest-building is not normally a steady or uninterrupted process. Birds may work actively at the nest for a few hours in the morning, then cease building for a time. Inclement weather may interrupt nest-building operations or cause desertion of partly built or even completed structures. Some nests, especially those with mud walls, are allowed to "set" a day or two before the lining is added; then one or more days may intervene before the first egg is laid. In many species (herons, hawks, doves), nest-building and incubation overlap, with fresh materials being brought to the nest more or less continuously. Wolf and Wolf (1971) suggest that female hummingbirds may save valuable energy during initial nest-building and egg production by postponing completion of the nest until incubation, when less expenditure of energy is required. A bird's attachment to its nest is not very strong at first, and it readily deserts if disturbed. Even if undisturbed, the female may decide on another nest site, sometimes deserting her mate and territory. Nest-building is apparently largely instinctive, for though a Northern Oriole may improve its technique with age and experience, many birds do as thorough a job with their first nest as with subsequent attempts.

Usually the female takes the initiative in nest construction, selecting the site and doing most or all the building. In many species the males help gather materials; these may be merely token presentations that the female as likely as not discards; or in other cases, the male may bring all or nearly all the material, leaving the female to work it into the nest. In still other cases the male and female work cooperatively, on a nearly equal basis (Fig. 10.10). In the Phainopepla, a handsome black and white passerine bird in southwestern United States, the male does most of the nest-building before he acquires the female who joins in on the finishing touches (Rand and Rand, 1943). In still other birds—the Ostrich and its allies, kiwis, tinamous, and some shorebirds—there is a nearly complete sex reversal in family affairs with the male constructing the nest if one is needed. In ducks, most gallinaceous birds, hummingbirds, and "arena" birds, however, the male usually has nothing to do with the nest. The onerous chore of chiseling out woodpecker holes or

FɪG. 10.10. *Avocets, unlike phalaropes (Fig. 10.16), but like many other shorebirds, work cooperatively at domestic chores. Both sexes join in group courtship activities, and share nest-building and incubation duties. The peculiar recurved bill is used with a sweeping motion to rake aquatic invertebrates from shallow waters.* [Photo by Allan D. Cruickshank.]

kingfisher tunnels is usually a joint enterprise, the male and female working in shifts. In the Black-capped Chickadee, the female does two thirds of the drilling and the pair may work together or in shifts, but apparently only the female carries nesting material (Odum, 1941). Mountfort (1957) reports that a pair of bee-eaters (Meropidae) in France alternated in swooping at a hard-baked earthen bank, striking it with their bills, then one sitting exhausted after 15–18 strikes while the other worked. Not all nest sites, however, were that difficult.

The Eggs

Gᴇɴᴇʀᴀʟ Cʜᴀʀᴀᴄᴛᴇʀɪsᴛɪᴄs

The eggs of birds vary in size, shape, color, and texture. Not surprisingly, the smallest eggs are laid by the smallest birds, the hummingbirds, and the largest eggs by the largest bird, the Ostrich (Fig. 10.11). Egg size or weight, however, is not always strictly correlated with the

FIG. 10.11. *Comparative egg sizes. Ostrich, Domestic Hen, and Ruby-throated Hummingbird.* [Photo by Philip G. Coleman, from specimens in Mich. State Univ. Mus.]

size or weight of the bird. The eggs of precocial birds, for example, which contain more nutritive material (yolk and albumen) and hatch out more fully developed young, are comparatively much larger than the eggs of altricial birds (see Fig. 10.18). The eggs of Old World cuckoos are peculiarly small for the size of the parent, as an adaptation for deposition in the nests of small host species. The Ruddy Duck lays larger eggs than does the Canvasback, which is three times the size of the smaller bird (Fig. 10.12). A 14-egg clutch of the Ruddy Duck, laid at the rate of nearly an egg a day, weighs about 3 pounds, though the female parent may weigh only 1 pound. The kiwis lay even larger eggs,

a.   b.

FIG. 10.12. a. *Canvasback female and egg.* b. *Ruddy Duck female and egg. The smaller bird lays a slightly larger egg.* [Drawings by Diane Pierce.]

but often only one, probably the largest egg known in relation to the size of the parent, although the total clutch of a Ruddy Duck would have a greater volume. The largest eggs known, probably the largest animal cells, are those of the extinct elephantbirds of Madagascar whose egg (in the larger species) measured 13″ × 9.5″, weighed about 18 pounds, and had a capacity of about 2 gallons. This has been computed to be equal in volume to 6 Ostrich eggs or 148 hen eggs.

The shape of a bird's egg (Fig. 10.13) necessarily conforms somewhat to the limitations of the oviduct in which it is molded, but may vary from nearly round, as in the nearly sedentary owls, to extremely elongate, as in fast-flying swallows and swifts. Thus the length and breadth of the egg have some correlation with the speed of flight, or more precisely with the diameter of the oviduct during passage. The pointed, pear-shaped egg of the murres and other alcids is believed to be an adaptation for nesting on precarious ledges; the shape of the egg and concentration of weight at the larger end cause it to pivot on its axis when disturbed rather than rolling off the cliff. The four-egg clutches of many species exhibit considerable tapering toward one end; obviously they fit in a nest better that way.

The colors and pigmentation in birds' eggs (Fig. 10.14) run nearly the gamut of possible hues and patterns, but the brighter colors are unusual. White and off-white shades predominate, with or without the brownish or darker maculations that are characteristic of eggs laid in exposed nests. White eggs may be so thinly peppered with spots that the markings are hardly visible, or so heavily pigmented that they appear all brown like the eggs of the Long-billed Marsh Wren. The configuration of markings is determined by the motions of the egg in the oviduct; spots result when the egg is stationary, and scrawls and scratches when the egg is in motion. Hyndman and Hyndman (1972) have pointed out that the apparent "pigmentation" on some Golden

F IG. 10.13. *Egg shapes.* Left to right, *Double-crested Cormorant, Common Murre, Bobwhite, Great Horned Owl,* and *a bee-eater* (*Meropidae*). [Photo by Philip G. Coleman from specimens in Mich. State Univ. Mus.]

Eagle eggs is caused by blood stains from ruptured blood vessels in the oviduct or cloaca during passage of the egg; the stains are reddish at first but darken with age. Heavier deposition of pigments on the larger end of eggs, often as an encircling wreath, is fairly characteristic, perhaps because the larger end often leads in the oviduct. Blue eggs, with or without spots, are characteristic of thrushes, while those of the Gray Catbird are more greenish. Red eggs are unusual, but a warbler (sylviid) in Japan lays a red egg, which is almost perfectly matched by the egg of an imposing cuckoo. The Namaqua Sandgrouse (*Pterocles namaqua*) in South Africa also lays red eggs occasionally, although green is the more common color (Maclean 1969). The tinamous in South America lay eggs of exceedingly variable colors. They may be blue, green, brown, yellow, or even purple. Many birds that normally lay colored or spotted eggs occasionally lay all white or "albino" eggs, sometimes a single egg in the clutch, sometimes a whole set (Gross, 1968).

Egg colors may change strikingly with age, beyond the mere tarnishing that goes with continued incubation. Rhea eggs, for instance, are said to be a rich green when laid, but later fade to yellow, then to blue, and finally to white. The Emu of Australia usually lays a dark green egg, but completely black eggs of this species have been collected. In general, white eggs are characteristic of birds that lay eggs in concealed places, such as in holes in trees or in burrows, whereas more protectively colored spotted eggs are found in exposed nests, but there are exceptions to both of these generalizations.

In texture the eggs of different species vary from those with thin, fragile, and transparent shells (which become more opaque with age) to those with thick-shelled chalky coverings. The eggs of the tinamous have a smooth glossy surface, like glazed porcelain (Fig. 10.14), while those of the Emu and of cassowaries are rough, and those of toucans

FIG. 10.14. *Pigmentation of eggs.* Left to right, *Domestic Hen, Silver Pheasant* (*buff*), *Royal Tern, tinamou* (*smooth and glazed*). [Photo by Philip G. Coleman from specimens in Mich. State Univ. Mus.]

and the Ostrich are pitted and grooved. Ducks and some other water birds have greasy, water-resistant eggs; those of grebes and boobies are chalky. More detailed data on egg characteristics may be found in publications by Kendeigh et al (1956), Preston (1968), and Romanoff and Romanoff (1949).

CLUTCH SIZE

Birds lay from 1 to about 20 eggs per clutch. A single egg is characteristic of some penguins and albatrosses, the California Condor, and cliff-dwelling alcids. Two-egg clutches are usual but not invariable in loons, eagles, pigeons, Great Horned Owls, Whip-poor-wills, nighthawks, and most hummingbirds. Boobies often lay two eggs but hatch only one; if both hatch, usually only one young is raised. Adelie Penguins normally raise only one young from their two-egg clutch; skuas dispose of unhatched eggs and neglected nestlings (Maher, 1966). Skuas usually hatch both of their eggs but sacrifice one of the young. Several authors suggest that such practices are evolutionary steps geared toward maximum production; if conditions are favorable, more than one young might be raised.

Many birds, perhaps the majority of our passerine species (flycatchers, swallows, thrushes, warblers, sparrows) as well as many non-passerine birds (e.g., shorebirds), lay from three to five eggs per clutch, with four as a common number. In fact, four-egg clutches are so typical of charadriiform birds that Maclean (1972) concludes that four is the ancestral number; the two-egg clutch in some species is a result of reduction perhaps representing "end-lines" out of the ancestral range. Similarly, most gulls lay three eggs—and have three brood patches (see page 274)—hence, Beer (1965) suggests that three is the optimum number and that less than that is a derived number.

Many insectivorous birds wintering in northern latitudes (titmice, nuthatches, creepers, and kinglets), as well as wrens and some woodpeckers, have larger sets of eggs with 6–10 or more per nest. Ducks and gallinaceous birds have even larger clutches of 8–15 or more eggs, but higher numbers in duck nests are apt to be the product of more than one female. Dreis and Hendrickson (1952) report a Wood Duck nest with 20 eggs laid in 15 days, obviously the product of more than one female. Fuller and Bolen (1963) found a 13-egg clutch of Wood Ducks being incubated by two females and found some "dump" nests with 20–40 eggs. Dump nests are fairly common among ducks and gallinaceous birds. Many birds occasionally drop an egg in the wrong nest, perhaps in an emergency because their own nest is not ready or has been destroyed.

Clutch size is roughly correlated with mortality. Birds that lay small clutches may be longer lived and thus have more time to replace themselves, or they may nest in more protected situations, as in isolated island colonies free from four-footed predators. Waterfowl and game birds, on the other hand, even before being hunted by man, have always faced many dangers; a young duck or a young grouse has many hungry enemies. Further exploration of survival in relation to egg number, however, reveals inconsistencies. The Great Horned Owl, a much-persecuted species, appears to do well with only two eggs per year, whereas the Barn Owl commonly lays six or more eggs per clutch and often raises two broods but in most regions is less common than its close relative.

In many species, clutch size increases with latitude. Tropical birds lay fewer eggs than their northern relatives, although some species (most sea birds, shorebirds, and pigeons) lay a constant number regardless of their geographic range. Two- and three-egg clutches are common to many birds in the tropics and subtropics, whereas the same or closely related species lay double or triple that number in more northern latitudes, presumably because of the greater hazards and less dependable food supply in northern regions. Indeed, this may be the critical determining factor in northern latitudes. Rough-legged Hawks and Snowy Owls lay larger clutches in good lemming years than in years of food shortage. Bell (1965) reports 25 cases of Long-tailed Jaegers not breeding at all in poor lemming years; peak numbers and full clutches coincided with lemming maxima. But food is apparently not a critical factor in the tropics, since birds there do not spend all their time foraging, even when feeding young, and often one parent can raise the brood.

In many books and papers Lack, Moreau, Skutch, Ryder, Wagner, Wynne-Edwards, and others contribute lively discussions of these problems. Lack favors a genetically fixed clutch size correlated with the food supply, but Skutch shows that this does not apply to neotropical birds. Wagner (1957) suggests that the upper limit of egg production may be genetically fixed, but obtainable only under certain conditions and governed by food reserves in the bird. That is, if a Snowy Owl uses 25 percent of its reserves to lay one egg in low lemming years, then four eggs would be its limit, but in good lemming years that number might be (and is) doubled. Ryder (1970) believes that clutch size in Arctic-nesting geese is determined by food reserves in the female *before* arriving on the breeding grounds, with enough diverted to the eggs to nourish the young until they can feed themselves. Tropical species store little or no fat and have low reserves.

LAYING SEQUENCES

Passerine birds, as well as many of the smaller nonpasserines, usually lay an egg a day until the clutch is complete, but irregularities do occur, especially in tropical birds. Often the egg is laid at a particular time of day, such as in the early morning. After an egg is laid, the female does not stay at the nest, unless incubation is started immediately, which is unusual. Many of the larger birds do not oviposit daily, though ducks at least approximate an egg per day. Pheasants and grouse require about 15 days to lay 10 eggs. Hawks and owls usually lay every other day, or in some cases may skip more than a day between layings.

The Slender-billed Shearwater (*Puffinus tenuirostris*) has a remarkably precise seasonal timetable for egg laying (Serventy, 1963). The birds return to their nesting grounds on islands off the Australian coast in clouds that darken the skies, start laying on November 20–21, reach a peak on November 25–26, and are all through laying December 1–3; 85 percent of the eggs are laid within three days of the mean and some marked females laid on the same dates in successive years. Probably other shearwaters and some other oceanic birds have similarly precise timetables.

Birds possess an unexplained control over egg production. If the complete clutch is four eggs, they cease laying with that number, partially formed ova in the ovary presumably being resorbed. But if the first clutch is destroyed or part of the eggs taken, more eggs are produced after a waiting period. The capacity to continue uninterrupted laying is well illustrated by the oft-cited example of a flicker that laid 71 eggs in 73 days when an egg a day was experimentally removed from the nest before the clutch was complete. Domestic fowl lay almost continuously when not setting; there are records of 351 and 352 eggs in a year. There is also a record of a duck that laid 363 eggs in 365 days.

On the other hand, the total clutch of the Herring Gull is three. The female cannot be induced to lay more than three eggs or less than three, when eggs are experimentally removed from or added to the nest (Davis, 1942). Experiments with Tri-colored Blackbirds (Emlen, 1941) and with Barn Swallows and Black-billed Magpies (Davis, 1955) yielded similar results. Such birds—perhaps the majority, though there are few experiments to prove it—are spoken of as *determinate* layers, as opposed to *indeterminate* layers such as flickers, ducks, and poultry.

Incubation

As soon as a clutch is complete, or in some cases before it is complete, one of the parent birds sits on the eggs (incubates them) with varying degrees of attentiveness until they hatch. During the breeding season, most incubating birds develop an *incubation patch* or *brood patch*, which is a featherless area on the ventral apterium containing an anastomosis of blood vessels that can be brought into close contact with the eggs.

FIG. 10.15. *Wilson's Plover settling on eggs. The abdominal feathers are parted to permit vascularized skin (incubation patch) to come in close contact with the eggs. Note the nest—merely a depression in the sand—and the extremely large eggs.* [Photo by Allan D. Cruickshank.]

When a bird settles upon its nest, it parts the abdominal feathers so that they surround or enclose the eggs (Fig. 10.15). Thus the heat of the body, through the brood patch, can be transferred more or less directly to the eggs. Ventral skin temperatures of incubating females are higher than those of the males (1.7°F higher in the House Wren).

Studies of the incubation patch (Bailey, 1952) show that it is produced by defeatherization by molting ventral down (not feather plucking) and vascularization (increase in size and number of blood vessels in the patch). In most birds it is a median ventral patch, but in charadriiform, gruiform, and galliform birds lateral patches (sometimes merging) develop. Schreiber (1970) demonstrated experimentally that Western Gulls, with three brood patches, cannot incubate four eggs successfully. The brood patch is absent in struthious and anseriform birds, occurs only in the males when only the males incubate, and develops in both sexes if both sexes incubate regularly. In most passerines the brood patch occurs only in the female, lending some doubt as to the efficacy of incubating males in this order. However, the Great Crested Flycatcher (Parkes, 1953), Clark's Nutcracker (Mewaldt, 1952), and some neotropical flycatchers have proved to be exceptions, with the males possessing well-developed incubation patches during the nesting season. Development of the patch is controlled by hormones—chiefly estradiol and prolactin (see pages 184–85). Selander (1960) could not induce brood-patch formation in the Brown-headed Cowbird (which has no need for it) by estrogen and prolactin treatments.

ATTENTIVE AND INATTENTIVE PERIODS (SESSIONS AND RECESSES)

Most birds incubate intermittently, with *attentive periods* (*sessions*) at the nest and *inattentive periods* (*recesses*) away (Skutch, 1962). This results in oscillating egg temperatures. Some birds bring water in their bills or on wet breast feathers and sprinkle the eggs, either for cooling or humidifying purposes. Mayhew (1955) reports that Mallards often shake out their wet feathers while standing over the eggs, and that hatching success is better in wet years and higher in incubators when the eggs are dipped or sprinkled. Open nests in hot exposed situations are sometimes shaded by a parent bird to prevent the hot sun from killing the embryos. This is particularly true of nighthawks on open rooftops and of birds with nests exposed to the hot sun. Most birds turn their eggs frequently or occasionally, which heats the eggs more uniformly and prevents the embryonic membranes from adhering to the shell, but this practice is not universal. Crested swifts (Hemiprocnidae) in India, for example, glue their single egg to the center of a nest so small and flimsy that the egg might roll off the structure if not glued to it. Most birds with open nests stand on the rim, sometimes several times an hour, and shift the position of the eggs or turn them with the bill. Rotary motions of the body serve a similar purpose.

Egg tempertures averaging in the nineties seem to be about optimum; tests with 37 species in 11 different orders gave an average egg temperature of 93.2°F, with not much fluctuation between the off (92.1°) and on (93.7°) periods (Huggins, 1941). Kessler (1960), using a self-recording potentiometer, found mean egg temperatures in pheasant nests varied from 94.0 to 98.3°F in 1955 and from 92.0 to 97.8°F in 1956. Highest egg temperatures were in the center of the nest, the lowest on the periphery. Graber (1955) kept incubator temperatures for six nongalliform birds somewhat higher (98–104°F); on three occasions, temperatures went up (accidentally) to 108, 112, and 120°F without harmful consequences. Embryonic development is said to stop at about 82°F, but air temperatures can drop much lower for short periods without killing the embryo.

In experiments at a state game farm in Michigan, pheasant eggs were subjected to temperatures as low as 45°F for periods up to 8 hours during the first two weeks of incubation with only slight lowering of hatching success, but exposures of 16 hours or more at that temperature (45°F) resulted in heavy losses during the last week of incubation (MacMullan and Eberhardt, 1953). Large eggs, of course, can stand more exposure than small ones.

Observations on attentive and inattentive periods, now standard procedure in many life-history studies, have been quite thoroughly documented in Kendeigh's *Parental Care and Its Evolution in Birds* (1952). Recesses are often lengthened at midday and shortened in the cooler parts of the day, and the nest is seldom left unattended during storms or rains. Song Sparrows (Nice, 1937) usually have attentive periods lasting from 20 to 30 minutes, followed by 6–8-minute intervals off the nest, a rhythm believed to be correlated with hunger. In Cactus Wrens (Anderson and Anderson, 1960), 28 attentive periods averaged 14.8 minutes, ranging from 1 to 28 minutes, and 29 recesses averaged 11.7 minutes, ranging from 2 to 26.5 minutes. These averages are remarkably close to those recorded for House Wrens (14.3 minutes on, 6 minutes off) by Kendeigh (1941).

Longer vigils are characteristic of some passerine birds, especially if the male feeds the female on the nest. A Red Crossbill has been known to spend 14 hours 34 minutes on the nest and only 36 minutes off, but was fed by the male three times during her long vigil (Snyder and Cassel, 1951). In the Cedar Waxwing, the female incubates 90 percent of the observed daylight time after the last egg is laid; attentive periods may last 2 hours or more, though commonly only 20–40 minutes, but the female is fed frequently at the nest by the male (Putnam, 1949).

Larger birds commonly have longer attentive and inattentive periods. Birds of prey are sometimes gone from their nest for hours without apparent reduction in hatchability of the eggs; but they also sit for long periods. A Sandhill Crane watched by Walkinshaw (1949) sat for 16 hours 10 minutes of continuous incubation. Long attentive periods also are characteristic of many sea birds (page 276) in which the male and female work in shifts. Richdale (1963), in a 17-year study of Sooty Shearwaters, recorded sitting sessions lasting from 1 to 16 days, but daily or weekly changeovers were much more common than the longer periods.

THE ROLE OF THE SEXES DURING INCUBATION

Of course, the female, in most species, is largely responsible for incubation; the absence of the brood patch in most male passerines precludes the possibility of any real incubation, although rapid cooling can be prevented by the male sitting on the eggs. The role of the male is primarily to guard the territory and to help care for the young when they hatch, but he may also assume varying degrees of responsibility at the nest at other times. Among geese and swans, the gander and the cob stand guard near the incubating female. Among songbirds, the male frequently visits the nest, inspects the eggs, and may stand guard while the female is away. Other males actually relieve the female at the nest, either occasionally, as in the case of the Rose-breasted Grosbeak, or regularly, as in the case of many nonpasserine birds.

Kingfishers work in shifts varying from a few hours in some species to a day and night (12-hour) shift in others, and a 24-hour shift in the Ringed Kingfisher of the tropics (Skutch, 1952, 1957). In the Mourning Dove the female incubates at night, but is relieved by the male for an approximately 8-hour spell of daytime incubation, a schedule somewhat similar to that of some other columbids. In many shorebirds the male and female share incubating duties about equally, in others it may be either sex that takes charge.

Some birds have special signals or elaborate nest-relieving ceremonies when changing places. In the Common Tern one adult may tempt its mate off the nest by offering a fish at changeover time; if unsuccessful, it may push the incubating bird off the nest (Palmer, 1941). In general, changing over is a more striking ceremony among birds that nest in open places (shorebirds) than at nests concealed in marshes, where one bird may slip on quickly as the other skulks away.

One of the most dramatic cases of alternation of incubatory duties is found in the Adelie Penguin, which sometimes nests long distances from its source of food in the sea. After a prolonged period of courtship and nest preparation (page 203), which may last nearly a month, the females leave for the sea while the males remain on the eggs for a two-week period, thus fasting for a month to six weeks before the females return to relieve them. Similarly, in the Rockhopper (Warham, 1963) the female takes the first regular shift of 14 days, is relieved by the male for 10 days, and then returns to finish the job.

Cases of partial or complete sex reversal in breeding behavior, in which the male does all or most of the incubation, include the struthious birds—the Ostrich, rheas, Emu, and cassowaries—as well as the kiwis, tinamous, button-quails, jaçanas, and phalaropes. Female Ostriches may resort to brief spells of diurnal incubation, but it is the male that keeps the eggs covered during the cool desert nights. In some

Fɪɢ. 10.16. *Male Wilson's Phalarope on nest. The female merely laid the eggs in a makeshift nest, probably prepared by the male. Then the male took over completely, deserted by the more colorful female.* [Photo by L. H. Walkinshaw.]

shorebirds, notably the phalaropes and several sandpipers, the females desert after oviposition and leave incubatory duties to the males (Fig. 10.16). The phalaropes have carried sex reversal so far that the females also wear the brighter nuptial plumage and initiate courtship. Of added interest in the many scolopacid breeding adaptations is the disclosure that in a few species—Old World stints, Sanderlings, and probably the Spotted Sandpiper—the female sometimes lays two sets of eggs, one set tended by the male, the other by the female, so that two broods are raised nearly simultaneously. (See Hayes, 1972, and Oring and Knudson, 1972, for discussion of these peculiarities.)

As already indicated, some males feed their mates at the nest, thus eliminating the need for the female to do much hunting. Particularly among the corvids (magpies and jays), titmice (chickadees), and some fringillids (crossbills, towhees, and goldfinches), mate feeding during incubation is a regular habit. Incidental mate feeding during incubation, however, may be a holdover from courtship feeding. Care of the female during nesting is carried to the extreme in the hornbills, in which the females are imprisoned by partially sealing off the nesting cavity with regurgitated castings and mud. The male then proceeds to feed her through an opening too small to permit her egress and cares for her during the whole incubation period as well as for both her and the young after hatching (Fig. 10.17). In some fruit-eating species, the period of confinement may last as much as four months, but in the insectivorous species the female breaks out when the young are half grown. In both cases, the female emerges in good flesh, with a new coat of feathers, whereas the male is worn and emaciated.

INCUBATION PERIODS

The incubation period, or time interval required for hatching an egg, varies from about 10 to 80 days and is not necessarily uniform within a species. Excessive periods of inattentiveness, too much disturbance, or unfavorable weather can retard hatching. Many ornithological references give specific incubation periods for all birds included in the book, but many of them are misleading or actually wrong; some of the erroneous periods have been traced by Nice (1954) back to Aristotle.

The incubatory period is correlated to some extent with egg size, the larger eggs generally requiring longer to hatch. But more fundamental is the stage of development of the young at hatching, the precocial birds usually requiring longer periods for incubation. The eggs of the precocial Killdeer take about 26 days for hatching, but the young are able to run from the nest within a few minutes of hatching. The much smaller eggs of the similar-sized altricial robin hatch in about 13 days, but the young remain in the nest for another 10–14 days (Fig. 10.18). Daniel (1957) points out that a domestic chick hatched at 21 or 22 days is at about the same stage as a 22-day-old Red-winged Blackbird leaving

FIG. 10.17. *A male hornbill* (Bucerotidae) *presents a food item to his mate in the partially sealed-in nest cavity. Although this is a mount in a museum exhibit, the frayed tail may be characteristic of hornbills in late season mate-feeding.* [Courtesy of Amer. Mus. Nat. Hist.]

a.   b.

FIG. 10.18. *Comparative egg sizes of altricial robin and precocial Killdeer. The larger egg has more storage material and requires twice as long to hatch.* [Drawings by Diane Pierce.]

the nest; the chick spent its 22 days in the egg, the Red-wing spent 12 days in the egg and 10 more in the nest.

Passerine birds in northern climates have incubation periods varying from about 11 to 20 days, with a few records of early 10-day hatchings. House Sparrows studied by Weaver (1943) at Ithaca, New York, hatched a few of their eggs in 10 days, but the average of 22 sets was 12 days. Berger (1957), at Ann Arbor, Michigan, reported one 10-day and one 13-day incubation period, but most sets hatched in 12 days. Case and Hewitt (1963) reported an 11-day incubation period for the Red-winged Blackbird, but *some* eggs hatched in 10 days and some in 12. Ward (1965) gave the incubation period of the Black-faced Dioch (*Quelea quelea*) in Nigeria as 10 days *or less*. Chipping Sparrow eggs will hatch in 11 days in warm weather, but average 12.3 days in colder weather (Walkinshaw, 1952). Robin eggs average 13 days, but some hatch in 12, while others take 14 or 15 days. Incubation periods are longer in the corvids; crows and magpies take about 18 days and most jays average 17, while ravens take 2 days longer or more. Some tropical passerines require considerably longer than their northern relatives; Skutch (1945) lists several flycatchers that require more than 20 days for incubation. Snow (1972) gives 26–27 days for the Calfbird (a passerine cotingid that "moos"). Lyrebirds (Menuridae) in Australia are reported to take up to 40 days to hatch their single egg, but one has been hatched under a domestic hen in 28 days (Austin, 1961).

Incubation periods for nonpasserine birds, as might be expected, are generally much longer, but there are a few exceptions. Some quail in captivity (*Coturnix*) and ground doves in the wild (Skutch, 1956) may hatch their eggs in 10–12 days. Two marked eggs of a Black-billed Cuckoo in northern Michigan hatched in 10 and 11 days, respectively (Spencer, 1943); possibly retention of the 10-day egg in the oviduct an extra day may have accounted for the shorter period (Nice, 1953). At the opposite extreme some sea birds, including penguins, petrels, shearwaters, and albatrosses, have exceptionally long periods of six weeks or more—up to a maximum of 80 days for the Royal Albatross. A pair of kiwis in captivity usually required 75 days for incubation (Robson, 1948); Cottrell (1955) gave the period as 75–77 days, or even 80. Emus, both captive and wild, have been checked at 58–61 days, much longer than the 42-day period assigned to the larger Ostrich.

A curiously delayed incubation has been reported for the European Goshawk (Holstein, 1942) with the male ineffectively covering the eggs for much of the first 16 days, then the female taking over for another 26 days, giving an extremely long total (for a hawk) of 42 days. The male, however, was observed to have no brood patch, and the temperature of the eggs, checked by thermocouple, was kept too low for embryonic development. Incubation of the first egg in the Common Pigeon is partial—it hatches a half day earlier than the second egg although laid 44 hours earlier. Many life-history studies indicate that incubation is partial or irregular in some birds until the clutch is complete.

Though a number of birds are known to take advantage of solar heat and decaying vegetation to aid in incubation, the megapodes of the Australasian region are the only ones known to do so entirely. Mound-building species scrape together a huge pile of vegetation and soil, sometimes 10–15 feet high, on which many eggs are deposited (18–24 in the Mallee Fowl, *Leipoa ocellata,* although large clutches may be communal affairs) in an amazingly symmetrical pattern with the pointed ends down, and then cover them. The adults, primarily the male, attend the nest for months, both before and during "incubation," and actually regulate the temperature by alternately opening and closing over the mound, testing it with bare head and neck (hence the name "thermometer birds") and keeping it at about 92°F (90–100°). Decomposition heat is important at first, but solar heat is more important toward the end of the cycle. The incubation period is very variable (50–96 days in *Leipoa*); then the young break out and fly away. Relatives of the Mallee Fowl, notably the Maleo (*Macrocephalon*) in the Celebes, bury their eggs in volcanically heated sand instead of mounds. (See Clark, 1964, for a comprehensive review of megapode life histories, based largely on work by Frith in Australia.)

Eggs lose from 10 to 20 or more percent of their weight during incubation, the rate of evaporation (which causes the weight loss) increasing with continued incubation and with rising temperatures. House Wren eggs lose 13.7 percent of their weight up to hatching time (Kendeigh, 1940), and the weight loss in gull eggs is even greater, roughly 20 percent.

Hatching

At the end of the period of incubation fertile eggs with live embryos hatch; unhatched eggs may be infertile or contain embryos that died at some stage of development. Hatching is preceded by "pipping" of the egg (Fig. 10.19), which is accomplished by the action of the hatching muscle (page 97) and the struggling chick within. Young birds are equipped with an "egg tooth," a horny protuberance on the tip of the upper mandible that is used to open the shell and disappears soon after birth. Sometimes eggs appear to hatch rather suddenly, but usually they are pipped for several hours to a day or more before the young finally emerge. Often the chick can be heard peeping within the shell, even a day or two before hatching. During its struggles it utilizes air in the air chamber of the egg. Eventually the crack in the weakened shell, initiated by pipping and completed by mechanical breakage, extends completely around the egg, usually around the larger end, and the chick kicks itself free. The parent bird may watch the proceedings, ostensibly with great interest, and may try to help out by picking at the shell, but too much parental (or human) interference may rupture the blood vessels in the lingering yolk sac.

Soon after hatching the parent removes the shell, often carrying it a considerable distance before dropping it. Shells and their associated membranes emit considerable odor, which might attract predators, so it is important that some disposal be made of them promptly. Some birds eat the shells, although there is probably little mineral value in them since the developing embryo extracted much of the lime for bone formation. Precocial birds that leave the nest promptly, such as ducks, pheasants, and grouse, may leave the shells at the nest (Fig. 10.19), but precocial birds that remain for a day or more in the nest, such as gulls and terns, generally remove the shells. Eggs that fail to hatch are usually left in the nest, and may even survive the tramplings of the young through nest life.

Fig. 10.19. *Hatching.* top *White Pelican nest scene in Montana. One young has just emerged, the other egg is pipped.* [Photo by L. H. Walkinshaw.] bottom *At a pheasant nest. Most pheasant eggs hatch more or less simultaneously, and the young soon depart, leaving the discarded shells.* [Photo by Mich. Dept. Nat. Res.]

Parent birds appear to be conscious of and even "excited" by hatching events. Probably they hear the peeping of the young or feel the pulsations within the egg. A female Gray-cheeked Thrush (Wallace, 1939) has been observed to hop to the rim of the nest repeatedly during the hatching process and to watch and poke at the eggs. Several times she sang while perched on the edge of the nest. Once she flew away and returned with a worm that she tried to feed to an unhatched egg, as if her feeding instincts were getting ahead of schedule ("anticipatory" feeding); but, as subsequent events disclosed, the egg was probably a day late in hatching. Putnam (1949) describes a similarly dramatic scene in the nest life of the Cedar Waxwing. At hatching time the female in some way communicated the event to the male who came and inspected the young. He then left but later returned with a mass of small caterpillars for the young, whereas previously he had brought only fruit for the female. The female Common Tern (Palmer, 1941) appears nervous at hatching time and often refuses to change over when her mate comes to relieve her; her restlessness to some extent is communicated to the male.

All the eggs in a nest may hatch more or less simultaneously (*synchronous hatching*) or at staggered intervals (*asynchronous hatching*). Among the ducks, gallinaceous birds, plovers, and sandpipers, which leave the nest soon after hatching, it is important that they all hatch at about the same time; sometimes, indeed, unhatched eggs with live embryos are abandoned. Among the gulls and terns, however, hatching of the two or three eggs may occur at intervals, even a day apart, but the young stay at or close to the nest site for several days after hatching. A common but far from universal sequence in passerine birds that lay four eggs is that three of the eggs hatch close together, with the fourth egg hatching several hours to a day later, indicating that incubation began with laying of the third egg.

In the hawks, owls, North American cuckoos, and Boat-tailed Grackles, it is common to find young of different ages in the nest, indicating asynchronous hatching (Fig. 10.20). As previously pointed out, this may result in cannibalism in some predatory birds, and may be a means of bringing family size into adjustment with the food supply. Ricklefs (1965) suggests that some passerines (e.g., thrashers) may be able to "evaluate" the potential food supply at laying time, lay eggs to meet "average" conditions, then sacrifice one or more young if the food supply proves inadequate. Some of the northern owls, notably the

FIG. 10.20. *These young Marsh Hawks hatched at staggered intervals. The one in the background is nearly ready to fly; the one at the far left is relatively helpless and is lucky to have survived trampling by his larger nest sibs.* [Photo by Edward M. Brigham, Jr.]

Great Horned, and the Gray Jay start nesting in winter and must keep their eggs covered quite constantly to prevent freezing; hence the young hatch at staggered intervals.

When eggs fail to hatch, incubating birds may prolong incubation for indefinite periods. Peterle (1953) reported that a Ruffed Grouse sat on infertile eggs for 70 days, nearly three times the normal incubation period. Kelly (1956) reported that an Anna Hummingbird sat faithfully on her nest for 95 days. A male Bobwhite (Jickling, 1940) sat for nearly three months. Vigils of 22–24 days, roughly double the normal incubation period, have been reported for several passerines (Berger, 1953a). Holcomb (1965) reported a Cardinal sitting on infertile eggs for 27 days.

Selected References follow Chapter 11.

11

The Annual Cycle: Postnatal Life

THIS final chapter on the annual cycle describes events in the life of young birds in and out of the nest, including some of their activities in late summer and fall and concluding with a brief account of winter habits. Several of these topics might well be expanded to a full chapter, but are here only briefly summarized to round out the picture of the annual cycle.

Early Growth and Development

Young birds, as mentioned earlier, are commonly divided into (1) *altricial:* birds born in a relatively helpless condition and wholly dependent on parental care, and (2) *precocial:* birds well developed and capable of locomotion at birth, covered with natal down, and able to feed themselves. Such a division, however, is not always definitive. Typical altricial birds are naked or nearly so at birth, but herons, hawks and owls, though altricial with respect to helplessness and need for parental care, are well covered with natal down when hatched. Some charadriiform birds (gulls, terns, and alcids) are well covered with down at birth, but remain at or near the nest and are cared for by their parents for some time. Hence, the following more restricted terms are often used in advanced work in ornithology (definitions—but not examples—from Van Tyne and Berger, 1959):

1. Nidicolous—Birds whose young remain in the nest for some time after hatching (includes herons, gulls, terns, alcids, hawks, and owls, as well as all birds born naked or nearly naked—Fig. 11.1 TOP).
2. Nidifugous—Birds whose young leave the nest shortly after hatching (loons, grebes, anatids, galliform and gruiform birds, plovers—Fig. 11.1 BOTTOM).
3. Psilopaedic—Naked at hatching, or having down only on the future

FIG. 11.1. TOP *Young Chimney Swifts are naked and helpless at birth and are cared for in the nest for about three weeks.* [Photo by Robert Knickmeyer from National Audubon Society.] BOTTOM *Piping Plovers, like ducklings and gallinaceous chicks, are fully clothed with natal down and scamper from the nest soon after hatching.* [Photo by Bertha Daubendiek; courtesy of *Jack-Pine Warbler.*]

pterylae (all passerines, woodpeckers, kingfishers, swifts and hummingbirds, cuckoos, parrots, pigeons).

4. Ptilopaedic — Completely clothed with down at hatching, the down covering the apteria as well as pterylae (particularly true of anatids and gallinaceous birds, but also includes many of the nidicolous types). The terms *nidicolous* and *nidifugous* are more commonly used in descriptions of young birds than *psilopaedic* and *ptilopaedic*.

At hatching time most altricial birds, such as passerines, are blind, relatively helpless, and nearly naked, with varying amounts of wet natal down sparsely distributed on some of the feather tracts. Some, such as Chimney Swifts (Fig. 11.1), House Sparrows, and Cedar Waxwings, are entirely naked; others, like flycatchers, may have a fairly dense coat of long down. The eyelids are completely closed; if pried open manually, they do not remain open. The young bird appears to be all head (to accommodate the large eyes) and abdomen (yolk sac); its relatively undeveloped limbs will not support it in its efforts to rise to beg for food. The body is of a pinkish hue due to suffusion of blood vessels beneath the thin skin through which the viscera can be seen. The interior of the mouth is usually a conspicuous red, orange, or yellow, providing a more visible target for the parents feeding young (Fig. 11.2). Essentially cold-blooded (*poikilothermous*) at this stage, the young have to be brooded quite constantly to keep them warm. Many pas-

FIG. 11.2. *These young Brown Thrashers raise their heads hungrily for food at the appearance of an adult. Light-reflecting papillae on the brightly colored palate aid the parents in hitting the target.* [Photo by Allan D. Cruickshank.]

serines weigh from 1 to 3 grams at birth, some even less, having lost about one third of the original weight of the egg from which they hatched.

Growth and development are rapid in all passerine species. Young Vesper Sparrows, for example, weigh about 2 grams at birth and average about 18 grams when they leave the nest at 9–10 days of age (Dawson and Evans, 1960). Young robins average 5.5 grams (4.1–6.7) at birth and 56.7 grams (52.2–63.2) at nest-leaving time at 13 days (Howell, 1942). The sparse natal down is carried out on the tips of the incoming juvenal feathers, which show as dark dots beneath the skin at first, but break through as pin feathers in a few days. Feathers of the juvenal plumage begin to unfold about midway through nest life (5–7 days in most passerines) and by nest-leaving time have expanded until they practically cover the apteria. The eyes usually open, at first a mere slit, some time during the first 3–5 days (a little earlier in a few, considerably later in others), but the young bird apparently does not see clearly at this stage.

About midway through nest life the young birds become conscious of things about them: they rear up in the nest with loud begging cries when the parents appear with food (Fig. 11.2); a sense of fear, perhaps correlated with improved vision, begins to develop so that they "freeze" in the presence of an intruder; they stretch, preen, and jostle each other; and just before departure are so active that the nest can hardly contain them. If unduly disturbed or startled by an intruder, they explode from the nest, often prematurely, in this way perhaps escaping a predator for the moment, but perhaps later falling prey to another predator because they cannot fly at this stage. Temperature control is established rather gradually, concomitant with feather development, and is quite unstable at first, but is fairly well established before nest-leaving time (see page 159 for development in Vesper Sparrows). Grasping ability also develops prior to nest leaving; this could have survival value in premature departures to prevent drowning, for instance, in marsh birds (Holcomb, 1966).

Weight is one of the best criteria of growth, but development of structural parts—bill, wings, tarsus, and tail—is also readily measurable. Growth, though rapid and fairly uniform (except for the first day or so), shows some fluctuations due to irregularities in feeding, health, and general welfare of the birds, and rapidity of feather development; a slowing down of the rate of increase in weight is characteristic during periods of rapid feather development. Some structural parts develop more rapidly than others; the bill and tarsus may have attained full length at nest-leaving age, but the wings and especially the characteristically stubby tail of all birds continue their growth for some time after departure from the nest (Fig. 11.3). The young bird is nearly as large as its parents at nest leaving (70–80 percent of the adult weight in those that leave in preflight stages, 100 percent or so for those that stay longer). It is a mistaken concept among many laymen that small individuals are young birds. In some nonpasserines, in fact, as in eagles, the immature are often larger than their parents during their first summer. Oilbirds (*Steatornis*) in South America, fed on fattening palm fruits, attain a weight of 600 grams—200 grams more than the adults—while in the nest, but trim down their body weight to that of the adults when they fledge (Lindblad, 1969).

This generalized account of the development of a young passerine does not apply to all birds. Even among the passerines the growth rate is slowed down in some hole-nesters; it is an advantage for birds in a safe nesting cavity *not* to develop too rapidly and not to leave the nest prematurely. Among the nonpasserines that are not precocial are some species that develop comparatively slowly. The Ruby-throated Hummingbird remains in the nest for about 20 days, compared to 9–14 days for most passerines. Chimney Swifts may stay in the nest for 20 days, then linger in the chimney another 8–10 days before venturing outside. Young woodpeckers are slow of growth; a flicker's eyes do not open until about the tenth day, and it may remain in the nest cavity nearly a month. The young of hawks and owls linger in the nest for 6 weeks or more, young eagles for 12 weeks. Some of the oceanic birds are notoriously slow of growth. Petrels stay in their nest burrows for 8 or more weeks; penguins require several months for growth; and young of the larger albatrosses remain at the nest site for 8–9 months, then spend 7 or more years at sea before returning to their natal isle to breed.

The development of nidicolous birds described above differs in many respects from that of nidifugous birds, which at birth are usually alert and lively, with open eyes and a full covering of natal down. Gallinaceous chicks run about soon after birth, and downy ducklings swim; both are led away by the mother and at least guided in their first quests for food. Their legs are well developed for locomotion, but their wings are retarded, and the juvenal plumage develops slowly compared to altricial passerines (Fig. 11.4). Though well proportioned, they are smaller in relation to the size of the parent (1–6 percent of adult female weight) than passerines (6–8 percent or more) both at birth and at the preflying stage. In anseriform birds flight feathers and flight are particularly slow to develop (see Weller, 1957, for remex development and first flights in ducks).

Temperature control in precocial birds is not fully developed at hatching time. Though in chickens and gulls the capacity to regulate temperature is known to be present to some extent before birth, regulatory ability after hatching is good only in ordinary air temperatures at

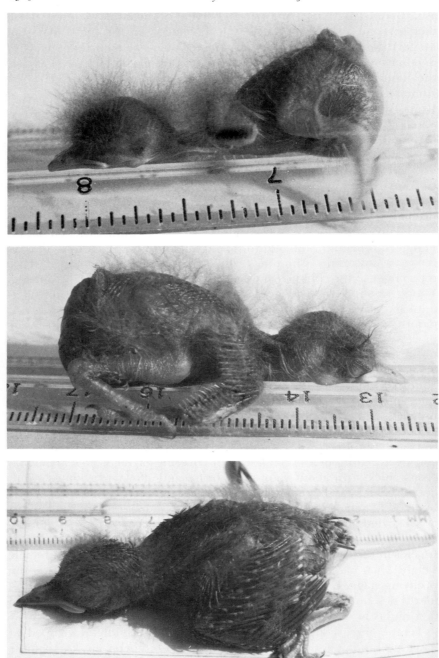

FIG. 11.3 *Five stages in the nest life of the Eastern Phoebe. Shown are 1-day (birth), 3-day, 5-day, 10-day, and 15-day (nest-leaving) stages.* [Photos by Harold D. Mahan.]

TABLE 11.1. COMPARATIVE NESTING DATA FOR REPRESENTATIVE SPECIES

Species	Family	Pair Bond	Number of Eggs per Clutch	Incubation Period	Sex Incubating	Fledgling Period (days)	Reference
Ostrich	Struthionidae	polygynous	16–23	39–42	male	0	Sauers, 1966
Kiwi	Apterygidae	monogamous	1–3	75	male	6	Robson, 1948
Boucard Tinamou	Tinamidae	polygamous	10	16	male	<20 hr	Lancaster, 1964
Emperor Penguin	Spheniscidae	monogamous	1	62–64	male	?	Rivolier, 1959
Common Loon	Gaviidae	monogamous	2	28–30	both	0	Kendeigh, 1952
Laysan Albatross	Diomedeidae	monogamous	1	55–72(65)	both	160	Fisher, 1971
American Bittern	Ardeidae	monogamous	4–6	25–26	female	14	Kendeigh, 1952
Whistling Swan	Anatidae	monogamous	2–7(4–5)	35–40	female	0	Kortright, 1942
Canada Goose	Anatidae	monogamous	4–10(5–6)	28–30	female	0	Kortright, 1942
Mallard	Anatidae	monogamous	6–15(8–10)	23–29(26)	female	0	Kortright, 1942
Bald Eagle	Accipitridae	monogamous	2	35	both	84	Broley, 1947
Mallee Fowl (*Leipoa*)	Megapodiidae	monogamous	18–24	57	neither	0	Clark, 1964
Ruffed Grouse	Tetraonidae	promiscuous(?)	11	23–24	female	0	Edminster, 1947
Sandhill Crane	Gruidae	monogamous	2	28–30	both	0	Walkinshaw, 1949
Killdeer	Charadriidae	monogamous	4	26	both	0	Kendeigh, 1952
American Woodcock	Scolopacidae	promiscuous	4	20–21	female	0	Sheldon, 1967
Northern Phalarope	Phalaropodidae	monogamous(?)	4	20–21	male	0	Kendeigh, 1952
Herring Gull	Laridae	monogamous	2–3	24–28(26)	both	0	Bent, 1921
Common Tern	Laridae	monogamous	1–5(3)	21–26(23)	both	2	Palmer, 1941
Mourning Dove	Columbidae	monogamous	2	14	both	13–15	McClure, 1943
Black-billed Cuckoo	Cuculidae	monogamous	2–4	10–14	both	7	Kendeigh, 1952
Great Horned Owl	Strigidae	monogamous	2	33–35	both	45	Hoffmeister et al. 1947
Nighthawk	Caprimulgidae	monogamous	2	18–19	female(m.?)	23(brooded)	Kendeigh, 1952
Chimney Swift	Apodidae	monogamous(?)	4	18–19	both	19–26	Kendeigh, 1952
Belted Kingfisher	Alcedinidae	monogamous	6–7	23–24	female(m.?)	28–32	Bent, 1940

Common Flicker	Picidae	6–8	11–12	both	25–28	monogamous	Bent, 1939
Eastern Phoebe	Tyrannidae	5	16	female	16–18	monogamous	Kendeigh, 1952
Horned Lark	Alaudidae	3–5	11–12	female	11–12	monogamous	Pickwell, 1931
Barn Swallow	Hirundinidae	4–6	14–16	female(m.?)	18–23	monogamous	Kendeigh, 1952
Common Crow	Corvidae	4–6	18	both	28–35	monogamous	Bent, 1946
Black-capped Chickadee	Paridae	5–8(6.7)	13	female	16	monogamous	Odum, 1941–42
Dipper	Cinclidae	4–5	16	female	24–25	monogamous	Bakus, 1959
House Wren	Troglodytidae	4–9(5.5)	13	female	15–16	polygynous	Kendeigh, 1952
Gray Catbird	Mimidae	3–5(4)	12–13	female	10	monogamous	Bent, 1948
American Robin	Turdidae	3–4	12–14	female	9–16(13)	monogamous	Howell, 1942
Ruby-crowned Kinglet	Regulidae	5–11	12(?)	female	12(?)	monogamous	Bent, 1949
Cedar Waxwing	Bombycillidae	4–5	12.2	female	15.9	monogamous	Putnam, 1949
Loggerhead Shrike	Laniidae	5–7(6)	16	female	20	monogamous	Miller, 1931
Starling	Sturnidae	4–6	11–14	both	14–21	monogamous	Bent, 1950
Red-eyed Vireo	Vireonidae	2–4(3.07)*	12–14	both	11–12	monogamous	Southern, 1958
Kirtland's Warbler	Parulidae	3–6(4.6)	13–16(14)	female	9–11(9.4)	monogamous	Mayfield, 1960
Ovenbird	Parulidae	3–6(4.7)	11–14(12)	female	8	monogamous	Hann, 1937
House Sparrow	Ploceidae	4–6(4.7)	10–14(12)	female	12–16(14.4)	monogamous	Weaver, 1943
Brewer's Blackbird	Icteridae	3–7	12–14(13)	female	13	polygynous	Williams, 1952
Scarlet Tanager	Thraupidae	3–5(4)	13	female	9–10	monogamous	Prescott, 1965
Cardinal	Fringillidae	2–4(3)	12 or 13	female	9–10	monogamous	Laskey, 1944
Song Sparrow	Fringillidae	3–5(4)	12 or 13	female	10	monogamous	Nice, 1937, 1943
Snow Bunting	Fringillidae	4–7	12.5–13	female	12–14	monogamous	Sutton et al. 1954b

* Clutch size may have been reduced by heavy cowbird parasitism.

F I G. 11.4 *Day-old Black Duck. Downy ducklings are alert and lively with well-developed legs for running or swimming, but their wings and flight feathers are slow to develop.* [Photo by Miles D. Pirnie.]

first and unstable in colder weather. Farner and Serventy (1959) found that Slender-billed Shearwater chicks have adequate thermoregulation for life in burrows within the first day of hatching, surpassing both galliform birds and gulls in this respect. Occupied burrows had higher temperatures than burrows without young birds.

Parental Care

BROODING

Brooding is an important function in early nest life. During their first days, altricial young are brooded fairly constantly, in a rhythm similar to incubation. Thereafter, brooding gradually decreases with feather development and may cease altogether during the late stages of nest life, except that the adult generally sleeps on the nest at night. Birds nesting in colder climates and in mountainous regions, understandably, brood more constantly than those in warmer areas. Brooding is likewise correlated with local weather conditions, the adult, usually the female, remaining at the nest through rains and storms while the male brings the food.

The more advanced development of precocial birds at hatching does not preclude the necessity for some brooding, for they lose body heat rapidly with exposure. Nearly all species brood their young until they leave the nest and thereafter resort to frequent off-nest hovering. Duck mothers assemble their ducklings for brief spells of brooding, and gallinaceous birds brood their young at night and at intervals during the day for many days. In fact, the necessity for increased brooding during spells of bad weather seriously curtails feeding time in gallinaceous birds and may cause losses.

FEEDING THE YOUNG

A major task in the care of nestlings is providing sufficient food for their proper growth and development. The appetites of young birds are enormous; they often consume their own weight in food per day, or even more, although perhaps about one fourth of the body weight is more usual. Finding and delivering this food is nearly a full-time occupation for one or both parents, especially in northern latitudes where larger broods are reared. Feeding problems are solved in many different ways by different species. Some of these variations are dealt with in the following sections.

ROLE OF THE SEXES. The role of the sexes in feeding programs varies greatly. The somewhat exceptional cases in which the male plays little or no part in family affairs, as well as instances of sex reversal in which the male does it all, have already been mentioned; these, in most cases, apply to precocial birds, which are soon able to forage for themselves. In most other birds, however, feeding is handled on a cooperative basis (Fig. 11.5). Frequently the female appears to take the initiative, the male helping out somewhat dilatorily; or, again perhaps somewhat exceptionally, the male may bring the bulk of the food while the female remains at the nest. In the Marsh Hawk, for instance, the male brings nearly all the prey for the first five days and transfers it to the female, often in midair, and she then portions it out to the young. Later the female also hunts prey, but the male continues to leave his contributions at the nest (Hecht, 1951). A similar arrangement prevails in the Goshawk (Schnell, 1958); the male brings 85 percent of the food to the female and she feeds it to the young. Schnell suggests that this functions in keeping the male from wasting time at the nest (the female is hostile to him) and forces him to do more hunting.

In the Black-capped Chickadee (Odum, 1942) the male does most of the feeding at first (two to four times as much) while the female broods, but thereafter they feed with about equal frequency. In the Cedar

F IG. 11.5. *The males of most passerine species cooperate faithfully in feeding the young. Here a male Hooded Warbler brings food to the nest.* [Photo by Hal Harrison; courtesy of *Jack-Pine Warbler.*]

Waxwing the male does all the feeding for the first three days, the female brooding about 90 percent of the time (Putnam, 1949).

At nests of the Gray-cheeked Thrush on mountains in Vermont (Wallace, 1939), the male and female shared feeding assignments about equally; the female was a little more regular in fair weather, but during the frequent mountain storms when she remained at the nest to brood, the male did double duty, bringing insects at about the same rate that both birds usually maintained. Once during a two-hour period of steady rain the female remained constantly at the nest while the male brought food, which she took from his bill and relayed to the young; or she stood aside while he fed them. Hole-nesting species, of course, do not face the problem of extra brooding during storms.

Evenden (1957) reports an unusual situation in which House Finches in California had two nests operating simultaneously. At hatching time the male took charge of one nest while the female started another. In one case the female left the first nest before the eggs hatched, but the male carried on successfully. Ames (1967) reports two cases of probably overlapping broods in Barn Owls. Curiously, the male Painted Bunting in Oklahoma helps feed the young only after they are fledged; in fact, the female does not renest *unless* the male takes over the first brood (Parmelee, 1964).

Usually, however, among birds fed for a period at the nest, feeding is a cooperative enterprise, and without the participation of both sexes the welfare of the young might be in jeopardy. An illuminating example is that the Yellow-eyed Penguin raises only one chick if one parent is lost, but two young if both parents survive (Richdale, 1951).

FEEDING SCHEDULES. Frequency of feedings varies greatly with different species, but also with the age of the young, the time of day, and the amount of food brought per trip. Usually feedings are more concentrated in the early morning and in late afternoon or evening, with a pronounced midday slackening. The young are fed less frequently—or, more rarely, not at all—for the first day or so, since they are still deriving some nourishment from the yolk-filled abdomen, but the feeding rate increases rapidly thereafter. Cuthbert (1954) noted that Black Terns increased their feeding rate from 1.2 times per hour on the first day to 16.8 times on the eighth day.

Some of the smaller birds have extremely rapid feeding rates. Tabler (1956) reports that a pair of Barn Swallows nesting on a light fixture on a porch fed their young about once a minute, or more often on the eighth day. Davis (1968) found Barn Swallows carrying 5–11 flies per trip to their young from the 10th to the 16th day. Assuming a uniform feeding rate from dawn to dark throughout the 16-day period (not strictly accurate—see preceding paragraphs), he calculated that a brood of five would consume 59,470 flies. When two broods were raised, the total would be 118,940 flies. Since there were 25–30 local nesting pairs, he had no problem with house or stable flies that summer! Colonial Bank Swallows feed their young at about half the hourly Barn Swallow rate (Peterson, 1955), presumably because they have to go farther afield for their prey. A Warbling Vireo has been reported making 45 trips per hour to a nest, in contrast to the slow feeding rate of 6.5 times per hour at two nests of the Red-eyed Vireo (Southern, 1958). Using daily rates, Kendeigh (1952) gives 491 feedings as the maximum of many observations on House Wrens and 845 times on the 12th day of nest life as the climax at a nest of Eastern Phoebes. The phoebes may have established a record, with 8,942 total visits to the nest recorded in 17 days or 2,275 per young bird.

Most of the birds cited above have small territories and make short trips with small insect prey. Thrushes make fewer trips, at 10–15-minute intervals, but usually bring larger loads. Cedar Waxwings (Putnam, 1949) average about three feedings per hour, but bring a whole cropful of berries, enough to go around the whole brood one or more times (Fig. 11.6). Goldfinches feed their young by regurgitation at half-hour intervals, but bring a gullet full of predigested seeds for each feeding (Stokes, 1950). A fairly typical feeding schedule for many passerines, with male and female cooperating, is described by Davis (1960) for the Rufous-sided Towhee in California (Fig. 11.7).

Larger birds feed their young less frequently. With predatory birds, the rate may vary with hunting success. The young may gorge them-

FIG. 11.6. *Cedar Waxwings feed their young less frequently than most passerines, but may bring a large load at one time — 5 large mulberries, or 7 chokecherries, or 9–13 elderberries in one trip.* [Photo by Edward M. Brigham, Jr.]

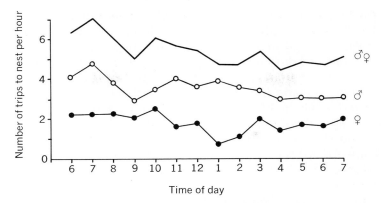

F IG . 11.7. *Feeding rates of male and female Rufous-sided Towhees and of the two combined. Note morning peak of male activity and midday slump of female.* [From J. Davis, *Condor,* 62, 1960.]

selves until their crops threaten to burst, if food is available, then fast for long periods if it is not. Hourly feedings are said to be characteristic of some hawks, but often a large carcass, such as a rabbit, may be left in the nest for the young to feed on intermittently. Herrick (1934) noted that Bald Eagles renewed the food supply at the nest from two to eight times per day. Young of the world's largest and most powerful eagles, the Crowned Eagle (*Stephanoaetus coronatus*) in Africa and the Harpy Eagle (*Harpia harpyja*) in South America, are fed much less frequently, usually at several-day intervals, and can go without food for 10–14 days (Brown, 1966; Fowler and Cope, 1964).

One of the most extraordinary feeding programs is that of a European swift (*Apus*), which has been well documented by Koskimies (1950). Under adverse weather conditions, aeroplankton (flying insects mainly) are not available and the swifts cease hunting. During such periods the nestlings become torpid and are able to fast for long intervals, up to a recorded maximum of 21 days, though the average survival time of young swifts subjected to starvation experiments was about 10 days. In good weather, however, the adults collect large food balls of massed insects (up to several hundred per ball), which they regurgitate to their young at infrequent intervals—hence, swifts are well fed in good weather but fast during bad weather.

Some fish-eating birds—gulls, terns, kingfishers, and the Osprey—bring their young whole fish, but others, like herons, cormorants, and pelicans, bring larger loads from more extended fishing excursions and feed their young by regurgitation. White Pelicans nesting on Gunnison Island in Great Salt Lake, where fish are not available, make long flights, apparently twice daily, to fishing waters that are 50–75 miles away (Low et al., 1950); hence, the young are presumably fed twice daily. Pink-backed Pelicans (*Pelecanus rufescens*) in Africa also feed and nest in different areas; they hunt fish in the early morning and return to the nesting colony to feed their young between 9:00 A.M. and 1 P.M. (Burke and Brown, 1970). At heronries closer to the feeding grounds young herons are fed more often, about four times a day at Great Blue Heron colonies in California (Pratt, 1970). Sea birds feed their young infrequently; procellariiform chicks (albatrosses and shearwaters) are fed at several-day intervals, or even less frequently toward the end of the nest life.

Long absences of such birds as albatrosses and petrels from their nest sites have given rise to a *starvation hypothesis,* namely, that the adults desert their young in late nest life for weeks or months (three months in the Wandering Albatross) until the starving young go out to sea and learn to forage for themselves. Richdale (1954) has shown that this is not always true, since the adults may come back unobserved at several-day intervals to feed their young; yet there appears to be no question about such a starvation period during late nest life of many of the Procellariiformes. In a later analysis Richdale (1963) states that four species of migratory Procellariiformes desert their young to migrate, and that five nonmigratory species do not. In many land birds less frequent feedings toward the end of nest life are thought by some to function as an incentive for young birds to leave the nest. In the Crowned Eagle previously referred to, Brown (1966) thought that long periods of food deprivation (up to 13 days) might stimulate the young to hunt food for themselves.

FEEDING TECHNIQUES. Most birds bring food items in their beak (predatory birds also use their talons) and feed directly by cramming the food deep down into the throats of the young. A special muscular mechanism in the throat facilitates swallowing, as many a novice has discovered in trying to rear nestlings, which appear unable to swallow food unless it is placed deep down in the throat. The oral cavity of many birds is brightly colored, in some cases even equipped with papillae or light-reflecting knobs that aid the adults in hitting the target quickly. The swollen rictal region or oral flanges (Clark, 1969) on the lower jaw of nestling woodpeckers and around the gape of passerines may provide a tactile stimulus that causes the young to gape for food. The red spot on the bills of several species of gulls apparently aids the young in spotting the source of food. Swaying movements of the head, special food cries, or vocal signals of either adults or young, and even the draft of air created by Chimney Swifts descending a chimney (Barton, 1958) provide other stimuli for feeding.

Some birds utilize the pharynx and the crop for carrying food. Such birds feed largely by regurgitation. Hummingbirds deliver predigested food (nectar and insects) by thrusting their long rapierlike bills deep into the gullets of the young until it appears that the nestlings would be stabbed to death. Young herons grasp their parent's beak at right angles in their own and shake it vigorously. This seems to initiate the release of the prey, which is then dropped in the nest for the young or else taken directly from the bill of the adult. (See Pratt, 1970, for a discussion of the controversy over these different techniques). Pelicans simply open their huge maw and let the hungry nestlings go exploring, which they do so vigorously that it sometimes injures the gullet of the adult (Burke and Brown, 1970).

Hunger reflexes help determine the rotation in which the young are fed. The hungriest bird is apt to reach higher or clamor more loudly; when satisfied, it drops back relaxed while another bird is fed. If the food placed in the mouth is now swallowed promptly, it is taken out and fed to another. Young cowbirds flourish, usually at the expense of rightful occupants of the nest, by being larger and able to obtain most of the food. Such a process of food distribution, to the one that appears hungriest, is called *automatic apportionment.*

The nature of the food fed to the young usually reflects the food habits of the species, already dealt with in more detail in Chapter 5. Young birds, however, even in primarily seed-eating species, are fed animal food at first, mainly insects in the smaller forms, and larger prey, such as fish and other vertebrates, in the larger forms. Cedar Waxwings feed their young on insects for the first three days, then shift largely to fruits. Many insectivorous birds—thrushes, vireos, warblers, even swallows and flycatchers to some extent—become partially frugivorous after fledging. A male Cardinal has been observed to visit a feeding station with his beak full of green worms, lay down the worms, crack and eat sunflower seeds, then pick up the worms and go to the nest site, presumably to feed the young (Van Tyne, 1951). The American Goldfinch, however, which postpones nesting until fresh pulpy seeds of thistle and other Compositae are available, feeds its young on predigested seeds from the beginning of nest life, a feeding method also characteristic of redpolls, crossbills, and some other fringillids. Pigeons, which are primarily grain-feeders, produce pigeons' milk (page 121) for the early nourishment of their young. Flamingos in Africa feed their young on a blood-red fluid, rich in carbohydrates and fats and containing 1 percent blood secreted from the lining of the throat (Kahl, 1972).

Penguins in particular, but also some other water birds, are noted for forming assemblages of postnest fledglings called "creches" ("pods" in pelicans). These are nurseries for young birds with "babysitters" guarding them while other adults gather food. It was long supposed that the food-gatherers fed the young in the creche at random, but it is

FIG. 11.8. *Nest sanitation.* ABOVE *An Old World babbler (Timaliidae) in India removes a fecal sac from the nest.* [Photo by R. M. Naik.] OPPOSITE *A Long-billed Marsh Wren disposes of a fecal sac.* [Photo by Edward M. Brigham, Jr.]

now known for many species that the adults single out and feed only their own young, both adults and young using vocalizations and visual clues to recognize each other.

NEST SANITATION

In species in which the young are cared for in the nest, some system of sanitation is almost imperative. Considering the large amounts of

food consumed, and the fact that a nestling's digestive apparatus does not function perfectly at first, it is natural that excreta are expelled frequently. Sanitation of the nest is usually carefully provided for by cleaning it at nearly every feeding. Excreta are voided by the young in a convenient fecal sac, which can be readily picked up by the adults (Fig. 11.8). Usually the young defecate immediately after being fed; frequently the adult waits at the nest after feeding, then picks the fecal sac off the elevated cloaca, sometimes even nudging the young to prompt it to perform.

Oddly enough, the parents often eat the feces, presumably getting some nourishment from them because of the incomplete digestion in the young at this stage. Eating the excreta is not universal, however, either as to species or individuals; often they are merely carried away and dropped at a distance. Wible (1960) noted that a Yellow-bellied Sapsucker in Montana used a bracket fungus on a tree for fecal disposal; the male regularly brought the sac from the nest hole to the fungus, ate the sac, and allowed the contents to dribble over the shelf to the ground. Kilham (1962), however, noted that young sapsuckers fed on sap and insects did not form fecal sacs; the adults mixed the feces with sawdust and carried them away to a definite spot. The feces are eaten less frequently as young birds mature. Laskey (1944) noted that the female Cardinal ate them the first five days, the male the first four days, and that thereafter both sexes carried them away. Artifacts, such as wads of paper or plasticene models of feces placed in or on the nest, are also removed (Smith, 1947). Sometimes adults try to remove shiny leg bands from nestlings, even dumping the young out of the nest.

There are numerous exceptions to the scrupulous sanitation described above. Barn Swallows are untidy about their nests, letting the droppings accumulate until they may become a distraction in and around farm buildings. A seed diet and lack of formation of convenient fecal sacs perhaps are the reason that the otherwise beautiful nest of the American Goldfinch may become much soiled toward the end of the nest cycle. The "white-washed" nests in heronries and "white-washed" ledges in sea-bird colonies are familiar scenes. In dry climates, where such accumulations are not subject to leaching, rich deposits of guano result (see Fig. 15.2). Kingfishers and woodpeckers apparently make no effort at nest sanitation. In kingfisher burrows in sandy soil, the liquid is quickly absorbed, but when drainage is poor it may ooze out of the entrance almost continually. Kingfisher tunnels reek with ammonia gases, and the adults often plunge into water on emergence and bathe (Skutch, 1952).

Nest Leaving

The days just prior to departure from the nest witness preparations for the event. Exercises, which actually began in a limited way before birth, now reach a new peak. The nestlings stretch and yawn, extending first one foot and then the other, elevate and flap their wings, and preen their feathers. They become increasingly interested in affairs outside the nest. Sometimes they snap at passing insects, but seldom catch one.

A nest with several well-developed young becomes so overcrowded that much pushing, pecking, and jostling take place. Two birds may stage a sparring match on the nest, perhaps standing on top of one or more nest mates; loss of a bird or two out of a tree nest because of such activities is not uncommon. When an adult appears with food they all make a lunge for it, the boldest perhaps even venturing out of the nest momentarily to meet its parent. Such exercises probably help prepare the young bird for its first flight.

These activities eventually culminate in departure from the nest (Fig. 11.9), but the manner of leaving varies considerably. Often, as previously mentioned, nestlings may be startled into a premature departure by an intruder. Birds leaving prematurely usually cannot fly but merely flutter over the ground to a hiding place; in fact, most ground-nesting passerines regularly leave the nest before they can fly, but tree-nesting species that stay the full term are capable of short flights at nest-leaving time. Young Dippers remain in their cozy streamside nest an unusually long time — 24 or 25 days — but flutter to a safe landing, sometimes across a stretch of water, when they leave.

When nest leaving is normal, the young may depart gradually, one at a time, although often the departure of the first seems to be the signal for all to leave. At a hawk nest it is not uncommon to see the least developed bird crouched in the nest, another perched precariously on the edge of the structure, and still another out in the branches of the nest tree. Herons commonly crawl out into the branches of the nest tree before flight (Fig. 11.10). Some birds seem so hesitant about leaving that they may have to be coaxed away by the adults, which appear to tempt them with food just outside the nest, though whether or not this is a deliberate device to promote nest leaving or merely a waning of the parent's feeding instinct has been a controversial question.

The extended nest life of hole-nesting species may be an important survival factor. Black-capped Chickadees, for instance, remain in the nest for about 16 days and Tufted Titmice for 17 or 18 days, compared to 10–14 days for other tree-dwelling passerines and 8–10 days for ground-nesting passerines.

Postnest Life

Details in the life of young birds are difficult to secure after fledging, but the parents usually accompany them and continue to care for them until they are able to shift for themselves. Nice (1943) found Song Sparrows being cared for by the parents for 18–20 days after nest leaving. Ovenbirds, which leave the nest at the early age of 8 days, are described by Hann (1937) as having a 3-day hopping stage (when 8–11

FIG. 11.9. *Two stages in the nest life of the Gray-cheeked Thrush.* TOP *Restless and getting ready to leave.* BOTTOM *Safely fledged.* [Courtesy of *National Geographic Magazine.*]

FIG. 11.10. *Young Black-crowned Night Herons, like young hawks, usually clamber about the branches of the nest tree or adjacent tress before taking flight. They show little resemblance in plumage to their parents.* [Photo by Edward M. Brigham, Jr.]

days of age), then a 10-day early-flying stage (11–20 days of age), an 11-day semidependent stage (20–30 days of age) when they do some feeding for themselves, and then the independent stage beyond 30 days when they leave their parents and shift for themselves. They apparently stay in or near the nest area by traveling in circles, rather than going long linear distances at first. Other birds may disperse more rapidly. Sutton and Parmelee (1954a) found a whole family (male, female, and four young) of Wheatears on Baffin Island a mile and a half from the nest 8 days after nest leaving.

Postnatal care of precocial birds follows a similar pattern, with an adult, usually the female, supervising their activities until they become independent. Young ducklings are led to the water, can swim and dive instinctively, and are quick to learn to glean surface insects from the water. Geis (1956) found that the young of Canada Geese in Montana left their island nest sites within 48 hours and traveled 2–10 miles to brood-rearing areas. The young of gallinaecous birds accompany their parent on their first explorations, and may remain associated with her until fall.

Edminister (1947) describes an imaginary but probably representative account of a Ruffed Grouse family after hatching. The mother led her 10 newly hatched chicks away from an 11-egg nest, deserting, in her haste to get away, an unhatched egg with a live embryo. She called them together and brooded them at intervals, taught them how to look for and find the proper kinds of food, how to freeze at a given signal, and how to scurry for cover at another. In spite of parental teachings, one of the 10 was lost to a plundering Sharp-shinned Hawk, another lost track of its mother one evening and perished of cold, and then, on a stormy night a week later, a blundering skunk accidentally broke up and scattered the family, so that three more failed to respond to the breakfast call. By the end of a month the five survivors were becoming more and more independent, until a Cooper's Hawk picked off an insubordinate youngster that failed to heed his mother's danger signal. Then the hunting season took another victim, the mother and another young failed to survive the winter, and another young was caught by a weasel in the spring, leaving only the original father and one young female to carry on.

Thus, most young birds go through a period of postnest training under parental supervision, but the extent of family ties varies greatly. At one extreme are the megapodes (see page 281), which hatch out in the sand or brush and go their way without any parental care at all; at the other extreme are the geese and swans, which maintain strong family ties through the fall and winter until another breeding season separates them (Fig. 11.11). California Condors are dependent on their parents so long that the adults usually nest only every other year, as is true of some albatrosses and the larger eagles. The young of some tropical birds remain in association with their parents for a long time, roosting and sleeping in family groups over winter and sometimes even helping the adults at the nest the following year. Ashmole and Tovar (1966) found adult Royal Terns still feeding juveniles on their wintering grounds in Peru six months after hatching. Aside from these somewhat exceptional cases, however, family ties are of short duration, and the young separate from their parents during their first summer.

Birds inherit many traits of behavior, such as feeding, singing, and bathing, without necessarily having to learn them from their parents. A young Gray-cheeked Thrush, kept in confinement after his 7th day, learned to pick up food for himself on his 15th day of age. Clumsy and inaccurate at first, he soon developed such a precise aim that on his 17th day he picked 30–40 leafhoppers (Cicadellidae) in quick succession off a window pane when held on the finger. He started to sing an

FIG. 11.11. *Geese maintain strong family ties — the gander guards the goose at the nest, both participate in rearing the young, and the family migrates as a unit, remaining together over winter.* [Photo by Allan D. Cruickshank.]

off-tune song at 24 days of age (another young thrush sang at 15 days) and when 25 days old plunged into his drinking receptacle and took a complete bath. Thereafter he bathed daily, but another young thrush was irregular about this feature (Fig. 11.12).

Three young Song Sparrows kept by Mrs. Nice started to sing at 13, 14, and 15 days, respectively. Two young Cardinals kept by Mrs. Laskey started warbling at three and four weeks, added two adult songs at two months, and sang songs indistinguishable from adults by late January and February. One young Cardinal picked up its own food 13 days after nest leaving, another at 20 days, but cracking sunflower seeds was not mastered until nine weeks. Young Song Sparrows attempted to shell seeds at 17 days of age, but were not successful until 25 days old. Young terns make many misses in their early attempts to catch fish.

Mortality and Nesting Success

One of the features of bird life that comes as a shock to most laymen is the apparent wastage in reproduction. Heavy losses occur among birds,

FIG. 11.12. *Young birds learn to bathe instinctively. This Gray-cheeked Thrush, separated from his parents when 14 days old, jumped into his drinking water one day for his
first bath. Although another young thrush kept in captivity bathed frequently, the one
shown here was less regular. Often he only spashed water over himself without jumping
in.* [Courtesy of *National Geographic Magazine.*]

not only during the nesting season but also throughout the first year of
life. Many eggs never hatch, either being infertile or containing defective embryos; storms destroy some nests, others are broken up by
predators; accidents of many kinds to one or both adults interrupt or
terminate the nesting cycle; prolonged cold rainy weather may cause
heavy loss of nestlings, particularly among the aerial insect feeders; and
ectoparasites and diseases exact their toll. Of course, when nests are
deserted or destroyed in early stages, the adults renest, often repeatedly, and thus have an opportunity of nullifying earlier failures. A detailed history of an individual Song Sparrow at Ann Arbor, Michigan,
revealed that she nested five times in 1949, with only two of the nests
successful, and four times in 1950, with none of the nests successful
(Berger, 1951b).

Many life-history studies now show that a nesting success of about 50 percent may be normal for passerines with open nests. Greater losses may represent a decreasing population for one season, but could be atoned for in the following season. Nice (1957) figures that success for 7,778 nests of altricial species with open nests ranged from 38 to 77 percent (average, 49). Fledging success for hole nesters (33 studies involving 94,400 eggs) ranged from 26 to 94 percent (average, 66).

Precocial birds may show a high hatching success at times, in spite of longer periods of vulnerability during incubation. Nests of Clapper Rails, in salt marshes in Virginia, in 1950 showed a hatching success of 94 percent in the absence of storm tides during the nesting season, but in 1951 storm tides lowered the hatching success to 45 percent (R. Stewart, 1951, 1952). Similarly, Geis (1956) found that 73 percent of her Canada Goose nests in Montana were successful in 1953, but only 51 percent in 1954. High water is often disastrous to the nests of aquatic birds (Fig. 11.13).

FIG. 11.13. *Nests and eggs of the Common Tern destroyed by high water in Saginaw Bay, Michigan.* [Photo by Mich. Dept. Nat. Res.]

Later losses, in fall and winter, result from further inroads by predators, particularly among the inexperienced immature birds; storms
during migration; and freeze-ups on the wintering grounds, which
sometimes produce severe mortality among insect-eaters in the
southern states. Additional, man-induced factors include hunting; road
kills; such obstructions as lighthouses, monuments, television towers,
and picture windows; pollution and poisonous pesticides; and loss of
habitat. Birds possess an amazing capacity to recover from or adjust to
the natural factors (e.g., weather and predators) and from some of the
man-induced factors (e.g., regulated hunting), but how well or how
long they can withstand increasing habitat restrictions is a moot question.

TABLE 11.2. MAXIMUM AGE RECORDS FOR
SOME BANDED WILD BIRDS

Species	Age in Years	Remarks	Reference
Laysan Albatross	40+	Eleven at 40 or more	Fisher (unpubl.)
Blue-faced Booby	22+	Three birds	Clapp et al., 1966
Frigatebird	34	Still living?	Clapp et al., 1969
Great Blue Heron	21	Oldest of 349	Owen, 1959
Glossy Ibis	20	USSR	Turček, 1958
Canada Goose	23	Shot, Swan Creek, Mich.	Douville et al., 1957
Mallard	16	USSR	Turček, 1958
Common Teal	20	USSR	Turček, 1958
Common Goldeneye	17	USSR	Turček, 1958
Red-shouldered Hawk	20+	Another 18	Schmid, 1963
Osprey	21	—	Hann, 1953
Oystercatcher	34	Three others 27–32	Bird Banding **33**:205
Eurasian Curlew	32	Shot in England	Austin, 1961
Herring Gull	31	—	IBB News, 45:178
Common Tern	25	England	Bergstrom, 1956
Arctic Tern	27	Killed by cat, Germany	Bergstrom, 1952
Sooty Tern	28	Two; two others 26	Clapp et al., 1966
Least Tern	21	Cotuit, Mass.	Bergstrom, 1952
Caspian Tern	26	—	Bergstrom, 1952
Chimney Swift	14	—	Dexter (pers. comm.)
Hairy Woodpecker	11	Oldest woodpecker?	(Fig. 11.14)
Blue Jay	14	L.I. to Md.	Beals, 1952
Barn Swallow	16	Europe	Thomson et al., 1952
Mockingbird	12	Another 10+, Calif.	Michener, 1951
Black-and-White Warbler	11+	Dead, hit window	Blake, 1969
House Sparrow (2)	13	Killed at feeder, Ohio	Dexter, 1959
Red-winged Blackbird	14	Shot	Frankhauser, 1967
Common Grackle	14+	—	Hann, 1953
Cardinal	13.5	—	Laskey, 1944

FIG. 11.14. *An 11-year-old Hairy Woodpecker. Banded at Pleasant Valley Sanctuary in Massachusetts on Jan. 19, 1939, this bird was observed nearly every winter through March of 1949, but did not reappear the following winter.* [Photo by Alvah Sanborn; courtesy of *Jack-Pine Warbler.*]

Longevity

The rapid death rate among wild birds means, of course, that their *average* or *mean natural longevity* (in the wild) is very low. Their *potential natural longevity* is much higher, and can be increased by confinement (*potential* as opposed to potential natural) in birds well adjusted to captivity (parrots, pigeons, and swans). There are now many records of banded wild passerines living 10–20 years and nonpasserines for more than 20 (Table 11.2), but computed averages usually are less than 2 years. From voluminous banding records Farner (1949) computed the mean longevity of 10 species of passerines in Europe and America that ranged from 1.1 years in the English Robin (the American Robin was given as 1.3 and 1.4 years) and several other European birds (Redstart, Starling, and Great Tit) to a maximum of 2.0 in Song Sparrows. Speirs (1963) calculated the longevity of color-banded Black-capped Chickadees at *less* than one year. Marshall (1947) computed the average natural longevity of Herring Gulls, a notoriously long-lived species, as only 1.75 after September 1 of their first year, thus not including high nesting losses.

Table 11.2 gives some maximum age records of wild banded birds. Intensive banding during the past half century has produced some reliable age records for the smaller short-lived birds, but banding records

TABLE 11.3. MAXIMUM AGE RECORDS FOR SOME CAPTIVE BIRDS

Species	Age in Years	Remarks	Reference
Ostrich	27	—	Hann, 1953
Australian Pelican	60+	Wellington, N.Z. Zoo	Jubb, 1966
Canada Goose	33	—	Hann, 1953
Yellow-billed Duck (2)	48	Blind but still alive in 1965	Jubb, 1966
Andean Condor	52	—	Hann, 1953
Golden Eagle	48	Three others over 40	Brown and Amadon, 1968
Bateleur Eagle	55	Oldest raptor?	Brown and Amadon, 1968
Sarus Crane	42	—	Hann, 1953
Siberian Crane	61.9	Washington Zoo	Davis, 1969
Common Pigeon	27	—	Hann, 1953
Sulphur-crested Cockatoo	56	—	Hann, 1953
Eur. Eagle Owl	68	—	Hann, 1953
Saw-whet Owl (2)	17+	Rec'd as injured adults	Schumacher, 1964
Mockingbird	15	Hand-raised	Laskey, 1962
White-eye (*Zosterops palpebrosa*)	21	Spec. in Wash. State Mus.	Schumacher, 1964
Honeycreeper (*Cyanerpes cyaneus*)	21	Two still alive in 1963	Schumacher, 1964
Western Tanager	15.4	Spec. in Wash. State Mus.	Schumacher, 1964

of some of the larger potentially long-lived birds, such as hawks and owls, have not yet yielded sufficient return or recovery records.

Many birds attain a much greater longevity in captivity. Parrots, swans, and eagles are often credited with attaining ages of 50–100 years, or even more, but many of these records are open to doubt. Hann (1953), citing records from Old World zoos, lists nine species that are known to have attained records from 27 to 68 years (Table 11.3). Only three of them exceeded 50 years of age.

Some Late Summer Activities

One of the major events in the late summer schedule is the postnuptial molt of the adults and the postjuvenal molt of the young, already described in Chapter 3. Both adults and young, except for those that postpone molting until after migration, thus get a complete or nearly complete new plumage before the fall journey. During these molts the birds are relatively quiet and inactive, thus making late summer an unrewarding time for bird watching. However, birds may be gathering in flocks in late July or August, preparatory to migration. Summer assemblages of swallows, shorebirds returning early from the Arctic, and early warbler movements prevent late summer from being a period of complete inactivity. Late summer and fall flocking has certain advantages for birds. Flocks serve as information centers and predator warning systems, and the many eyes help find new sources of food.

Another late summer feature is random wandering of many species. Banding of young birds shows that they may travel long distances from their birthplace, not necessarily southward at first. Every summer witnesses invasions of herons and egrets into the northern states. Bald Eagles hatched in Florida invariably travel northward, even up into Canada, during their first summer (Broley, 1947). Possibly such movements are correlated with the development of the gonads in the young until unfavorable climatic conditions in the north reverse the trend of the gonads and the direction of travel.

The Fall Journey

The journey south is the complement of the trip north in the spring, but differs in several important features. Many of the participants in the fall are birds of the year that have never made such a journey, yet they seem to know when, how, and where to go, without adult guidance. Though some families (e.g., geese) migrate together, possibly so that the adults can guide the young, in some species the adults and the young go separately. Analysis of fall specimens recovered from TV towers often, but not invariably, discloses different flight schedules of males and females and of adults and young.

The fall journey is usually more leisurely than that of spring. Though some birds make long spectacular flights (page 341), many progress more slowly, lingering for days or weeks in favorable localities. The span of the fall migration period may last from late June to December. Many shorebirds, mainly the "unemployed" sex and unsuccessful breeders, leave their Arctic nesting grounds in late June and July, and reappear in the northern states. Most of the swallows, the Bobolink, and some flycatchers and warblers start moving out of the northern states in July. Fall migration probably reaches a peak in the northern states in mid-September; in spite of swollen numbers due to numerous young birds, the spectacular waves characteristic of spring are largely lacking because static late summer weather is not conducive to waves, and by the time cyclonic weather is established many birds have gone. However, many hardy species, mainly seed-eaters and water birds but also some insectivorous types, linger until virtually "frozen out" in November or December.

Wintering Habits

This brief account of the activities of birds in winter is appended to round out the picture of the annual cycle. Most individual birds have a rather specific winter home, although the range of the species may cover half a continent or more. The Common Crow's winter range includes most of the United States, and that of the Sanderling extends from Cape Cod to Cape Horn; but the Whistling Swans, which breed in the far northwest, winter largely in the Chesapeake Bay area or along the Atlantic Coast, and the Blue Geese winter almost entirely on the Gulf Coast. The American Woodcock, coming from a large breeding area in eastern North America, winters mainly in the lower Mississippi Valley. A few birds have divided winter ranges; some ducks and geese winter in part on the Atlantic Coast and in part in the Gulf of Mexico. The Lesser Snow Goose divides its wintering population between the Gulf of Mexico and Lower California.

The wintering grounds of birds, though often far removed from the breeding grounds, nevertheless cover practically all habitable portions of the globe, Ptarmigans remain in the snow-bound Arctic, providing sustenance for the few raptors that also remain; but some northern nesters go across the equator into nearly all parts of the southern hemisphere. A few Antarctic species seek winter homes *north* of their breeding grounds, but many are essentially nonmigratory.

Not only does a particular species have a fairly specific winter range, but banding now discloses that individuals may return each winter to the same locality. Three Tree Sparrows, banded by Dr. Allen at Ithaca, New York, in the winter of 1921, returned in 1922; two of them returned in 1923; and the sole survivor returned in 1924. Another Tree Sparrow apparently returned for eight successive winters to a sanctuary in Lenox, Massachusetts (Wallace, 1942). Though not recaptured every winter, it at least spent its fifth, sixth, seventh, and eighth winters in the same place, usually returning in early November and leaving in April. Of eight color-banded Tree Sparrows present in the winter of 1939 seven (87.5 percent) returned the following winter, perhaps the highest recovery percentage for any migratory bird. Snow and Snow (1960) report a banded Northern Waterthrush that returned for two successive winters to Trinidad, an island off the coast of Venezuela—an indication that migrants wintering in the tropics may have a similarly strong attachment to a specific place. Downer (1972) had two male Indigo Buntings returning to Jamaica for *eight* successive winters.

Some birds are influenced by the distribution of winter food supplies. Some hawks will linger in the northern states in a relatively snow-free winter, but go on to more southern feeding grounds if heavy snows persist, or congregate in regions where rodent outbreaks prevail. Ground-feeding seed-eaters are similarly affected by snow cover and may do considerable roving in quest of new weed fields. Winter finches are erratic in their distribution, for they depend on a particular food supply whose local availability varies from year to year.

Recent literature on birds, not only on their northern wintering grounds but also in the tropics, abounds with studies on the winter habits of birds. In the northern states the shortened winter day, which curtails feeding time, and low temperatures, which necessitate a good food supply to maintain body heat, make it imperative for birds to spend the day foraging. Then at the end of the short day they seek shelter in a cavity of a tree, a protective grove of evergreens, or tangle of vines, usually sleeping with their heads tucked under the wings or buried in the scapular feathers to conserve heat.

In the selection of a nightly roost, many birds show a preference for the type of place in which they were born: woodpeckers drill new cavities in trees, House Sparrows resort to crevices and vines, and Red-winged Blackbirds, though foraging over fields by day, congregate in the marshes at night. Many birds tend to roost at low elevations on winter nights, where there is often good shelter with radiation-produced warmth, but crows, hawks, and owls do not. Ptarmigans and sometimes other grouse burrow in the snow. Many birds, especially tropical species, build special nests or dormitories for winter. Central American wrens (at least 11 species of 22 found in Costa Rica) construct sleeping quarters; some sleep singly, some in pairs, some in family groups (Skutch, 1940). Cactus Wrens in the southwest and Long-billed Marsh Wrens in the northwest also construct special sleeping or roosting quarters for winter, the male and female in separate nests (Anderson and Anderson, 1957; Verner, 1965). Winter Wrens and Pygmy Nuthatches in the western states also roost in cavities or nest boxes, sometimes in large groups (Phillips and Black, 1956; Knorr, 1957). Smith (1972) found 29 Common Bushtits roosting in a *Crataegus* tree on the University of Washington campus in Seattle near the northern extremity of their range during January and February, usually "correctly spaced" two inches apart but more closely huddled on the coldest nights.

The narrowly circumscribed winter ranges characteristic of some birds are well illustrated by the Black-capped Chickadee. Flocks of a half dozen to a dozen individuals assemble in the fall and remain together in close association until spring. If one is transferred to an adjacent territory or taken several miles away, it promptly returns to its former territory and associates. In three winters of observations, in Massachusetts, most individuals did not cross between feeding stations that were a half mile apart (Wallace, 1941). White-breasted Nuthatches and Downy and Hairy Woodpeckers appear to cover only slightly larger areas, but Blue Jays and Tree Sparrows may range considerably farther.

Apparently, individual bonds or social ties exist among certain birds that stay together during the winter. In the Black-capped Chickadee the same individuals can be found together in close association day after day and, if survival permits, year after year. At Lenox, Massachusetts, in the winter of 1937–38, ten chickadees, color-banded for individual recognition, were often observed at a woodland feeding stand. The following winter nine of the ten birds were back at the same station. Sabine (1955) found winter flocks of Oregon (Dark-eyed) Juncos quite stable, but in Tree Sparrows (Heydweiller, 1935; Sargent, 1959) flock organization is less stable. Similar group associations exist in migratory Slate-colored (Dark-eyed) Juncos and White-throated Sparrows wintering in the states, and in Indigo Buntings wintering in Guatemala (Van Tyne, 1932).

Such winter associations terminate with the coming of another breeding season, when individuals segregate, establish territories, and mate for another annual cycle.

SELECTED REFERENCES
PERTAINING TO
THE ANNUAL CYCLE

(Others listed in literature cited, pages 490–528)

ALLEN, A. A. 1914. The Red-winged Blackbird: a study in the ecology of a cat-tail marsh. Abst. Proc. Linn. Soc. New York, 24–25:43–128.

———. 1961. *The Book of Bird Life.* Princeton, N.J.: Van Nostrand. Chapters 9 and 10.

ANDERSON, A. H., and A. ANDERSON. 1957–1963. Life history of the Cactus Wren. Parts I–VI, Condor, Vols. 59–65.

BENT, A. C. 1919–1968. Life histories of North American birds. U. S. Nat'l. Mus. Bull., 107–237. Twenty-three volumes covering loons through fringillids; complete.

BLANCHARD, B. D., and M. M. ERICKSON. 1949. The cycle in the Gambel Sparrow. Univ. Calif. Publ. in Zool., 47:255–318. Refers also to earlier comprehensive paper on White-crowned Sparrows.

BROWN, L., and D. AMADON. 1968. *Eagles, Hawks and Falcons of the World.* 2 vols. New York: McGraw-Hill. Life histories of all diurnal birds of prey.

BUMP, G., et al. 1947. The Ruffed Grouse: Life history, propagation, management. Albany, N.Y.: N. Y. State Cons. Dept.

CLARK, G. A., JR. 1964. Life histories and the evolution of megapodes. Living Bird, 3:149–167.

DAVIS, J. 1960. Nesting behavior of the Rufous-sided Towhee. Condor. 62:434–456. See Literature Cited at end of article for other towhee papers by Davis.

EDMINSTER, F. C. 1947. *The Ruffed Grouse, Its Life Story, Ecology, and Management.* New York: Macmillan.

FISHER, H. I. 1971. The Laysan Albatross: its incubation, hatching, and associated behaviors. Living Bird, 10:19–78.

HANN, H. W. 1937. Life history of the Oven-bird in southern Michigan. Wilson Bull. 49:145–237.

HERRICK, F. H. 1911. Nests and nest building in birds. Jour. Animal Behavior. 1:159–192, 244–277, 336–373.

———. 1934. *The American Eagle.* New York: Appleton.

HOCHBAUM, H. A. 1944. *The Canvasback on a Prairie Marsh.* Washington, D.C.: Amer. Wildl. Inst.

HOWARD, H. E. 1920. *Territory in Bird Life.* London: J. Murray.

HOWELL, J. C. 1942. Notes on the nesting habits of the American Robin (*Turdus migratorius* L.). Amer. Midl. Nat., 28:529–603.

KENDEIGH, S. C. 1941. Territorial and mating behavior of the House Wren. Ill. Biol. Monographs, 18(3):1–120.

———. 1952. Parental care and its evolution in birds. Ill. Biol. Monographs, 22(1–3):1–356.

KESSEL, B. 1957. A study of the breeding biology of the European Starling (*Sturnus vulgaris* L.) in North America. Amer. Midl. Nat., 58:257–331.

KORTRIGHT, F. H. 1942. *The Ducks, Geese, and Swans of North America.* Washington, D.C.: Amer. Wildl. Inst. Data on waterfowl life histories.

KOSKIMIES, J. 1950. The life of the swift, *Micropus apus* (L), in relation to the weather. Ann. Acad. Scient. Fennicae, 4, Helsinki, 15:1–151.

LACK, D. 1943. *The Life of the Robin.* London: Witherby.

LANYON, W. E. 1957. The comparative biology of the meadowlarks (*Sturnella*) in Wisconsin. Publ. Nuttall Ornith. Club, 1:1–67.

LASKEY, A. R. 1944. A study of the Cardinal in Tennessee. Wilson Bull., 56:27–44.

LINSDALE, J. M. 1937. The natural history of the magpies. Pacific Coast Avifauna, No. 25, Cooper Ornith. Club, Berkeley, Calif.

MAUNDER, J. E., and W. THRELFALL. 1972. The breeding biology of the Black-legged Kittiwake in Newfoundland. Auk, 89: 789–816.

MAYFIELD, H. F. 1960. The Kirtland's Warbler. Cranbrook Inst. Sci. Bull., 40. Bloomfield Hills, Mich.

MEANLEY, B. 1971. Natural history of the Swainson's Warbler. No. Amer. Fauna, 69:1–96.

MICHENER, H., and J. R. MICHENER. 1935. Mockingbirds, their territories and individualities. Condor, 37:97–140.

MILLER, A. H. 1931. Systematic revision and natural history of the American shrikes (*Lanius*). Univ. Calif. Publ. Zool., 38:11–242.

MUMFORD, R. E. 1964. The breeding biology of the Acadian Flycatcher. Univ. Mich. Mus. Zool. Misc. Publ. 125:1–50.

NELSON, B. 1968. *Galapagos: Islands of Birds.* New York: William Morrow. Detailed life-histories of boobies, gannets, and other sea birds.

NICE, M. M. 1937, 1943. Studies in the life history of the Song Sparrow, I and II. Trans. Linn. Soc. New York.

ODUM, E. P. 1941, 1942. Annual cycle of the Black-capped Chickadee, Auk, 58:314–333, 518–535; 59:499–531.

PALMER, R. S. 1941. A behavior study of the Common Tern (*Sterna hirundo hirundo* L.). Proc. Boston Soc. Nat. Hist., 42:1–119.

PETERSON, A. J. 1955. The breeding cycle of the Bank Swallow. Wilson Bull., 67:235–286.

PETTINGILL, O. S., JR. 1936. The American Woodcock, *Philohela minor* (Gmelin). Mem. Boston Soc. Nat. Hist., 9:167–391.

PICKWELL, G. B. 1931. The Prairie Horned Lark. Trans. Acad. Sci. St. Louis, 27:1–153.

PRESCOTT, K. W. 1965. Studies in the life history of the Scarlet Tanager, *Piranga olivacea.* New Jersey State Mus. Invest., 2.

PUTNAM, L. S. 1949. The life history of the Cedar Waxwing. Wilson Bull., 61:141–182.

RICHDALE, L. E. 1951. *Sexual Behavior in Penguins.* Lawrence, Kans.: Univ. of Kans. Press.

SAUER, E. G. F., and E. M. SAUER. 1966. The behavior and ecology of the South African Ostrich. Living Bird, 5:45–76.

SHELDON, W. G. 1967. *The Book of the American Woodcock.* Amherst: Univ. Mass. Press.

SKUTCH, A. F. 1945. Incubation and nestling periods of Central American Birds. Auk, 62:8–37.

———. 1954, 1960. Life histories of Central American birds. Pacific Coast Avifauna, No. 31 and No. 34, Cooper Ornith. Soc. Berkeley, Calif.

SMITH, S. 1947. *How to Study Birds.* London: Collins. Chapters 1–7.

SOUTHERN, W. 1958. Nesting of the Red-eyed Vireo in the Douglas Lake Region, Michigan. Jack-Pine Warbler, 36:105–130, 185–207.

STODDARD, H. L. 1931. *The Bob-white Quail: Its Habits, Preservation and Increase.* New York: Scribner.

TINBERGEN, N. 1939. The behavior of the Snow Bunting in spring. Trans. Linn. Soc. New York, 5:1–94.

VAN TYNE, J., and A. J. BERGER. 1959. *Fundamentals of Ornithology.* New York: Wiley.

WALKINSHAW, L. H. 1949. The Sandhill Cranes. Cranbrook Inst. Sci. Bull., 29, Bloomfield Hills, Mich.

WALLACE, G. J. 1939. Bicknell's Thrush, its taxonomy, distribution, and life history. Proc. Boston Soc. Nat. Hist., 41:211–402.

WILLIAMS, L. 1952. Breeding behavior of the Brewer Blackbird. Condor, 54:3–47.

12

~~~~~

Migration

ONE OF the most spectacular events in the animal world is the migration of birds. It has intrigued mankind for many centuries; in ancient times it led to observations designed to explain the phenomenon (Aristotle, Job, Jeremiah), but also to wild speculations, such as the persistent views that swallows hibernated in the mud and that birds flew to the moon for winter. Now after more than 2,000 years of observations, climaxed by a half century of experimental studies, we are still in the dark regarding many aspects of migration. Though we know in detail the breeding range, wintering grounds, and often the migration routes of practically every North American bird, we still cannot always explain why birds go where they do at the time they do, and their manner of navigation is only partially solved.

Birds are by far the best known examples of migrants, but of course many other animals have migratory habits. Many marine invertebrates have definite but often not well understood movements in relation to tides and seasons; aeroplankton, perhaps unwittingly as in the case of wind-blown ballooning spiders, take long journeys through space; certain insects, particularly butterflies, undertake regular migrations, some going south for the winter, others merely taking suicidal journeys out to sea. Sharp periodicities in the breeding cycle of most amphibians involve short journeys to and from water, and some reptiles travel far for suitable winter quarters. Among the fishes the amazing migrations of the eel and salmon are the most publicized, and among the mammals the cyclic outbreaks of the lemmings are classic. Many other animals—reindeer and caribou on their grazing grounds, desert creatures seeking a new food supply, montane forms making altitudinal adjustments—respond to seasonal and climatic changes by undertaking regular or irregular migrations.

Definitions and Migration Patterns

Migration in a broad sense is simply moving from one place to another. A more complete definition would include some correlation with the seasons and would limit its use to two-way journeys of animals under their own power, thus excluding unidirectional movements and those in which animals are helplessly carried by some other agency. Thus, a more precise definition, condensed from Dorst (1962), would be: a series of periodic round trips — usually annual — between a breeding area and a nonbreeding or wintering area.

With regard to frequency, movements may be (1) *daily*, as of birds passing to and from a roost (crows, Starlings) or rookery (herons); (2) *lunar*, as in the case of many marine organisms under the influence of tides, and Sooty Terns and perhaps some other birds on oceanic islands, which have a 10-month breeding cycle corresponding to the lunar rather than calendar year (Chapin and Wing, 1959); (3) *seasonal*, which includes the regular, as well as some of the irregular, migrations of birds; and (4) *cyclic* (*emigration*), movements correlated with some cycle of longer duration than, or not associated with, the seasons. Cyclic and seasonal migrations are of particular importance in bird life and merit further analysis.

CYCLIC MIGRATIONS (EMIGRATIONS)

Irruptions of animals, typically, are characterized by periodic population explosions and consequent food shortages that force an exodus of surplus individuals out of the normal range. Irruptions may be *sporadic* (irregular) or *rhythmic* (correlated with a cycle of definite length). They are often unidirectional, and thus not considered a true migration by some authorities, but among insects sometimes another generation makes the return trip. A classic example is the lemming, a mouselike Arctic rodent, which periodically (every three to five years) reaches population peaks apparently compelling it to undertake a suicidal journey overland, sometimes to the sea where individuals surviving the march plunge in and drown. Perhaps the concept of drowning has been overemphasized, however, since only the animals living near the coast, as in Scandinavia or Labrador, would ever reach the sea. Curry-Lindahl (1963) vividly describes a suicidal lemming march in Norway in 1960–61 and suggests a new theory: "an endocrine malfunction may cause lemming psychosis."

Such emigrations are not unknown among birds. The Pallas Sandgrouse (*Syrrhaptes parodoxus*), a pigeonlike bird of central Asia, has staged several such irruptions in the past, spilling over Europe in large numbers. In the flight of 1863, some 700 individuals reached Great Britain. A few pairs remained to breed in western European countries, but failed, as on subsequent invasions, to establish themselves. Cade and Buckley (1953) report a similar irruption following a buildup of Sharp-tailed Grouse in the Tanana Valley, Alaska. A large flock, two to three miles long and half a mile wide, took off en masse on a day in

October and disappeared. Periodic population peaks (the 11-year cycle) are well known for the Ruffed Grouse. The fall shuffle and dispersal of Bobwhites may become irruptive, though not necessarily suicidal, when populations are high (Agee, 1957). The presumed function of such outbreaks is periodic reduction, or perhaps in some cases merely more effective scattering, of rapidly breeding herbivorous forms that might otherwise devegetate their range. Höhn (1967) supports the theory of an endocrine imbalance as a cause of the 10-year cycle in ptarmigans: overpopulation stress results in decreased gonadotropic secretion, decreased reproduction, adrenalin exhaustion, and slow recovery over a 10-year period.

Irruptions of birds are usually correlated with fluctuations in the food supply. These may be in tune with known cycles, as in the case of the three-to-five-year movements of the Snowy Owl, Rough-legged Hawk, and Northern Shrike, or more irregular, as in the case of cross-bills and other wandering seed-eaters. The Snowy Owl (Fig. 12.1) in particular is affected by the three-to-five-year lemming cycle. When these rodents decline, the birds are forced to seek new feeding grounds or starve. Many of them move southward from Canada, spilling over into the northern states in large numbers. In the heavy flight of the winter of 1945–46, 13,502 reports were received by an international committee collecting records on the invasion. Presumably the largest flights are correlated, not alone with lows in the lemming cycle, but also with population peaks in the owls, so that if a summer of high production is followed by a winter of scarcity of lemmings, a large southward flight will occur.

The crossbill movements are much more irregular, and not necessarily southward in direction nor confined to winter. Crossbills are dependent on seed production in northern conifers and wander widely in quest of a new cone supply. Even their breeding season seems to be correlated with this peculiar situation, since the birds are very irregular as to time and place of nesting. Svärdson (1957) correlates "invasion" migrations of certain species in Sweden with high fruiting of trees every three or four years; redpolls raise two broods in good years, one in March on seeds, one in June on insects. Evans (1969) reported that redpolls, invading southern England in winter, moved on to the continent if the birch seed crop on which they depended was poor; on the continent they found the advantages of longer days, shorter nights, and an almost untouched birch seed supply. Bock and Lepthien (1972) proposed three theories for Red-breasted Nuthatch irruptions: (1) an abundant winter food supply in the south, (2) cone failures in the north, (3) and overpopulation pressure. They favored the second view. Davis and Williams (1957, 1964) describe similar irruptions of Clark's Nutcracker in California; population buildups during years of heavy

F I G . 12.1. *The periodic winter invasions of Snowy Owls into the United States arouse great interest. Trophy collectors, even in states according legal protection to these birds, practically exterminate the visitors before they can return to their Arctic home. Unlike most owls, this species is largely diurnal, living on lemmings and other vertebrate life of the tundra.* [Photo by Allan D. Cruickshank.]

pine-cone production were usually followed by years of pine-cone failures and a consequent exodus of the nutcrackers. They concluded these movements were not true migrations, since they were not triggered by hormonal and physiological changes.

Vagaries in weather cycles are responsible for some of the irregular movements of birds. In Australia, for instance, there appears to be no particular incentive for regular northward and southward migrations, for winter and summer as we know them here are not a conspicuous feature on that continent. Consequently, bird movements in Australia may be associated with the highly irregular dry and rainy spells, with seed-eating as well as insectivorous and predatory birds following the rains to new feeding grounds, and raising their young during the staggered periods of maximum food production. In Africa, the Black-faced Dioch or Quelea (*Quelea quelea*), perhaps the most abundant (and destructive) bird in the world, performs mass movements correlated with periodic rains and consequent availability of seeds and grain. The direction, distance, and frequency of their movements depend on rains and they may breed several times a year in different areas (Ward, 1965, 1971).

Seasonal or Annual Migrations

Some of the seasonal movements of birds are quite irregular and include random wanderings in quest of new food supplies and the summer dispersal of young birds in all directions from their birthplace. These are called *vagrant* migrations by Lincoln (1950), but might better be called *postbreeding wanderings*. They nearly always involve a subsequent return to the regular wintering and breeding grounds. The most noteworthy of these vagrant movements are the northward exodus of herons from southern breeding grounds (see page 317) and the random dispersal of gulls, terns, Barn Owls, and other species far from their nest sites before migrating southward. Banding data and other late summer records indicate that Yellow-breasted Chats, unlike other warblers (and indeed the chat may not be a warbler), have a northward dispersal after nesting (Dennis, 1967b).

Tropical and subtropical countries witness a type of seasonal migration not well known in northern latitudes. These movements, termed *moisture rhythm* migrations by Woodbury (1941), are like those described above for Australia and Africa except that they are regular migrations of the seasonal type. In the spacious savanna lands of South America and Africa, which are broad belts characterized by sharply divided wet and dry seasons, the breeding season of birds coincides with the rainy season, or period of maximum food production, and their migratory movements follow the seasonal rains. Often there is no hard and fast line between irregular movements in response to rains, as in Australia and parts of Africa, and the more regular movements in savanna lands where the rainy seasons are quite regular.

The regular seasonal migrations of birds in northern latitudes may be divided into (1) *altitudinal,* as up and down a mountain (the term *vertical* migration is perhaps more properly restricted to aquatic organisms); (2) *longitudinal* (east-west migrations); and (3) *latitudinal* (north-south).

Altitudinal migrations are well-known phenomena in western United States, where jays, chickadees, nuthatches, kinglets, and some fringillids avoid long north-south migrations by merely changing altitude with the seasons. Some of these species are much less migratory in the western mountains than in the eastern parts of their range. Most western species exhibiting altitudinal migrations nest at the higher altitudes and seek more favorable lower levels for winter, but the Blue Grouse and Rosy Finch reverse this trend and go to higher elevations in late summer (Mussehl, 1960; Johnson, 1965), a movement that coincides with the fruiting and seeding of certain plants. The Clark's Nutcracker also seeks higher levels after nesting, but like the Rosy Finches does not remain there over winter.

East-west migrations are well illustrated by the peculiar route taken by Redheads (ducks) nesting in the Bear River marshes in Utah and wintering on the Atlantic Coast, in about the same latitude as where they were reared. The fall migration route takes them on a sweeping curve, north and northeastward at first, then southeastward to the coast. Evening Grosbeaks (Fig. 12.2) and California Gulls are also good examples of east-west migrants. The former are believed to have followed the fruiting of box elders, a favorite food, from the midwest to the eastern states in winter until it became an established route. California Gulls nest in the interior but winter on the Pacific coast. Some western races of the Red Crossbill appear in the eastern states in large numbers on occasions, but these might more properly be termed "invasion migrations."

However, most journeys of birds in the northern hemisphere take them between a northern summer (breeding) and more southern winter home. Migration is best developed in the northern hemisphere, where the largest land masses, and thus the greatest extremes of winter and summer, occur. The less extensive land areas in the southern hemisphere, with their more equitable climate, witness some interesting migrations, but they are not comparable to those of northern hemisphere birds. Many southern hemisphere birds are essentially non-migratory. Others merely move short distances toward the equator, and a few actually cross the equator. Perhaps the most spectacular example of the latter (excluding oceanic birds) is a race of the Blue and White Swallow (*Atticora cyanoleuca*), which nests in southern South America,

FIG. 12.2. *Evening Grosbeaks, which formerly bred exclusively in the northwest, now visit eastern feeding stations in winter in large numbers, often dominating the feeders and consuming large quantities of expensive sunflower seeds. Now they breed in limited numbers in the Adirondacks, northern New England, and eastern Canada.* [Photo by Alvah W. Sanborn.]

and has been taken in Costa Rica, 2,200 miles north of its northernmost breeding limits (Howell, 1955).

In local field work birds are commonly divided into *status groups* on the basis of their seasonal distribution. Terms commonly used and definitions for them are as follows:

Permanent Residents—species that are present the year round in a given locality.
    a. Nonmigratory—game birds (grouse, pheasants, quail), some owls (Great Horned, Barred, Screech), some woodpeckers, some titmice, House Sparrows, Cardinal.
    b. Migratory (species present in the same locality throughout the year but certain individuals at least are migratory)—some ducks, some hawks, some crows and jays, Cedar Waxwing, Starling, Song Sparrow.
Summer Residents—species that come in the spring, remain for summer (breeding), and leave in the fall. Includes nearly all summer birds that are not permanent residents.

Transients—birds that pass through a given region in spring en route to more northern breeding grounds and again in fall for more southern wintering grounds. Examples (in the United States) are Arctic breeding ducks, shorebirds, and northern warblers.

Winter Residents (or Visitants)—somewhat loosely defined as birds that remain (or appear) in a given locality in winter. Some, like the Tree Sparrow, may come in the fall and remain in a small area all winter (see page 319); others, often called erratic finches (redpolls, crossbills, grosbeaks), may appear some winters for short or long periods and not show up at all in other winters.

Often these status groups are further qualified by such terms as abundant, common, uncommon, and rare, but such terms are hard to define accurately and are not used in exactly the same meaning by all observers. Widely wandering or displaced birds, such as tropical species in the north, sea birds blown inland, or Old World birds appearing in North America are called "stragglers" or "accidentals."

The annual migration of birds, from a summer to a winter home and back again, may be extremely short or extremely long. The Wrentit (Chamaeidae) in California is nonmigratory; certain individuals at least probably never range over more than a few acres within their lifetime. Gallinaceous birds, in general, are also essentially nonmigratory. The Bobwhite, except for unusual irruptions that may take it up to 26 miles (Agee, 1957), is perhaps the least mobile of American game birds; some spend their entire lives within one quarter of a mile of their birthplace, and most others stay within the square mile where they are hatched (Murphy and Baskett, 1952). Some of the owls (page 330) are nonmigratory (Fig. 12.3); others (Barn, Snowy, Short-eared) are very irregular in this respect. Most of the titmice, in this country at least, are quite sedentary, although they may move to more sheltered locations in winter. White-breasted Nuthatches, except for unusual flights (see Heintzelman and MacClay, 1971), are essentially nonmigratory, but Red-breasted Nuthatches are frequent invasion migrants.

In some cases adult birds, established on their breeding grounds are nonmigratory, whereas their young go long distances during their first year or two. Herring Gulls hatched in New England coastal colonies may go to Florida for their first winter, to the middle Atlantic Coast their second winter, and then remain in northern waters after their first breeding season. A pair of Barn Owls in East Lansing, Michigan, apparently occupied the same barn almost continuously for three years, but one of their banded young was taken in Alabama during its first winter. Mueller and Berger (1959) report that two Barn Owls banded as nestlings in Wisconsin on July 1 were taken in Florida, over 1,200 miles distant. Yet there are records of migrating adult Barn Owls also,

FIG. 12.3. *This nonmigratory Screech Owl may inhabit the same cavity throughout its lifetime, though the chances are that it had to move from its natal nest site, if one of its parents still occupied it. Some other owls are highly migratory, at least on occasions.* [Photo by Allan D. Cruickshank.]

up to 850 miles in one case (P. Stewart, 1952). Many Blue Jays banded in the northern states have been taken in the southern states in winter; often they can be seen traveling in large flocks during migration, yet many Blue Jays are quite sedentary.

Widely ranging young of sedentary parents are not uncommon; there are records of banded broods of titmice in England, some of which migrated and some did not. Among Mockingbird broods in California, the Micheners (1935) found some that migrated and some that did not. Starlings are quite irregular; some migrate and some do not, some migrate some years and not other years, and range expansion is accomplished by young birds (not adults) invading new regions (Kessel, 1953). In a sense all of these are examples of dispersal rather than true migration.

To add another quirk to the peculiarity of migration patterns, Ludwig (1960) reports that of two young Mourning Doves banded in the same nest in Michigan on May 24, one was recovered in Georgia on December 11, the other in Texas on the following day. Mason (1961) reports a banded dove that spent the winter of 1958 in Massachusetts, but was shot in January 1961 in Florida. Sharp (1971) describes a transcontinental migrant, banded in New York in August 1969 and shot in California in September 1970. He suggests that the young dove may have become "emotionally involved" with a western dove on their wintering grounds and therefore went to California in the spring.

Why certain individuals in a highly migratory species sometimes fail to migrate is not known. Every winter produces records of thrushes, warblers, orioles, and other birds remaining in the northern states when normally they would have gone south. In some cases, physical injury may have prevented migrating at the proper time, and then the urge to go subsided. Again some metabolic disturbance or hormonal imbalance may have been the underlying cause.

It has long been known that in birds with permanent pair bonds, such as geese, the adults and their young travel together, and that in birds with one-season pair bonds the males, females, and young often migrate separately, at different times, sometimes by different routes to different wintering grounds. Both banding evidence and TV tower kills indicate that migrating adults and young may appear at different times. Some shorebirds (e.g., Golden Plovers) leave their young on their northern breeding grounds and start their long southward journey ahead of the young. In some cases (e.g., Song Sparrows) the females are more migratory than the males; we have already seen that during spring migration the males arrive before the females, either having started earlier or having a shorter distance to go.

It has already been pointed out that migratory species go variable distances for suitable wintering grounds. In many species the northern part of the winter range overlaps, broadly or narrowly, the southern part of the breeding range. In other species the summer and winter ranges are widely separated. Many northern birds winter in the southern states; still others go to the West Indies, Central America, or northern South America; and a few make even more spectacular journeys across the equator. An exciting example of this last group is a Peregrine Falcon banded as a nestling in Northwest Territories, Canada, on July 29, 1965, and recovered in Argentina in January 1966.

It covered the intervening 9,000 miles in about four months (from fledging) at the average rate of 2,250 miles per month (Kuyt, 1967).

REVERSE MIGRATION

Observations of birds apparently flying in the wrong direction in both spring and fall are common, but the explanation for such paradoxical phenomena is somewhat speculative. On spring trips to Point Pelee, a peninsula jutting down into Lake Erie from the mainland of Ontario, observers often see flocks of birds, such as Blue Jays and blackbirds, moving *south* along the peninsula. Some banded birds have been recovered back in Ohio after having landed on the Point. Reverse migrations are common along the New England coast and the Canadian Maritime Provinces. Observers interpret these movements as migrants blown or strayed off course and trying to regain their bearings and get back on the preferred route. In some cases birds seem to start out over the water from New England or Nova Scotia; then, if they meet favorable flying conditions, they continue on to the West Indies or even South America, but if they encounter adverse winds or storms, they reverse their direction of travel and come back to land.

Origin and Causes of Migration

FOOD AND WEATHER

Though it is generally assumed, and rightly so, that cold weather and the consequent scarcity of food, particularly insects, cause birds to migrate, it is well known that this is an inadequate explanation in many cases. While it is true that a flycatcher or swallow cannot find enough flying insects in the north in winter, it is not clear why the Eastern Phoebe winters in the states when other flycatchers go to Central America or beyond, or why the Tree Swallow can make a go of it in New Jersey when the Cliff Swallow travels across the equator. Some ducks winter as far north as open waters permit, but the Blue-winged Teal and Pintail extend their winter range into South America. The Red-tailed Hawk winters in the northern states or southern Canada, but the Broad-winged Hawk reaches South America. Similarly, among the wrens, warblers, blackbirds (icterids), and fringillids, there are hardy forms readily weathering northern winters, while others go to the tropics or subtropics.

Another serious objection to the view that food supplies and cold weather are the underlying causes of migration is that fact that so many birds leave their northern breeding grounds in midsummer, long before food supplies are exhausted or even before they have reached a peak. Thus some shorebirds desert their breeding grounds in the Arctic in late June or July, with insects swarming in abundance. Most of the swallows, some flycatchers and warblers, and even seed-eating Bobolinks terminate nesting activities in July and start southward. Conversely, American Robins and Eastern Bluebirds often leave early from their wintering grounds, where food is usually adequate, and hurry northward into a region of questionable food supplies.

Often, however, a reasonable explanation can be found for some of these distributional peculiarities. The Eastern Phoebe has learned to exploit a special food supply—aquatic winter insects. All flycatchers could not be supported in this way. Tree Swallows wintering in New Jersey subsist in part on bayberries or other fruit, but Tree Swallows in Florida do not and sometimes starve during prolonged freezes while their northern relatives survive. Shorebirds that leave their Arctic nesting grounds early may do so to leave more food for their young. The Louisiana Waterthrush is one of the earliest warblers to arrive in the northern states in spring and perhaps is the first nester; Eaton (1958) notes that rearing of the young coincides with the heavy hatches of stoneflies and mayflies in the streams where the birds feed. Thus there may well be explanations for the enigmas of the timing of migrations, but we have yet to find the right answers to many of them.

NORTHERN ANCESTRAL HOME

Several theories have been advanced to explain the origin and perpetuation of the migratory habit in birds. One of these, the northern ancestral home (ice-sheet or glaciation) theory, postulates that in remote geologic times, nonmigratory birds inhabited the northern hemisphere, and as a mild climate prevailed, with palm trees and elephants occurring in Alaska, there was no incentive to migrate. Pleistocene glaciation ended this phase in the earth's history, and birds were forced by the first great ice sheet to go south or perish. As the icecap gradually receded, the displaced birds moved back to their ancestral nesting grounds, thus avoiding the more crowded tropics. Subsequent glacial periods and the gradual establishment of sharp winter and summer periodicities introduced a migratory rhythm in birds that became a hereditary habit.

This view of the possible origin of migration is now largely discredited; in fact, the migratory habit in birds probably preceded Pleistocene glaciation by many millions of years. However, glaciation has profoundly affected present-day distribution and migration routes of northern birds, both in North America, where persistent glaciation in the western mountains separated eastern and western forms, and in Europe, where ice-bound mountain ranges split populations into eastern and western subspecies. Hence, glaciation in the late Tertiary had a profound effect on the details of distribution, speciation, and migration routes, but probably had little or nothing to do with the origin of the migratory habit.

SOUTHERN HOME THEORY

The second hypothesis, in some respects, is diametrically opposed to the northern home theory. The southern home theory presupposes that the ancestral home of birds was in the tropics, which eventually became so congested that the more venturesome birds spread northward to seek isolation during the breeding season and then returned to their home in the tropics as soon as nesting was over. Such a view helps explain why some long-range migrants leave their breeding grounds as soon as nesting is over, long before the food supply is a critical factor. Hummingbirds, tanagers, and flycatchers, all of South American origin, may simply be returning home. Moreover, by moving northward in the spring, birds have a longer day (16–20 or more hours of daylight compared with 12 near the equator) for gathering food for their young. However, many northern species, perhaps a majority, are not of southern origin; hence their migrations cannot be explained by the southern home theory.

CONTINENTAL DRIFT

This theory, first proposed by Richard Owen in 1857, advanced by Wegener in 1915, and championed by Wolfson (1948) to explain some present-day distributional patterns, postulates that the continents were once fitted together into one great land mass. Eventually they separated into the present-day continents, carrying with them their existing plant and animal life. One aspect of the theory is that a tropical or subtropical Antarctica (Gondwanaland) was the place of origin of animal life, which subsequently spread northward into South America, Australia, and Africa and on into North America and Eurasia. This theory has often been discredited in the past, not only because of its seeming improbability, but because some geologists believe the separation took place before the origin of migratory habits in birds, probably in the Jurassic. However, new geological evidence now strongly supports the former existence of clustered continents and theories concerning it are being revived. It helps explain some puzzling plant distributions, the occurrence of chelid turtles in South America and Australia, the distribution of some fossils, and pantropical species of birds (see page 368). Glenny (1954) believes it explains the occurrence of relict forms in New Zealand and Australia and of related South American and Australasian birds.

Photoperiodism

A more modern and realistic approach toward explaining migration involves photoperiodism, which accounts for recrudescence of the gonads in the spring with increasing day length and regression with decreasing light in the fall. Photoperiodism is no longer theory, as far as its influence on the state of the gonads is concerned, but its exact relationship to migration is.

The first detailed experimental work demonstrating the effect of the daily photoperiod on birds was carried out by Rowan (1931 and earlier papers) in the Province of Alberta, Canada. He kept juncos and crows in outdoor aviaries in subzero fall and winter temperatures but with artificial lighting simulating spring. Birds thus exposed to increasing daily photoperiods showed enlarged gonads (the controls did not) and experimental releases in winter apparently "migrated," some of them northward, but the evidence was not very conclusive. Later Bissonette (1937, 1939) conducted similar experiments with Starlings and Blue Jays with similar results, but thought that increased light stimulated activity of the anterior pituitary, which in turn caused recrudescence of the gonads. Rowan, as well as subsequent workers, felt that increased wakefulness and increased exercise associated with the longer day were important factors in the development of the gonads.

More recently many workers have conducted extensive experiments with caged birds, trying to measure the effects of various photoperiods on the physiological state of a bird. One of the results has been the discovery of a postbreeding *refractory period*—a quiescent or preparatory period into which the gonads relapse before they can be stimulated by increased photoperiods to start another cycle. However, experimental results have been very variable. Some birds show no refractory period, some respond quickly, others more slowly or at different times. Annan (1963) found that Hermit Thrushes had a typical refractory period in the fall, and did not respond to artificial long days at that time. Juncos, White-throated Sparrows, and other fringillids have a refractory period that ends in November or December before they respond to light treatments (Shank, 1959); then they can be brought into breeding condition several months early by increasing photoperiods. Wiese (1962) kept several fringillid species on short photoperiods (nine hours) for 18 months without producing much recrudescence of the gonads. He concluded that a nine-hour day is too short for a full gonadal response. Apparently increased photoperiods are necessary for recrudescence of the gonads in northern hemisphere birds.

Another discovery from observations on caged birds is the occurrence of nocturnal restlessness (*Zugunruhe*) in migratory birds—that is, birds beating against the sides of the cages as if trying to migrate. Eyster (1954), experimenting with nonmigratory House Sparrows and migratory fringillids, found that the migratory species exhibited nocturnal unrest but that the House Sparrows did not. However, both migratory and nonmigratory races of the White-crowned Sparrow on the Pacific Coast exhibit *Zugunruhe*. Smith et al. (1969) interpreted this as a case of *atavism*—a "hangover" of ancestral migratory behavior. Some tropical species show no refractory period. Thapliyal and Saxena (1964) maintained Common Weaver Birds (*Ploceus philippinus*) in peak breeding condition continuously on 15-hour days. The Andean Sparrow, which begins breeding before it is a year old, has no post-juvenal refractory period.

Various shortcomings are evident when one tries to correlate migration with photoperiodism per se. Photoperiodism operates principally in birds that breed in northern latitudes, less so in tropical species because changing photoperiods are largely lacking in the tropics. Moreover, increased light could not initiate migration in transequatorial migrants, as they would have to start out on their journey under conditions of uniform or even decreasing light. According to Griscom (1945) such migrants may be well on their way up into the southern states before there is any perceptible increase in their gonads. Zimmerman and Morrison (1972) found that the testes of Dickcissels wintering in Panama started a slow recrudescence in winter but that full development was not attained until they reached their breeding grounds. Hence, it may be that the development of the gonads is coincident to, rather than a cause of, migration. East-west migrations seem to have no correlation with length of day. Heerman's Gulls on the Pacific Coast actually move northward in the fall and winter.

The breeding season of many birds does not conform to annual light cycles. The Barn Owl breeds irregularly; nests with eggs and/or young have been found in every month in both northern and southern states. Though most tropical birds have a specific time for breeding, it is not necessarily in the spring. The Andean Sparrow has two breeding seasons annually (Miller, 1959) and the Quelea may breed several times a year. Some tropical oceanic birds (Sooty Terns on Ascension Island and Gull-billed Terns in the Galapagos Islands) have 9–10-month breeding cycles. Well known also are some two-year breeding cycles (albatrosses, California Condor) in which the young mature so slowly that the adults usually breed every other year. Changes in or postponement of the breeding season in response to special environmental conditions, such as the Short-billed Marsh Wren in Arkansas delaying nesting until dense growth of rice fields is available in late summer (Meanley, 1952), Rufous-winged Sparrows waiting until summer rains make food more plentiful for their young (Phillips, 1951), and Piñon Jays breeding whenever the piñon crop is most plentiful (Ligon, 1971), are also interesting contradictions to the correlation of breeding with maximum light.

Hence, in summary, it is evident that the factors governing migration and breeding are exceedingly complex and that none can explain either phenomenon in all birds. The annual stimulus for migration, as outlined by several authors (see particularly Farner, 1955; and Marshall, 1961), is the development of a particular metabolic or physiological condition, called *Zugdisposition* by some Europeans, prior to migration. This may be reflected in the state of the gonads in some species but not in others, and in fat accumulations (see below) in some but not others. Weather conditions as external factors may be important in early spring migrants ("weather" migrants), less so, if at all, in later long-distance ("instinct") migrants. Experiments on photoperiodism have emphasized the importance of day length, but warm temperatures and other favorable vernal conditions supplement light in stimulating recrudescence of the gonads. The importance of the anterior pituitary, whose secretions stimulate gonadal development when exposed to light, has already been mentioned (page 184).

Some Mechanics of Migration

FAT RESERVES

It is now well known that accumulation and storage of fat are vital for long migratory flights. In general, nonmigratory birds do not accumulate fat reserves, leisurely migrants like Mourning Doves store very little (Hanson and Kossack, 1957), and birds making short trips, or on portions of their journeys not requiring long hops, have limited reserves. But long-range migrants accumulate adipose tissue up to 50 percent or more of their body weight, providing enough fuel to take them more than a 1,000 miles nonstop (Odum et al., 1961).

Conditions vary greatly. Caldwell et al. (1963) found fall warblers, thrushes, and vireos in Florida fatter than those in Michigan, presumably having lingered in the southern states to accumulate fat in preparation for trans-Gulf flights into Central and South America. But White-throated Sparrows were fatter in Michigan than in Florida in the fall, presumably because they were near the end of their migration in Florida and had used up their reserves. However, many intracontinental migrants in spring, such as Dark-eyed Juncos, Tree Sparrows, and White-throated Sparrows, start migration with low reserves for short flights, or for early stages of their journeys (Helms and Smythe, 1969; Johnston, 1966); later they may accumulate fat for longer stages of their migration. Nisbet et al. (1963) found Blackpoll Warblers stopping in southern New England for several weeks to fatten up before taking off overseas for South America. Some killed at a lighthouse in Bermuda, en route, were still fat. Blackpolls weighing 26 grams could fly for 105–115 hours, long enough to reach South America without stopping in Bermuda. Rogers (1965) found that migrants grounded in Panama in fall still had sufficient fat to carry them on into South America, although they had used up about half of their reserves getting to Panama from the United States. In spring he found birds arriving in Panama thin and fat-free, but those farther north in British Honduras were fat, in preparation for their trans-Gulf flight (Rogers and Odum, 1966). West et al. (1968) found Lapland Longspurs depositing fat en route to Alaska. Curiously, Irving (1960a) found male Water Pipits arriving on their Arctic breeding grounds, after migrating from a probable winter home in Texas, fat and in good conditon; then they lost weight during courtship when they spent much of their time singing instead of eating.

Nocturnal and Diurnal Migrants

Migrants are divided into nocturnal and diurnal types, with some birds (notably waterfowl, most shorebirds, and alcids) that migrate either by day or night. Most of the smaller insectivorous species migrate at night, resting and feeding by day, with prolonged stopovers at certain stations, either for replenishing depleted fat supplies or waiting for more favorable flight conditions. Swallows and swifts, however, which feed on the wing, migrate by day. Hawks are also conspicuous diurnal migrants and often have specific routes of travel, along coasts or mountain chains where they take advantage of thermals or updrafts created by mountain slopes. One such strategic and well-known route is at Hawk Mountain, Pennsylvania, formerly a rendezvous where hunters gathered to shoot the birds as they passed by at close range, but now a sanctuary where thousands of observers assemble to watch the miracle of migrating hawks (Broun, 1949). A less-publicized station, known to hunters for a long time, is at Duluth, Minnesota, where in the fall thousands of hawks converge from flight lanes along the shores of Lake Superior (Hofslund, 1954, 1966). Coastlines are favorable points for observing diurnal migrations, not only of hawks, but of shorebirds and waterbirds. Choate (1972) observed a spectacular flight of American Kestrels (6,000 per hour until noon, 25,000 for the day) on October 16, 1970, at Cape May, New Jersey, a station long famous for its concentrations of migratory birds.

Altitude and Speed

Most migratory flights at night take place at about 3,000 feet, but there are great variations under different weather conditions, in different terrains, and for different species. Birds fly higher on clear nights, but not necessarily at the same height for all stages of the flight. Able (1970), in a radar study of nocturnal migrants, reported that 90 percent of the birds were below 5,000 feet and 75 percent below 3,000 feet. Nisbet (1963), also using radar, got similar measurements (90 percent below 5,000 feet), but recorded occasional shorebirds up to 20,000 feet. Birds below 600 feet he thought might be nonmigratory nocturnal birds. But birds do fly dangerously low on cloudy nights with a low ceiling and often strike high obstacles, such as monuments, tall buildings, and television towers (see page 454).

Some birds fly at much greater heights, characteristically or under special conditions, such as crossing mountain barriers. Lapwings and cranes have been recorded at high altitudes in migration, the former up to 8,500 feet, the latter at 15,000. Swan (1970) describes the amazing migration of Bar-headed Geese (*Anser indicus*) over the Himalayas at about 30,000 feet, perhaps the highest known altitude record. He speculates that geese traditionally flew over the area before the mountains uplifted, then adjusted their altitude of flight over the same course when the mountains rose. Another curious altitude record is of a Mallard striking a plane at 21,000 feet; passengers heard the thud, and after landing safely, the pilot recovered a primary and down from the dent in the plane (Manville, 1963).

Apparently nocturnal migrants use flight calls to maintain proper position and contact with other members of a flock. Such calls increase in frequency in the early morning, probably to facilitate predawn landings, or to notify traveling companions of the descent (Ball, 1952; Hamilton, 1962).

Though the daily rate of migration for most birds is relatively slow because of rest stops (see page 222), some birds make long nonstop flights over all or parts of their routes. The eastern form of the American Golden Plover (see Fig. 12.10) makes its 2,400-mile trip from Nova Scotia to South America in about 48 hours, although Johnston and McFarlane (1967) maintain that the birds may make island-hopping stops in the West Indies. They also state that the plovers could make the trip in 37 hours instead of 48 by flying 65 miles per hour and that the fattest Pacific Golden Plovers (see Fig. 12.10) could fly 6,200 miles, although most of them do not and could not unless extremely fat. Blue Geese are believed to make long flights between their Arctic breeding grounds and wintering areas in Louisiana and Texas (Fig. 12.4). Apparently a part of the population assembles at James Bay in the fall, then takes off on a 1,700-mile journey for Louisiana, often without stopping en route (Cooch, 1955). Another segment of the population takes off from York Factory, Hudson Bay, and goes more or less directly to Texas (2,100 miles), sometimes stopping at Sand Lake, South Dakota, sometimes not.

FIG. 12.4. *Blue Geese nest in large colonies in a few places in the Arctic (e.g., Baffin Island and Southampton Island) and winter chiefly in a coastal strip bordering the Gulf of Mexico. They make the 3,000-mile journey with few stopovers en route.* [Photo by the Mich. Dept. Nat. Res.]

One of the most dramatic records of migratory flights is of thrushes radio-tagged at Urbana, Illinois, and followed by plane (Graber, 1965; Cochran et al., 1967). One of them, a Gray-cheeked Thrush, flew overland for 140 miles, over Chicago (where the plane lost track of it temporarily in the traffic din), and out over Lake Michigan for 250 more miles. It ran into a storm but went on undaunted, presumably landing in northern Wisconsin or Michigan at dawn, about 400 miles north of Urbana. Another thrush flew northwest, at a ground speed of 43 miles per hour for 353 miles, then landed north of Rochester, Minnesota, at 4:00 A.M. It deviated only three miles from a straight-line course.

WIND AND WEATHER FACTORS

The timing of migrating flights in relation to wind and weather has already been mentioned (page 221). In both spring and fall, birds travel chiefly with a tail wind and stop when they meet adverse conditions, such as a cold front ahead. Bennett's (1952) studies of fall migration at Chicago disclosed birds coming in chiefly with northwest or northerly winds. Static weather, common in late summer, produced few or no waves, but a cold front from the north, then or later in the fall, brought in heavy flights. Possibly vagaries of the weather explain why some migrants, such as White-throated Sparrows, are seldom if even retaken at banding stations along their migration routes, for their exact paths and stopover places vary so much from year to year.

There have been many stimulating discussions regarding the influence of weather on migration (see Lack, 1960, for a more than six page bibliography on the subject). Lack summarizes it as follows: "There is more migration in fine weather than in rain, more with clear than cloudy skies, and more with light than strong winds. . ."; and that there is "more migration in spring with warmth and in autumn with cold."

Studies on trans-Gulf migration in Louisiana by Lowery (1945, 1946) have disclosed a peculiar situation. Trans-Gulf migrants of certain species are abundant along the Louisiana coast in spring *only* when forced down by meeting a cold front from the north or northwest; in good weather they go on to northern Louisiana or Tennessee before coming down, creating a "coastal hiatus" (gap) described by Cooke in 1915. Williams (1950, 1952), however, maintains that the coastal hiatus results from converging lines of flight from Texas and Florida coming together north of the Gulf. Recent data from Forsyth and James (1971) tend to support Williams' view of a circum-Gulf route and that offshore winds contribute to the well-known heavy concentrations of spring birds at Rockport and other areas in east Texas. Stevenson (1957) also thinks that some of the migrants in both Florida and coastal Texas might be displaced trans-Gulf migrants. But there is no doubt that many birds, including hummingbirds, make the trans-Gulf crossing directly.

Orientation

One of the great unsolved problems relating to migration is a bird's means of orientation, or direction-finding ability. By some little understood method, birds are able to find their way through thousands of miles of space, usually flying at night, the young often separate from the adults, over routes they have never traveled before. Yet, they arrive at the place where they are supposed to go—perhaps a remote island in the Pacific, a specific nest site on the northern tundra, or a winter home in the Amazon rain forest. And if they survive, they may return to the same place year after year.

Many homing experiments now demonstrate this remarkable direction-finding sense in birds. Among the earliest are those of Watson and Lashley (1915) in which marked Noddy and Sooty Terns were shipped in various directions away from their breeding colonies on the Dry Tortugas Islands, off the southwest coast of Florida. Some of the transported birds returned promptly to their nests on the Dry Tortugas, within a few hours from short distances (55 miles), or within a few days from points hundreds of miles north of their regular range.

Similar experiments conducted by Griffin (1940) with Leach's Petrels taken from their nesting burrows on Kent Island, near Nova Scotia, demonstrated this homing ability even more dramatically. The majority of the 220 petrels involved in these trials returned to their burrows, some of them over 360 miles of relatively featureless ocean and 470 miles from their nests. Later, Billings (1968) performed similar experiments with this species, releasing the birds at points varying from 0.2 to 2,980 miles from home. Two petrels taken to England returned in 13.7 days, having traveled 2,980 miles at the rate of 217 miles per day. Birds taken inland returned readily over unfamiliar land areas and used overland routes in preference to overwater routes, if they were shorter.

Homing experiments with Manx Shearwaters in Europe (Fig. 12.5), which previously indicated that they would return from long distances with remarkable promptness, reached a climax when one was flown to Boston after being taken from its nest burrow on Skokholm, an island off the southwest coast of Wales. The bird was back in its nest burrow $12\frac{1}{2}$ days after its release at the Boston harbor, having traveled over 3,200 miles at an average rate of more than 250 miles per day (Mazzeo, 1953). Equally astonishing are homing experiments with Laysan Albatrosses (Kenyon and Rice, 1958). Eighteen adults were shipped by air from their nest sites on Midway Atoll to six different localities. Fourteen birds, representing all six localities, subsequently returned, the farthest from the Philippines, some 4,120 miles, in 32 days; the fastest from Whidby Island, Washington—3,200 miles in 10.1 days or at the average rate of 317 miles per day.

Fisher (1971), experimenting with fledgling Laysan Albatrosses, found that homing ability improved with age; one-month-old chicks had not acquired the homing capacity of one-to-five-month-old chicks; juveniles without visual clues during transport homed as well as those that could see land, sea, and sky during transport. Apparently land

F IG . 12.5. *Long-winged, skillful aerialists, the shearwaters perform some of the most amazing migrations known. The Manx Shearwater, the species shown above, has been used in several homing experiments.* [Courtesy of Amer. Mus. Nat. Hist.]

birds have similar homing ability. Cowbirds trapped at Waukegan, Illinois, by W. I. Lyon (Hann, 1953) were shipped to Denver (925 miles), New Orleans (875 miles), and Washington, D.C. (620 miles). Some birds from each of these three localities found their way back to Waukegan, though some of them were not retaken until the following year and others were never seen again. A cowbird, released in Ithaca, New York (Wharton, 1959), on June 15, headed east and was back in Groton, Massachusetts, on June 29, in spite of intervening spells of cold rainy weather. Much faster flights were recorded by Southern (1959) for Purple Martins. Sixteen birds released at various distances from their nest boxes at Douglas Lake, Michigan, all returned—the farthest from a 234-mile displacement to Ann Arbor, Michigan. It returned overnight, in 8.58 hours! Later, Southern (1968) concluded that the homing ability of martins was not as remarkable as his earlier experiments indicated. Adult males had higher return percentages than adult females, and percentages for subadults were still lower. Juveniles, as might be expected since they had less incentive to return, were poor homers; only two returned from short distances, none from long distances. Homing experiments with Bank Swallows in Wisconsin (Sargent, 1962) and in Minnesota (Mayhew, 1963) yielded somewhat indifferent success for such long-range migrants. They seemed not to have a true navigational sense and presumably homed by random search and use of landmarks.

But homing experiments with Golden-crowned and White-crowned Sparrows (Fig. 12.6) wintering in California are even more astonishing

FIG. 12.6. *White-crowned Sparrows, and related* Zonotrichia, *are used extensively in experimental studies, such as homing. Wintering White-crowned and Golden-crowned Sparrows displaced from California to Louisiana and Maryland returned to California, but those sent to Korea failed to return.* [Photo by William K. Kirsher.]

(Mewaldt, 1964). Of 411 birds shipped to Louisiana, 26 were back the following winter, and 6 of them returned to California again after being shipped to Laurel, Maryland, the following winter. Presumably they had gone to nesting grounds in Alaska or the Northwest before returning to California.

Apparently flightless birds also can home with great precision. Displaced penguins taken long distances to release points in all directions from their rookeries returned, although those taken inland usually headed for the nearest water, then presumably swam home (Emlen and Penny, 1966; Penny and Emlen, 1967).

None of the several theories that attempt to explain this direction-finding ability in birds gives a wholly satisfactory answer. Of course some diurnal birds can and do follow landmarks; some will actually deviate from a direct flight line in order to follow a coast, but will strike out across the open sea when the coastline takes them too far off course. River valleys and coastlines also offer much-used avenues of travel for diurnal migrants, less so or not at all for nocturnal migrants (Graber, 1968; Lowery and Newman, 1966). But landmarks are presumably of little help (though not necessarily invisible) to nocturnal migrants or to oceanic birds. Some think that landmarks, or familiarity with a region, are used by short-distance migrants, and that long-distance migrants might utilize such features at the beginning and the end but not over the main course of their journeys. Blake (1959) tries to distinguish "terminal" migrants, which proceed from landmark to landmark by visual clues, and "transmigrants," which use a true navigational sense.

Several largely discredited theories attempting to explain this navigational sense in birds have been advanced from time to time. One theory is that a bird might be able to judge its latitude by the *Coriolis force* produced by the rotation of the earth, but most authorities estimate that the effect of Coriolis force is so small, especially toward the equator (it is stronger toward the poles), that its effect on a bird would be negligible. Yeagley (1947, 1951), however, after extensive experiments with homing pigeons, feels that a combination of Coriolis force and a bird's sensitivity to a magnetic field might set up a grid pattern so that a bird could judge both its latitudinal and longitudinal field. Experiments by Southern (1972) with Ring-billed Gulls and by Keeton (1971) with pigeons showed that the birds were affected (disoriented) by the earth's magnetic field. Southern's gulls, ranging in age from 2 to 20 days, had an innate ability to select the appropriate migrational direction, but were disrupted under high but not low magnetic disturbances. Keeton believed that young, inexperienced pigeons used both magnetic field and sun for homing but that the more experienced adults used sun and landmarks and could get along without magnetic clues. He rated the sun first, magnetic field second, and landmarks third as orientation clues.

During the past two decades work has progressed remarkably on apparently much more fruitful lines of research. Kramer (1961) disclosed that Starlings in aviaries in Germany could orient themselves quite unerringly, southwest in the fall and northwest in the spring, by the sun's position. Homing pigeons, Herring Gulls, and sylviids behaved in a similar manner. They lost this directional sense when the sun was obscured, and they could be fooled by an artificial sun. Bellrose (1958) tested these hypotheses by releasing marked Mallards in fall, winter, and spring, both day and night, and in both cloudy and clear skies. He concluded that Mallards oriented by sun in the day (even under 70 percent cloud cover if the sun was visible) and by stars or constellations at night, but the birds scattered at random under cloudy skies. Later experiments (Bellrose, 1963, 1966, 1967) with Mallards and other waterfowl tended to confirm the results of his earlier tests, but with great variations. Often the birds "oriented" in the wrong direction, especially northward in the fall. They used landscape features for guidelines by day and celestial clues at night. Blue-winged Teal had a better navigational sense than Mallards.

Significant also are observations by Hamilton and Hammond (1960) in which pinioned geese, when released or escaping from pens in spring, headed true north and walked overland, covering about one-half mile per day for a period longer than the regular migration period of wild geese. Captive birds also persistently crowded the north fence in spring. The penguins referred to on page 346 walked at a much faster rate than the geese, up to 60 kilometers (37.5 miles) per day.

Although sun navigation is useful to diurnal migrants, it may also be an aid to nocturnal birds, which can set their compass direction correctly at or near sundown. However, additional information would be needed to guide birds throughout the night. Sauer (1957) demonstrated that three European sylviids in rotatable cages could orient themselves at night by the moon, if present, or by the stars. In planetaria they reoriented themselves correctly when the heavens were rotated and the position of the stars changed. S. T. Emlen, in several papers (see Emlen, 1969, for a bibliography of his and other work on celestial orientation), reported on his extensive work with Indigo Buntings in orientation cages in planetaria. Apparently the buntings "obtain directional information from the starry sky." They maintained normal migration direction when exposed to skies advanced or retarded from 3 to 12 hours. Northern stars and constellations appeared to be the most useful to them, but when these were not visible they could use other parts of the sky. They lost this ability, or failed to perform, when the sky was completely obscured.

In most experimental work, birds apparently lose their sense of direction when the skies (sun, moon, or stars) are obscured, yet it is well known that in nature there are often heavy migrations on cloudy nights. In such cases, Graber and Cochran (1960) suggest that birds might be able to utilize wind direction to enable them to fly along a proper course, a clue also believed to be used by ducks under cloudy skies (Bellrose, 1966). Other possible clues are sounds from the ground below, which according to balloonists' reports can be heard as high as 3,000 meters (D'Arms and Griffin, 1972). Can birds get directional clues from the sounds of waves on shores, rushing streams, or the wind in trees?

In spite of these remarkable discoveries on celestial navigation, there are still unanswered questions. Yet it would appear, from the evidence now available, that birds use a variety of devices to find their way: random wandering to regain their position when lost; use of landmarks in familiar territory; and a true navigational sense on long flights over unfamiliar territory.

Routes of Migration

Although many studies on migration indicate that birds migrate over a broad front, and perhaps spread out pretty much over the whole continent of North America in migration, there are nevertheless some special routes and preferred lanes of travel. As implied previously, birds use coastlines, river valleys, mountain chains, and even highways at times for lanes of travel. Birds that pass between North and South America have their choice narrowed down to several possible exchange corridors (see Fig. 12.8) over land or water. This produces a funneling effect, which is just the opposite of the situation in Europe where birds flare out over broad Mediterranean routes. Casement (1966), using radar observations, found a fairly uniform front across the Mediterranean except for the paucity of birds across the Ionian Sea. However, it has been known for a long time that White Storks (*Ciconia ciconia*) en route to Africa from Europe split into two great arms, a western arm across the Strait of Gibraltar, and an eastern one across the Bosphorus and the Red Sea. Other "soaring birds," including many thousands of raptors, use the eastern arm as a lane of travel (Porter and Willis, 1968; Safriel, 1968) where thermal currents would give birds with heavy wing loading considerable lift. Migration to and from Africa also involves a hazardous Sahara crossing; most birds fly across nonstop, but a few, especially if grounded, seek out the scarce wadis for water.

North American waterfowl are believed by some (Lincoln, 1950) to sort themselves in migration into particular flyways, so much so that the flyways have become the basis of the administration of regulations and policies in the Fish and Wildlife Service. The accompanying maps (Fig. 12.7) show the four major flyways defined by the Fish and Wildlife Service, and some of the detailed paths that contribute to the main networks. These special routes, if they actually exist (see Phillips, 1951a), are used not alone by waterfowl, but by many of the smaller birds as well. Bellrose (1968) does not discredit Lincoln's flyways, but thinks they need revision and redefinition. He prefers to use the term *corridor* for a collection of individual routes, with many corridors contributing to a major flyway.

FIG. 12.7 [FOLLOWING PAGES]. *Four of the flyways used by the U. S. Fish and Wildlife Service in the administration of regulations and policies on waterfowl.* [Drawings by Robert W. Hines, Bureau of Sport Fisheries and Wildlife.]

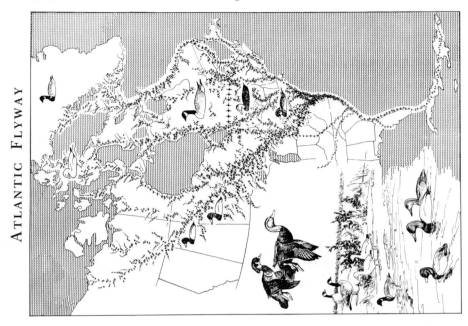

ATLANTIC FLYWAY

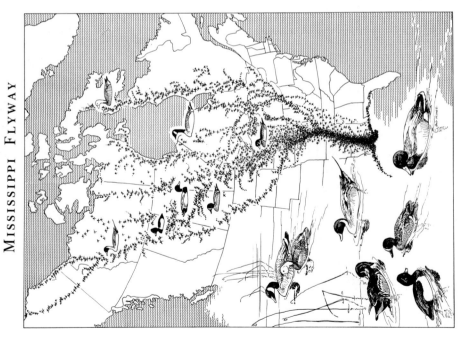

MISSISSIPPI FLYWAY

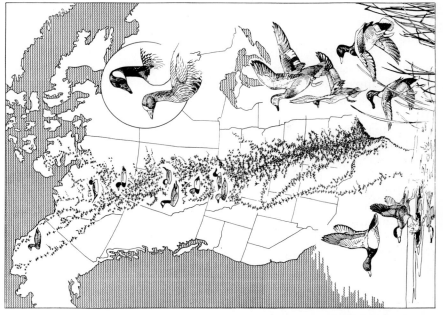

CENTRAL FLYWAY

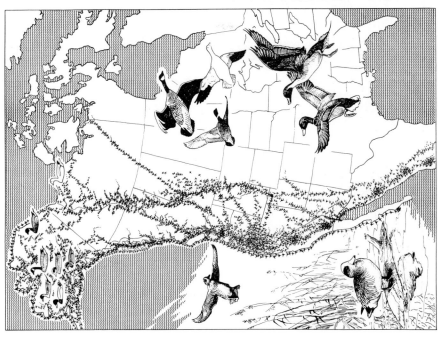

PACIFIC FLYWAY

Some birds use different routes in the spring and fall; Connecticut Warblers and the western race of the Palm Warbler, for instance, commonly detour through New England in the fall but use the Mississippi flyway in the spring. Young gulls hatched in the Great Lakes island colonies drift eastward along the St. Lawrence waterway in late summer and fall, apparently traveling with the prevailing winds, then proceed southward along the Atlantic Coast (Smith, 1959; Hofslund, 1959). Among waterfowl it is not uncommon for a part of the population to go by one route to one destination (Atlantic, Pacific, or Gulf Coast) and another segment of the population to go by a different route to another destination. Burtt and Giltz (1966) noted that recoveries of Starlings banded at a roost in Columbus, Ohio, indicated a northeast-southwest line of travel rather than strictly north-south. They suggested that this was a holdover from their ancestral northeast-southwest migration routes in Europe.

Figure 12.8 shows the various routes used by birds passing south of the United States. Route 1, for instance, from Nova Scotia to South

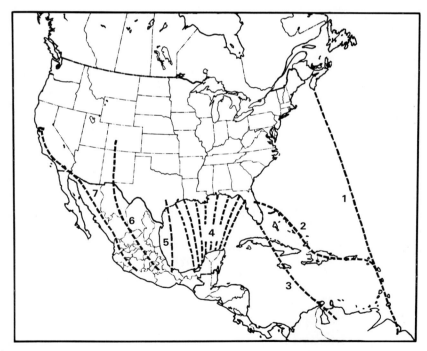

Fɪɢ. 12.8. *Principal migration routes used by birds passing between North America and their wintering grounds in the south.* [Drawing by Robert W. Hines, Bureau of Sport Fisheries and Wildlife.]

America, is used by adult Golden Plovers in the fall. Some Blackpoll Warblers use a similar route, flying nonstop from New England to South America (Nisbet et al., 1963); other Blackpolls move more leisurely along the Atlantic Coast (Murray, 1965), those reaching Bermuda being birds carried to the island by offshore winds. Route 2, from Florida to the West Indies, as well as Route 3, serves many North American warblers and other birds that winter in or cross these islands. Route 4, passing directly over the Gulf and involving a hazardous nonstop 500-mile flight for many land birds, is believed to be used extensively (Cooke, 1915; Lincoln, 1950; Lowery, 1946), but some birds take the more circuitous land routes (5 to 6) around the Gulf. Route 7, through western Mexico, is taken by Pacific Coast birds.

The following three illustrations (Figs. 12.9–12.11) show, and the legends explain, the migration routes of three outstanding travelers—the Arctic Tern, the Golden Plover (Atlantic and Pacific forms), and the Bobolink. The peculiar routes of the Arctic Tern and the (Atlantic) Golden Plover are extraordinary, duplicated by few if any other birds, but quite a few North American land birds make migrations similar to that of the Bobolink.

Hamilton (1962b) describes a fantastic experience with a Bobolink. An adult female, taken from its breeding ground in North Dakota in early August, was kept in an orientation cage on the roof of the Science Building at Berkeley, California, where Bobolinks do not occur. It was tested for orientation on three nights: on August 29 it oriented toward its home in North Dakota; on the next night it oriented toward the east, as if to join the migration path of other Bobolinks en route to South America; and on the third night it directed its attention toward the southeast, the shortest route to South America. Then it escaped, but (miraculously) was recaptured on its nesting ground in North Dakota the following spring. It is unlikely that it could have survived the winter in the states, even in California. It must have found its way by an unknown route (straight south would have taken it far out into the Pacific Ocean and certain death) to its wintering ground in southern South America and back to North Dakota with its fellow travelers the following spring.

SELECTED REFERENCES

BALL, S. C. 1952. Fall bird migration on the Gaspé Peninsula. Peabody Mus. Nat. Hist. Bull. 7. New Haven, Conn: Yale Univ. Press.

BELLROSE, F. C. 1967. Radar in orientation research. Proc. xiv Inter. Ornith. Congress: 281–309.

DORST, J. 1962. *The Migrations of Birds.* London: Heinemann.

DRURY, W. H., JR., and I. C. T. NISBET. 1964. Radar studies of orienta-

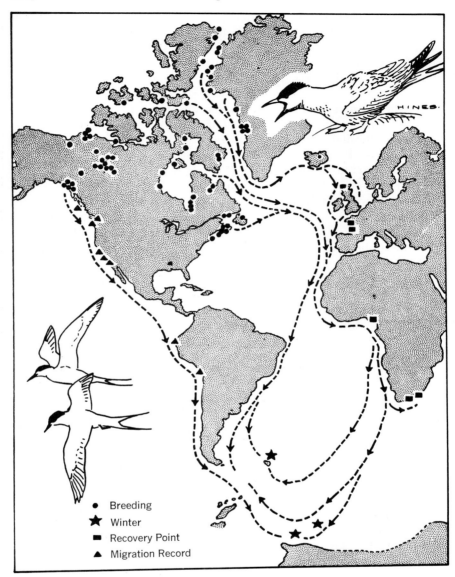

Fig. 12.9. *Distribution and migration of the Arctic Terns of North America. The route indicated for these birds is unique, as no other species is known to breed abundantly in North America and to cross the Atlantic Ocean to and from the Old World. The extreme summer and winter homes are 11,000 miles apart, and the route taken is circuitous; these terns probably fly at least 25,000 miles each year.* [Drawing by Robert W. Hines, Bureau of Sport Fisheries and Wildlife.]

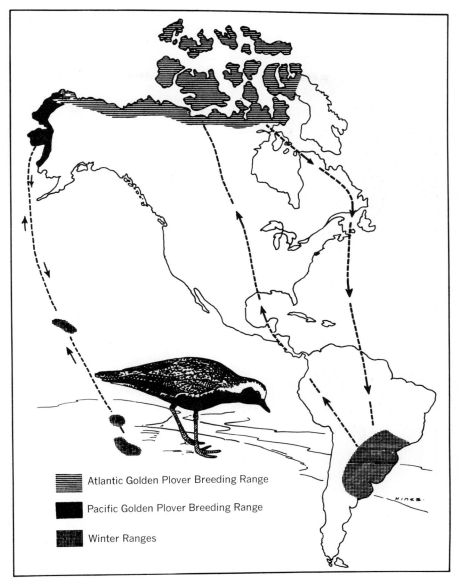

FIG. 12.10. *Distribution and migration of the Golden Plovers. Adults of the eastern form migrate across northeastern Canada and then by a nonstop flight reach South America. In spring they return by way of the Mississippi Valley. Their entire route is therefore in the form of a great ellipse with a major axis of 8,000 miles and a minor axis of about 2,000 miles. The Pacific Golden Plover, which breeds in Alaska, apparently makes a nonstop flight across the ocean to Hawaii, the Marquesas Islands, and the Low Archipelago, returning in spring over the same route.* [Drawing by Robert W. Hines, Bureau of Sport Fisheries and Wildlife.]

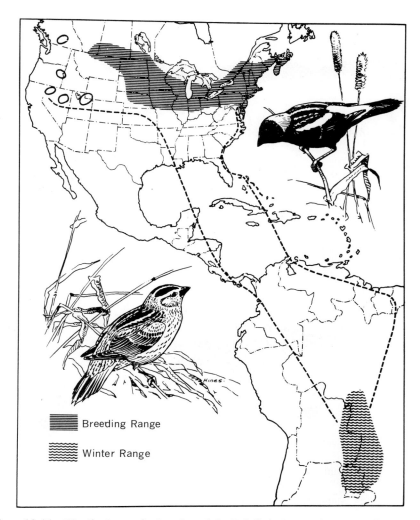

F IG . 12.11. *Distribution and migration of the Bobolink. In crossing to South America most Bobolinks use Route 3 (see Fig. 12.8), making the flight from Jamaica across an islandless stretch of ocean. Colonies of these birds have become established in western areas, but in migration they usually adhere to the ancestral flyways, not taking the short-cut across Arizona, New Mexico, and Texas.* [Drawing by Robert W. Hines, Bureau of Sport Fisheries and Wildlife.]

tion of songbird migrants in southeastern New England. Bird-Banding, 35:69–119.

EASTWOOD, E. 1967. *Radar Ornithology.* London: Methuen.

FARNER, D. S. 1955. The annual stimulus for migration: experimental and physiologic aspects. Chapter 7 in A. Wolfson (ed.), *Recent Studies in Avian Biology.* Urbana, Ill.: Univ. of Ill. Press.

GRIFFIN, D. R. 1964. *Bird Migration.* New York: Nat. Hist. Press.

HOCHBAUM, H. A. 1955. *Travels and Traditions of Waterfowl.* Minneapolis: Univ. Minn. Press.

KRAMER, G. 1961. Long distance orientation. Chapter 22 in A. J. Marshall's *Biology and Comparative Physiology of Birds.* New York: Academic.

LACK, D. 1960. The influence of weather on passerine migration: a review. Auk, 77:171–209.

LINCOLN, F. C. 1939. *The Migration of American Birds.* New York: Doubleday.

————. 1950. Migration of birds. U. S. Fish and Wildl. Service Circ. 16:1–102.

LOWERY, G. H., JR., 1951. A quantitative study of the nocturnal migration of birds. Univ. Kans. Publ. Mus. Nat. Hist. 3:361–472.

————. and R. J. NEWMAN. 1966. A continentwide view of bird migration on four nights in October. Auk, 83:547–586. Analyzes data of 1,391 moon-watchers at 265 stations; 15 full-page maps.

MARSHALL, A. J. 1961. Breeding seasons and migration. Chapter XXI in A. J. Marshall's *Biology and Comparative Physiology of Birds.* New York: Academic.

MATTHEWS, G. V. T. 1955. *Bird Navigation.* New York: Cambridge Univ. Press.

ROWAN, W. 1931. *The Riddle of Migration.* Baltimore, Md.: Williams & Wilkins.

WETMORE, A. 1930. *The Migration of Birds.* Cambridge, Mass.: Harvard Univ. Press.

13

⌒⌒⌒⌒⌒

Distribution

THE DISTRIBUTION of birds, particularly with respect to its geographical or spatial arrangements, has long engaged the attention of ornithologists. As a result, the summer and winter ranges and migration routes of birds are now fairly well known, several classifications of regions and zones have been presented, and special habitats have been studied in great detail. But when we ask why birds are where they are today, how they got there, or why some birds are restricted to a single island or a particular habitat while others are world-wide, we do not always have the answer.

Distribution may be divided, somewhat arbitrarily, into three kinds: (1) distribution in time (geological), (2) distribution in space (geographical), and (3) distribution in different habitats (ecological). The geological record was examined briefly in Chapter 2; it is chiefly the geographical and ecological patterns that need to be discussed here.

Every student has some concept of both geographical and ecological distribution. Nearly everyone associates penguins with the Antarctic, ptarmigans with the Arctic, and hummingbirds, for the most part, with tropical or subtropical regions. Similarly with respect to habitat, we associate ducks with water, woodpeckers with trees, and field birds with open spaces.

Origin and Dispersal

Distributional patterns involve *a center of origin* or *birthplace* of the species or group, then its subsequent *dispersal* from this center, and, lastly, its degree of success in the new regions invaded. *Spreading* is a term sometimes used to denote dispersal or increase in the range of a species, as opposed to migration, which is fundamentally a two-way movement not necessarily increasing the permanent range. (See Berndt and Sternberg, 1968, for additional definitions of terms.) Matthew (1939) postulates a Holarctic (northern hemisphere) origin for land vertebrates, but Serventy (1960), in part following Darlington (1957), broadens this to "the great Eurasian-African land masses as the differentiating center for main vertebrate evolution" with the tropical parts of Africa and the Orient as "the great reservoirs and apparent main dispersal centers of the vertebrates." The theory of continental drift (see page 336) also favors tropical centers of origin, which offers a solution to the problem of pantropical species (see page 368). Glenny (1954) even postulates a tropical or subtropical Antarctica as a center of origin of birds, which subsequently spread northward into the Americas and the Old World continents, although this does not explain the occurrence of the nearly flightless *Archeopteryx,* the first known bird, in Germany.

Some species, or groups of species, particularly those of recent origin and those handicapped by lack of mobility, may still be at or near their place of origin, never having succeeded in occupying new regions. Other species, with greater adaptability, better means of dispersal, or a higher reproductive or genetic potential, which creates population pressure and peripheral spreading, have successfully invaded new areas. Some of them, such as the hawks and owls, gulls and terns, swallows and swifts, and many water birds are now virtually worldwide. Still other forms may have moved into new homes and died out completely from their birthplace, leaving little or no trace of their origin. The rheas of South America are far removed from their Old World Ostrich relatives, which points to the probability that they spread to the New World before Pleistocene glaciation and have survived only in South America, where they have undergone further evolution. Ecological changes, such as the development of a more favorable climate in a previously inhospitable region, encourage range expansion (Serventy, 1960); then an unfavorable climatic change may restrict further spreading.

The Limpkin, a rail-like bird of the subtropics, is a good example of a changing range. Except for a relatively rare North American form, the Limpkins are now confined to tropical America, which might well be assumed to be their original home were it not for fossil evidence to the contrary. If the North American form dies out in the near future, only history, recent and fossil, will show that South America is probably the secondary home of a family of birds (Aramidae) that originated in North America. Apparently the New World vultures (Cathartidae) are undergoing a similar extension southward, and gradually disappearing from North America. Seven or more fossil species are known from the United States, and only three living forms, of which one (the California Condor) is nearly extinct. Curiously the Old World vultures (Accipitridae) once inhabited North America, and the New World vultures (Cathartidae) once inhabited Europe, but the two groups have reversed their former geographic positions; or more probably each has died out from a part of its former range.

Land and water barriers influence the dispersal of birds, but not always in the same way. A barrier to one may be an avenue of travel for another. A mountain range may effectively block further spreading of a plains species, but it is a means of further dispersion for others. The dippers, inhabitants of mountain streams, extend from Alaska to the Andes in the western mountains and from Scandinavia to North Africa in the Old World, but are presumably prevented from occupying suitable range in the Appalachians because of the formidable barrier interposed by the great plains, and similarly blocked from mountains in South Africa by the inhospitable Sahara.

Some species extend their range by "island hopping." A recent example believed to be attributable to this mode of travel is the Cattle Egret, which appeared in northern South America in the late 1930's, presumably from Africa or Europe, and spread rapidly up the coast of North America to Newfoundland and sporadically into the interior, reaching California in 1964 (Crosby, 1972). Though other explanations of its origin in South America are conceivable (transportation on cattle boats, escape from zoos), the recovery in Trinidad of a Little Egret banded in Spain (Downs, 1959) lends strong support to the island-hopping theory. Favorable winds also may have aided in such one-way migrations.

Transplanted birds, whether by natural means or introduced by man, are sometimes eminently successful in a new region. Lowe-McConnell (1967) reports on the breeding success of the Cattle Egret in Guyana, South America. Though not competitive with other herons for food (it is largely insectivorous), it usurps the best nesting sites by starting nesting earlier and progressing at a faster rate. Growth of the cattle industries also aided in its spread.

Changing ranges are an important part of distributional patterns. The spread of the Cardinal into the Great Lakes states in the early 1900's and the more recent spread of the Cardinal, Tufted Titmouse, and Mockingbird into New England have been well documented. Reasons for the range extension include climatic changes, tolerance and acceptance of man-made changes (farming practices, deforestation and reforestation, operation of feeding stations), and natural succession (Beddall, 1963; Boyd, 1962, Boyd and Nunneley, 1964). Northern Orioles, formerly rare in the United States in winter, now are of frequent occurrence. Many hundreds were banded in North Carolina in the 1960's, and many of them returned for successive winters (Erickson, 1969). Curiously, the operation of baited traps for blackbird-control studies in Ohio has apparently caused some birds that normally would have migrated to remain in the vicinity of the traps over winter (Burt and Giltz, 1971).

WIDESPREAD VS. RESTRICTED SPECIES

Several interesting dispersal problems are raised by a comparison of widespread and successful species with those that are restricted to a small area. Among the hawks, for instance, the Osprey (Fig. 13.1) occurs on every continent and on most of the larger islands throughout the world, whereas another hawk, the northern race of the Everglade Kite (see Fig. 16.9), is confined to a critically small area of southern Florida. In this case the food habits of the two species appear to provide a ready explanation, for the Osprey lives on a never-failing supply of fish, while the kite lives exclusively on a single species of snail (*Pomacea paludosa*). Similarly the success of the Marsh Hawks (harriers), both in numbers and distribution over nearly the whole earth, might well be attributed to their adaptability in diet (page 139) and the

FIG. 13.1. *The Osprey is nearly world-wide in distribution, and its chief source of food appears to be unlimited. Now, however, it is declining rapidly in the eastern states, owing to loss of nesting sites and reproductive failure from feeding on chemically contaminated fish.* [Photo by Allan D. Cruickshank.]

widespread availability of marsh habitats. Another factor, however, is the higher reproductive potential of the Marsh Hawk, which may lay a clutch of six or more eggs, whereas the Everglade Kite usually lays two or three.

Among the sparrows, to take another well-known example, the Song Sparrow is a widespread and abundant North American bird that occurs from coast to coast and from Alaska to southern Mexico, but the Ipswich race of the Savannah Sparrow is confined in the breeding season to Sable Island, Nova Scotia. There appears to be no ready explanation for this distributional enigma other than to generalize on the Song Sparrow's adaptability and genetic potential and Ipswich Sparrow's lack of these traits. Another enigma: the House Sparrow spread rapidly over most of North America after its introduction into Brooklyn, New York, in 1850, but the congeneric European Tree Sparrow is still confined to a small area around St. Louis, Missouri, more than 100 years after its importation in 1870.

Among the warblers, the Kirtland's Warbler (Fig. 13.2) breeds only

FIG. 13.2 *Kirtland's Warbler. This attractive warbler—limited to about 200 pairs— breeds only in a few counties of jack-pine country in north-central Michigan. It winters in the Bahamas.* [Photo by William Coates.]

FIG. 13.3. *The zone of stunted evergreens extending from the 3,000-foot contour line to the timberline on Mt. Mansfield, Vermont, harbors a race of the Gray-cheeked Thrush that is largely restricted to this zone in the breeding season.* [Courtesy of *National Geographic Magazine.*]

in the jack-pine country of a few counties in the north-central part of the Lower Peninsula of Michigan (about 200 pairs in 1972), whereas many woodland warblers range over nearly all of the deciduous forests or coniferous forests of eastern North America. Apparently the Kirtland's Warbler has been evolved to fit in only a particular stage in the jack-pine forest development, in nearly homogeneous stands of small jack pines, 6–18 feet high with branches close to the ground. Why it does not spread to available, and to our eyes nearly identical, jack-pine ranges include the nearly extinct Bachman's Warbler (see page 439) not readily answered. Other warblers with dangerously restricted ranges include the nearly extinct Bachman's Warbler (see page 439) and the Golden-cheeked Warbler in the juniper-oak scrub areas of south-central Texas.

In New England a race of the Gray-cheeked Thrush occupies the spruce-balsam zone between 3,000 and 4,000 feet on certain mountains (Fig. 13.3). It does not range much above or below this level, apparently because the conifers run into timberline above and deciduous forest below. But the White-throated Sparrow, a common nesting companion of the rarer thrush, extends across the timberline above and into the deciduous woods below, not only throughout New England, but over most of the forests of northern United States and Canada as well. In this case the White-throat is more adaptable than the thrush in its environmental requirements; but, in addition, a greater reproductive capacity—the White-throat lays a larger clutch of eggs and more often raises a second brood—may have prompted the sparrow to invade new and somewhat different habitats.

Among passerines the Barn Swallow is perhaps the most widely distributed species in the world. In North America it breeds from Alaska to Labrador and south to central Mexico, and winters from Panama to southern South America. In the Old World a related race breeds throughout Eurasia, wintering in southern Asia and south to South Africa. Australia is the only continent without breeding or wintering Barn (European) Swallows.

In western North America, where more abrupt climatic, ecological, and altitudinal differences create special niches, there are many birds, particularly subspecies, with restricted ranges. Often somewhat generalized answers (adaptability, habitat preferences, reproductive and genetic potential) can be found for these distributional enigmas, but such "solutions" leave much to be desired.

Zoological (Faunal) Regions and Subregions

Ever since Alfred Russel Wallace (1876) advanced his concept (largely adapted from Sclater) of the geographical distribution of animals, animal geographers have strongly adhered to the classical zoogeography of the 19th century and attempted to fit animals in zoological regions with fixed boundaries based largely on climatic conditions and land and water barriers. Though this concept of fixed regions (static zoogeography) as opposed to fluid faunas (dynamic zoogeography) has been questioned from time to time (Mayr, 1946), these regions and the terminology based on them are still widely used, and on a strictly geographical basis perhaps no better land divisions could be devised. Hence, Sclater's and Wallace's original regions, after many suggested modifications and criticisms, have been largely adopted by Darlington (1957) in one of the latest reviews of the situation. However, Neill (1969) divides the world into three great "Faunal Realms" (the Eurasian-African-North American, the South American, and the Australian), each with its Regions, Subregions, and Transitions, which in many cases do not correspond closely with the divisions used here.

A detailed analysis of the zoological regions of the world is beyond the scope of this book, but the major divisions, with a few birds characteristic of each, are described, and the Nearctic (North American) subregion is discussed in some detail, particularly with respect to the origins of its present-day bird life. Figure 13.4 shows these major regions and their approximate boundaries.

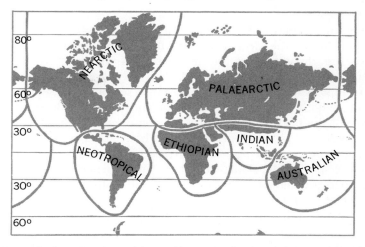

F IG. 13.4. *The faunal regions of the world. New Zealand, considered a separate faunal region by some, is here included in the Australian region.* [Map prepared by R. H. Manville from original by Sclater.]

H OLARCTIC (P ALAEARCTIC AND N EARCTIC) R EGION

Usually most of the northern hemisphere, which comprises a disproportionate part of the land area (but not species of birds) of the world, is included in one vast region, the *Holarctic*. This in turn is divided into two subregions, a Eurasian (*Palaearctic*) and a North American (*Nearctic*). The justification for treating most of the northern hemisphere as one major region is apparent when we consider the birds occupying it. At the polar extremes of the Holarctic are many aquatic birds—loons, alcids, and many waterfowl—that are circumpolar. In some cases the same species occur throughout the area, in other cases the Old and New World forms are only subspecifically distinct. South of the polar seas the tundras of both Eurasia and North America are occupied by the same, or closely related, species of plovers, sandpipers, ptarmigans, some hawks and owls, and even more recent passerines such as the Snow Bunting and Wheatear.

The extensive coniferous forest south of the tundra likewise forms a more or less continuous belt across Canada into Siberia and northern Europe, and we find such forest birds as Brown Creepers and Golden-crowned Kinglets, at best only subspecifically separable, ranging throughout the area. Each subregion has developed a few species of its own, such as the Snow Goose and several sandpipers in the American tundra, and the Red-breasted Nuthatch and the Black-backed Three-toed Woodpecker in the Canadian coniferous forests, but many species are common to the whole region.

Southward, in the deciduous woodlands, the grasslands, and the deserts, the dissimilarities in the avifaunas of the Palaearctic and Nearctic naturally become greater. Udvardy (1958) describes these relationships as follows: 32 species of Arctic sea birds (86.5 percent) are common to both subregions (Palaearctic and Nearctic), 36 species (75.0 percent) of the birds of the tundra are common to both, 28 coniferous forest species (43.8 percent) occur in both subregions, but only 7 species (3.1 percent) of deciduous forest birds are found in both. He lists only 2 species, the Short-eared Owl and Horned Lark, as common to the open grasslands of both regions, and both of these species are members of a circumpolar Arctic fauna that have spread far south of their probable port of entry into North America, the Horned Lark even reaching northern South America.

ORIGIN OF NORTH AMERICAN AVIFAUNA. Obviously, then, not all North American bird groups are strictly endemic; some had their roots on other continents. North American bird life is derived from three known sources: (1) part of it, at least 17 families and subfamilies according to Mayr (1946), is endemic, having undergone its main evolutionary development on this continent at an early date, probably in the Tertiary; (2) an equally large element, at least 18 families and subfamilies, is of Old World origin, with most representatives still over there but with smaller or more recent segments established in the New World; (3) still other family groups, principally the hummingbirds (Trochilidae), New World flycatchers (Tyrannidae), and tanagers (Thraupidae), spread northward from South America.

In addition to these known sources of our avifauna, many of the larger and older birds—oceanic forms, waterfowl, hawks, and shore-birds, as well as the more mobile smaller birds (swifts and swallows)—have been distributed over the whole earth so long that their place of origin is obscure. The following list presents some of the better known North American groups of birds and their known, or postulated, source (condensed from Mayr's *History of the North American Bird Fauna*).

Unanalyzed Element—now so old or widespread that their origins are
  largely unknown:
    Oceanic birds—8 families
    Shorebirds—5 families
    Freshwater birds—9 families
    Land birds—7 families
Old World Element—groups originating in the Old World, but now
  represented in North America:

Phasianids (Phasianidae)—in part (New World quail endemic in North America)
Pigeons (Columbidae)
Cuckoos (Cuculidae)
Kingfishers (Alcedinidae)
Larks (Alaudidae)
Titmice (Paridae)
Nuthatches (Sittidae)
Thrushes (Turdidae)—in part (some endemic in New World)
Shrikes (Laniidae)

North American Element—endemic, largely confined to or originating in North America:

New World Vultures (Cathartidae)—now entirely New World
Grouse (Tetraonidae)—have spread over Eurasia, apparently from American sources
Turkeys (Meleagrididae)—entirely New World
Limpkins (Aramidae)—have spread into South America
Wrens (Troglodytidae)—have spread into Eurasia and North Africa
Mockingbirds (Mimidae)—entirely New World
Vireos (Vireonidae)—entirely New World
Wood Warblers (Parulidae)—entirely New World

South American Element—primarily South American but represented in North America:

Hummingbirds (Trochilidae)—320 species, only one reaching eastern United States (20 in western United States)
Tyrant Flycatchers (Tyrannidae)—365 species, 33 north of Mexico
Tanagers (Thraupidae)—197 species, 4 north of Mexico

NEOTROPICAL OR SOUTH AMERICAN REGION

This is perhaps the most easily defined of the world regions, geographically because of its near separation from all other land areas, and avifaunally because it has so many birds (as well as other animals) restricted to it. Mayr (1946) gives South America as the probable home (origin) of 27 families, most of them still there. These include such well-known types as the rheas, the tinamous, the Hoatzin (see Fig. 4.8), and the toucans (see Fig. 5.8), as well as many passerine families (14) little known to the layman.

Ornithologically South America is one of the richest regions in the world. Lush tropical forests with heavy rainfall (exceeding 275 inches per year on parts of the Pacific Coast) and coastal deserts with little or none; broad savanna lands; a temperate plateau stretching deep into the southern hemisphere; and the high Andes with four "life zones" at different altitudes provide for a great variety of birds. Tropical regions, moreover, are noted for the development of large numbers of species;

85 percent of the bird species of the world, according to Griscom (1945), are found in the tropics. Colombia, for instance, with a land area approximately equal to Texas and Oklahoma, has 1,532 species (2,558 forms) of birds (de Schauensee, 1948–1952), which is roughly three times the number of birds recorded in Texas or California, and more than double the number that are of regular occurrence in North America north of Mexico.

Many tropical species are severely restricted in distribution, however, and the large flocks characteristic of northern latitudes are almost unknown. A rare grebe (*Podilymbus gigas*) is confined to a single lake in Guatemala; another (*Centropelma micropterum*) occurs only on Lake Titicaca in the high Andes. Griscom (1945) remarks further that a certain flycatcher is found only in a particular kind of palm tree, that a seed-eater dwells only in the reed beds of certain rivers in eastern Nicaragua, and that in the Amazonian basin many species are confined to the intertributary areas of wide rivers and will not cross the water. In the tall rain forest there are six strata of bird life from the forest floor to the lofty crowns of the trees. Hence, the New World tropics have an amazing diversity of bird life, but usually not large aggregations of any particular species.

Another distinctive feature of the Neotropical region is the *absence* of certain widespread groups. The cranes (Gruidae), titmice (Paridae), nuthatches (Sittidae), creepers (Certhiidae), and shrikes (Laniidae), for example, occur on all continents *except* South America and, in some cases, Australia.

In addition to its endemic bird life and the several groups that it shares with North America, the Neotropical region has eight families of birds that also occur in the tropics of the Old World. Such birds are known as *pantropical* (common to the tropics of both the Old and New World) and are oft-cited examples of discontinuous distribution. Best known of these are the parrots (Psittacidae), barbets (Capitonidae; Fig. 13.5), and trogons (Trogonidae) among the land birds and the snake-birds (Anhingidae) and skimmers (Rynchopidae) among the water birds. Peculiarly, a warm ocean current (the Gulf Stream) has enabled the Black Skimmer to extend its range up the Atlantic Coast to Massachusetts, just as a cold current (Humboldt's) has permitted a penguin from the Antarctic to reach the Galapagos Islands at the equator. The explanation for pantropical occurrences has long been a controversial issue, but the prevailing belief, supported by some fossil evidence, is that the species involved once had a more or less continuous distribution over the whole northern hemisphere, until increasing aridity and cold, followed by glaciation, eliminated tropical life from the north and left segments stranded in the warmer parts of Africa, Asia, and South America. A less generally accepted view, as already indicated (page

FIG. 13.5. *Crimson-breasted Barbet* (Megalaema haemacephalia). *Barbets (72 species) are pantropical, occurring in the tropics of both the New and Old World. They are mostly gaudily colored, rather sluggish fruit-eaters, with their bright colors not always in good harmony.* [From Delacour and Mayr, *Birds of the Philippines*, copyright 1946 by Jean Delacour and Ernst Mayr, Macmillan Publishing Co., Inc., New York.]

336), is that pantropical species originated in a tropical or subtropical Antarctica and spread northward into Africa, Australia, and South America before separation of the continents.

ETHIOPIAN REGION

The Ethiopian region includes all of Africa except the extreme northwest Mediterranean coast (portions of Morocco, Algeria, and Tunisia), which belongs with Mediterranean Europe. It also includes the portions of the Syrian and Arabian deserts that are more or less continuous with the Sahara and Madagascar. Some features of particular interest on this continent are the immense Sahara desert, broad savanna lands (including the veld and scrub thorn) with sharply divided wet and dry seasons, broad equatorial as well as montane rain forests, and the isolated subregion of Madagascar, which has a peculiar bird life of its own. Extensive wetlands—marshes, rivers and lakes— provide for both variety and abundance of aquatic birds.

The Ostrich (*Struthio camelus*) is a characteristic bird of the Ethiopian region, the northern form (there are 6 subspecies) formerly ranging into Arabia, but probably exterminated in World War II. The Secretarybird (Sagittariidae), which is a snake-eating falconiform bird, two odd storks (the Whale-headed Stork and Hammerhead—currently considered as monotypic families), and several other storks (Fig. 13.6) are African peculiarities that occur nowhere else. The guineafowl (Numididae) and plantain-eaters (Musophagidae), the latter noted especially for unique color pigments (see page 77), are likewise confined to the Ethiopian region. Five families, little known to the layman, are

FIG. 13.6. *The Wood Ibis or Yellow-billed Stork* (Ibis ibis) *is a characteristic bird of Africa, belonging to a fairly widespread family (Ciconiidae).* [From a Kodachrome by Roger T. Peterson.]

restricted to Madagascar. Still other families are shared with the adjacent Oriental region, and with the Palaearctic, which provides many winter visitors to Africa.

Africa, perhaps more than any other continent, has witnessed destructive exploitation of its animal resources, particularly of its big game animals, both from poaching and from government-sponsored programs to eliminate wild grazing mammals to make way for domestic livestock ill suited to the land (Grzimek, 1961; Mossman, 1966). But birds are also severely affected, not only by destruction of habitat from overgrazing but also by widespread use of insecticides to "eradicate" locusts, the tsetse fly (to protect cattle), and other insect vectors of diseases. In a *National Geographic* article, the Rodgers (1960) dramatically described the plight of Africa's vanishing wildlife as "The Last Great Animal Kingdom," which appears to be as true today as it was several decades ago.

On the more positive side of the ledger, however, is the serious effort of conservationists, both in and out of Africa, to try to save the continent's vanishing wildlife. Two of the world's largest national parks—the Serengeti, made famous by the Grzimeks' book *The Serengeti Shall Not Die,* and Tsavo—as well as numerous smaller parks, are ostensibly dedicated to the preservation of wildlife. Although they suffer from inadequate financing, the almost hopeless task of adequately patrolling such vast areas, too many visitors, constant pressures for agricultural and commercial developments, and the demand for animal resources such as ivory, the parks nevertheless are serving their purported mission of making their fabulous animal life available to appreciative visitors. East Africa in particular—Kenya, Uganda, and Tanzania—is justly famous for its wealth of bird life. Nowhere else in the world can so many birds be seen with so little effort. Kenya alone has recorded more than a thousand species of birds, and the alkaline lakes in the Rift Valley harbor concentrations of flamingos (see Fig. 4.3) that Peter Scott has called "probably the most striking ornithological spectacle in the world."

ORIENTAL OR INDIAN REGION

The Indian region comprises tropical Asia, south of the effective Himalayan barrier, as well as such islands as the Philippines and part of the East Indies. A fairly sharp line of demarcation separates the western islands of the East Indies from New Guinea and Australia which, with its surrounding islands, constitutes another region. Because of its continuity with the Australian region on the one hand and the Ethiopian on the other, the Indian region has few birds peculiar to it. The leafbirds or fairy bluebirds (Irenidae) are restricted to this region, and the broadbills (Eurylaimidae), a rather primitive passerine family (Fig.

FIG. 13.7. *Black and Yellow Broadbill* (Eurylaimus ochromalus) *from Malaysia. Broadbills (14 species) are primitive passerines of southern Asia and the Malayan states, but a few forms occur in Africa.* [From Delacour, *Birds of Malaysia,* copyright 1947 by Jean Delacour, Macmillan Publishing Co., Inc., New York.]

13.7), are chiefly Oriental but also occur in Africa. The region is also rich in phasianids, including *Gallus* (the ancestors of domestic fowl) and *Pavo* (peafowl), as well as in many familiar smaller birds (pigeons, parrots, kingfishers) and some less familiar birds that it shares with other regions. Among the latter are glittering sunbirds (Nectariniidae), the counterparts of our neotropical hummingbirds, colorful bee-eaters (Meropidae), grotesque storks (Ciconiidae), and hornbills (Bucerotidae). Its close affinity to adjacent regions is indicated by the fact that it shares 70 families (80.5 percent) with the Ethiopian region and 64 families (74.4 percent) with the Australian (Barden, 1941).

Like Africa, the Oriental Region has both assets and liabilities with respect to its bird life. Liabilities include war-torn and defoliated Vietnam, the apathy of most of the natives toward wildlife, and exploitation by outsiders. On the positive side are many wildlife sanctuaries, parks, and refuges, particularly in India where a veneration for all life affords protection to birds. Many predatory birds—kites, hawks, and eagles—exist even in the larger cities, prompting one American commentator to remark that India ought to be sending missionaries to the United States to teach us how to save our dwindling birds of prey. One of India's more notable sanctuaries is Keoladeo Ghana near Delhi, which includes a banding station and great concentrations of waterfowl, cranes, and colorful songbirds—this in spite of overgrazing by cattle and semidomesticated water buffalos.

AUSTRALIAN REGION

Probably because of its long period of isolation from mainland Asia, this region, comprising the Australian continent and its surrounding islands (Tasmania, New Guinea, and the subregion of New Zealand), has developed a remarkable bird life of its own, including many groups restricted to it. Best known of these are the cassowaries (Fig. 13.8) and emus (Casuariiformes), mound builders (Megapodiidae), the lyrebirds (Menuridae; Fig. 13.8), and birds of paradise (Paradiseidae) in New Guinea. Still other birds, notably pigeons, parrots, and kingfishers, reach their highest development or diversity in the Australian Region, which may well have been their center of origin. Tropical islands and a greatly varied climate, with examples of all the world's climates except Arctic and Antarctic extremes, account for a notably rich and varied avifauna—perhaps exceeded only by the Neotropical. New Guinea may vie with Colombia for the great variety (and inaccessibility) of its bird life. Like South America, however, Australia lacks quite a few birds that have spread over most other continents. Curiously, the great fringillid group has not yet reached Australia.

New Zealand, land of animal oddities and paradox of lush meadows in juxtaposition to barren hillsides and overgrazed pastures, has been separated from all other land areas so long, and has developed so striking a fauna, that it is sometimes regarded as an independent region (but otherwise included in the Australian). New Zealand has no native land mammals except bats, which reached the islands by flying, and its bird life is quite unique. Exclusive types are the moas, extinct ostrichlike birds that lived on the islands several centuries ago, and the odd kiwis (Fig. 13.9). The New Zealand wrens (Acanthisittidae) also seem to have developed only in New Zealand. As might be expected, however, the relatively small land area of the islands is not well stocked with native land birds, though of course many wide-ranging sea birds have colonized there.

New Zealand, like Hawaii and other islands, suffers from large-scale introduction of exotics—25 mammals, 24 birds, 14 fishes, and 600 plants (Cahalane, 1955). Cahalane calls New Zealand an "object lesson in biological mismanagement," but Williams (1953) notes that 13 British passerines introduced to the island appear not to be seriously competitive with native species and perhaps fill a void because of the scarcity of endemic songbirds.

Life Zones and Biotic Communities

Various attempts have been made in the past to divide large continental areas, such as the Nearctic or North American, into smaller

FIG. 13.9. *New Zealand habitat group in the American Museum, featuring a giant moa prominently in the foreground with a kiwi (see Fig. 7.2) foraging under tall fern fronds in lower right corner.* [Courtesy of Amer. Mus. Nat. Hist.]

FIG. 13.8 [OPPOSITE]. TOP *The lyrebirds (2 species) are versatile Australian song-birds noted for their extraordinary tails (see Fig. 3.7) and spectacular courtship displays.* BOTTOM *The cassowaries (about 30 closely related forms) are rather widely distributed through New Guinea and surrounding islands, including a portion of Australia. They are ferocious forest birds.* [Courtesy of Australian Information Service.]

workable units composed of relatively homogeneous elements. One of the earliest and best known of these in this country was the *Life Zone Theory* developed in the 1890's by C. Hart Merriam (1894, 1898), former chief of the Division (later Bureau) of Biological Survey. He divided North America into broad transcontinental belts, shown in Fig. 13.10, bounded on the north and south by definite isotherms, or lines of equal temperature, with each zone characterized by assemblages of similar plants and animals. In brief, the Life Zone Theory stated that (1) "animals and plants are restricted in northward distribution by the total quantity of heat during the season of growth and reproduction" and that (2) "animals and plants are restricted in southward distribution by the mean temperature of a brief period covering the hottest part of the year."

This life zone concept has had a wide following in the past, but has also been much discussed and severely criticized (Kendeigh, 1932; Shelford, 1932). Among its shortcomings is the fact that temperatures (erroneously computed) were used almost exclusively in setting the limits of the zones, although other factors, such as length of day in the north, aridity and humidity, and habitat (vegetation), are often equally important. In spite of flaws in the life zone theory, however, its terminology is still widely employed and such names as Boreal, Hudsonian, Canadian, Austral, Sonoral, and Tropical are commonly applied to birds considered characteristic of these zones. Figure 13.10 portrays these life zones, and their geographic limits and the names applied to each.

Of the various theories developed in later years (see also Dice's *Biotic Provinces*) to replace Merriam's perhaps outmoded life zone concept, one that is now widely employed by ornithologists is that of *biomes* or *biotic communities*, an idea advanced chiefly by Clements and Shelford (1939); its application to bird distribution has been worked out largely by Pitelka (1941). Biomes are major landscape features, such as *tundra, coniferous forest, deciduous forest,* and *grassland.* These are determined and characterized by the dominant life forms, chiefly plants, that make up the unit. The early successional stages in the development of a biome are called *seral communities,* the later, more persistent stages are *subclimax communities,* and the final stage, which is self-perpetuating and cannot be replaced by natural processes, is the *climax* or *major biotic community.* Transitional areas, where two major biotic communities meet and overlap, as where elements of the deciduous forest intermingle with the coniferous forest, are called *ecotones.*

Figure 13.11 portrays the major biotic communities of North America and their connecting ecotones. The ones in eastern North America are fairly well defined, and quite comparable to Merriam's life zones and Dice's biotic provinces, but have the advantage of a fairly

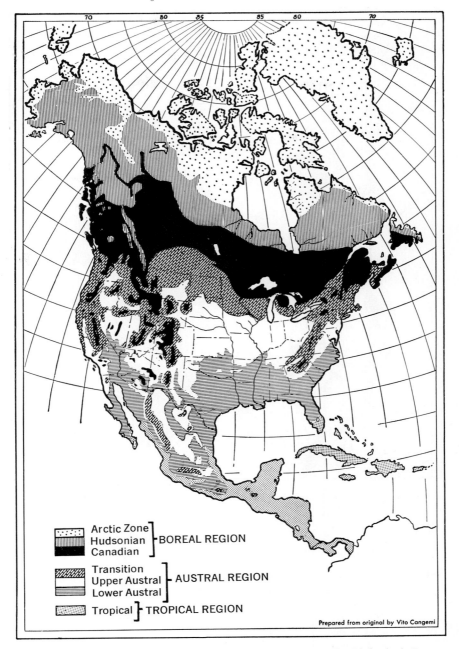

Arctic Zone ⎤
Hudsonian ⎬ BOREAL REGION
Canadian ⎦

Transition ⎤
Upper Austral ⎬ AUSTRAL REGION
Lower Austral ⎦

Tropical ⎱ TROPICAL REGION

Prepared from original by Vito Cangemi

FIG. 13.10. *Life Zone Map of North America.* [From U.S. Biological Survey Fourth Provisional Zone Map, by C. Hart Merriam, Vernon Bailey, E. W. Nelson, and E. A. Preble, 1910. By permission of Fish and Wildlife Service.]

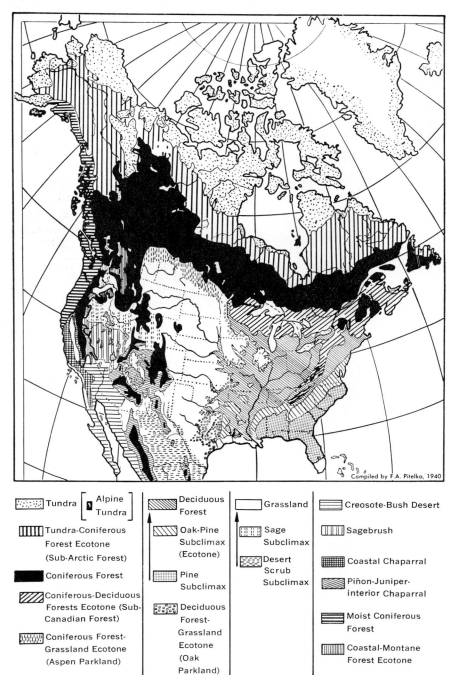

Compiled by F.A. Pitelka, 1940

Tundra [Alpine Tundra]	Deciduous Forest	Grassland	Creosote-Bush Desert
Tundra-Coniferous Forest Ecotone (Sub-Arctic Forest)	Oak-Pine Subclimax (Ecotone)	Sage Subclimax	Sagebrush
Coniferous Forest	Pine Subclimax	Desert Scrub Subclimax	Coastal Chaparral
Coniferous-Deciduous Forests Ecotone (Sub-Canadian Forest)	Deciduous Forest-Grassland Ecotone (Oak Parkland)		Piñon-Juniper-interior Chaparral
Coniferous Forest-Grassland Ecotone (Aspen Parkland)			Moist Coniferous Forest
			Coastal-Montane Forest Ecotone

simple and probably self-explanatory terminology in common usage by botanists and zoologists, as well as geographers. They also appear to fit the facts of animal distribution more satisfactorily, for the dominant plant type of an area so often determines what animals live there. The western communities, because of more abrupt climatic and vegetational transitions, necessarily present a more complicated pattern; it is not certain that those shown should be considered on an equal basis with the other biomes.

A biome is not uniform throughout, but is composed of different developmental stages with different plants and animals. Thus a meadow in the eastern states is not a part of the grassland biome, but merely a developmental stage of the deciduous (or coniferous) forest biome. If abandoned, the meadow would return, through several successional steps, to a climax forest, and the birds (and other animals) inhabiting it would change with each succession. Some birds are restricted to a particular developmental stage and others are not. A Bobolink, for example, inhabits the area only in the meadow stage, a Gray Catbird only in the shrub stage, whereas the more versatile American Robin or Common Crow might at least utilize the area during all of its successional changes.

North American Biotic Communities

The five major biotic communities in North America—tundra, coniferous forest, deciduous forest, grassland, and desert—are briefly characterized below, and some of the birds typical of each enumerated. The ranges of some birds coincide almost exactly with major biotic communities; some are more characteristic of the ecotones or transitional areas, some spread over several or even all the North American biomes, while still others are restricted to special niches within a biome.

TUNDRA. The treeless tundra is the northernmost habitable land, merging with the permanent icecaps poleward and with the coniferous forest southward. It has a permanently frozen subsoil, but thaws out on the surface during the short summer, giving rise to much surface water and myriads of insects. Lichens (reindeer moss), low grasses, sedges, flowers in season, and some brush at its southern extremities characterize the tundra.

Birds whose breeding range coincides quite closely with this biome, in both the Old and New World (Holarctic), are the Rock and Willow Ptarmigans, the Snowy Owl, and Snow Bunting. The latter, predominantly a seed-eater, is highly migratory in winter; the Snowy Owl

FIG. 13.11. *Major Biotic Communities of North America.* [Redrawn by Vito Cangemi from map by F. A. Pitelka; courtesy of *American Midland Naturalist.*]

migrates when its food supply fails (page 326); and the less migratory ptarmigans wander widely in winter, often invading the tundra–coniferous forest ecotone in search of berries and herbage. It is a mistaken concept that the tundra has a heavy snow cover in winter, for precipitation is slight. Gyrfalcons hunt extensively over the tundra, but are not confined to it.

Northern waterfowl (Whistling Swans, geese of several species, and ducks) and many shorebirds (Fig. 13.12) utilize the tundra for nesting and rearing young but desert it completely in winter. Gray-cheeked Thrushes inhabit the willow-fringed streams, as do Tree Sparrows, redpolls, and Northern Shrikes, but these four are more characteristic of the ecotone that meets the coniferous forest southward. Brooks (1968) discusses the adaptations of redpolls for survival in their Arctic environment: an effective insulating plumage, a croplike diverticulum for storing seeds before retirement, selection of high-calorie foods (birch

Fig. 13.12. *The Black-bellied Plover, here shown on its nest in Alaska, is one of the many shorebirds that breed on the tundra when myriads of insects are available for feeding young. It winters from southern United States into South America, and in the Old World from the Mediterranean to South Africa.* [Photo by L. H. Walkinshaw.]

seeds), ability to use low light intensities for foraging, and digestive efficiency at low temperatures.

CONIFEROUS FOREST (ALSO CALLED TAIGA). This occupies a broad belt around the world; the largest coniferous forest development in the world is in Siberia. It is composed predominantly of spruces and balsam fir, stunted on the north where it merged with the tundra, but forming fine stands of valuable though often inaccessible timber southward.

Strictly, or almost strictly, coniferous forest birds are the Spruce Grouse (Fig. 13.13), Northern and Black-backed Three-toed Woodpeckers, Gray Jay, Brown Creeper, Golden-crowned and Ruby-crowned Kinglets, several warblers, and several cone- or catkin-feeding fringillids. The Yellow-bellied Flycatcher inhabits the more boggy or swampy portions, and the Winter Wren favors the brushy undergrowth, particularly along streams. The Yellow-bellied Sapsucker and Black-throated Green Warbler are good examples of birds that extend their range southward into hardwood forests.

DECIDUOUS FOREST. This is the climax forest type that prevails over most of the eastern states, but of course it is greatly altered by cleared lands. Mature forests are chiefly oak-hickory, but a broad ecotone (sub-Canadian forest), primarily of beech-maple-hemlock, merges with the coniferous forest in the north, and a pine subclimax characterizes the southern states. The various pines (both northern and southern) and the beech-maple have their special birds, more or less distinct from those typical of the oak-hickory climax. Those characteristic of oak-hickory woods include such so-called Carolinian or Austral birds as the Red-bellied Woodpecker, Tufted Titmouse, Wood Thrush, and Yellow-throated Vireo, whose ranges are fairly sharply delineated northward by the beech-maple ecotone. Many other birds, however, range over both the oak-hickory and beech-maple areas, and still others occupy particular habitats within one or several biomes. The Song Sparrow, for example, occupies a type of habitat (primarily edge) that is widely available in or around all North American biomes.

In general, deciduous woodlots have a greater diversity and density of bird life than the other North American biomes (see page 392), but this varies greatly with the stage of succession, amount of dispersion (mixed stands usually have higher populations than uniform stands), presence of water, special food supplies, and other factors. Bond (1957), in a study of breeding birds in 64 upland hardwood stands in 17 counties of southern Wisconsin, found the greatest density in intermediate (mixed) stands and decreased numbers toward the climax stages.

GRASSLAND. The grassland biome occupies a broad strip of the central plains, paralleling the 100th meridian on both sides, from the

FIG. 13.13. *The Spruce Grouse is a characteristic coniferous forest bird. Here a displaying male shows off in a Michigan jack-pine area.* [Photo by Edward M. Brigham, Jr.; from Wood's *Birds of Michigan*, Univ. of Mich. Mus. Zool. Misc. Publ. 75.]

Rio Grande northward into the Canadian prairie provinces. Annual rainfall is probably inadequate for forest growth; presumably the western plains never were covered with trees, although perhaps repeated fires prevented such growth. Plantings for windbreaks and about homesteads now provide some cover for birds, but otherwise open country prevails. Particularly notable developments of similar grasslands outside this country include the pampas of Argentina, the steppes of Europe and Asia, and the parklike savannas (veld) of Africa.

Grassland birds in this country include the Burrowing Owl, prairie chickens, Horned Larks, the Western Meadowlark, Dickcissel, and several open-country sparrows. Most of these, though originally true grassland birds, have invaded cleared lands to the east, even reaching the Atlantic Coast. A possible explanation of both the occurrence and extinction of the Heath Hen (a form of the Greater Prairie Chicken) on the East Coast is that it represented a relict form from a former wider extension of grasslands. Many other species, not typically prairie forms, have taken advantage of special niches provided by man or nature—swallows about buildings and bridges, doves in the windbreaks, marsh and water birds about pot holes or sloughs—but in general the grasslands have a low density of breeding birds.

DESERTS. Deserts (Fig. 13.14) are characterized by deficiency of rainfall, which is low and erratically distributed. Permanent streams cannot arise in a desert, although they may flow through if they have a source in outlying humid regions. Permanent ground water is not available, so only *xerophytes* (drought-resistant plants) can survive. From the standpoint of plant and animal life deserts are inhospitable regions—day and night temperatures have sharp extremes, winds are often severe, food and cover are scarce and unevenly distributed—hence animals are highly specialized to exist under desert conditions. Many are cursorial and cover a wide range in foraging; others are nocturnal and many live in burrows to escape the heat of the day, wind-blown sand, and predators. One of the critical factors faced

FIG. 13.14. *Deserts vary from sandflats with little or no vegetation and practically no bird life to those that show quite a diversity, though low density, of both plants and birds.* [Photo by George J. Wallace.]

by desert creatures is the scarcity of water (see pages 163–64). Some, like doves and sandgrouse, visit distant waterholes periodically, many get along with preformed water from their food (succulent vegetation and insects or vertebrates), and a few subsist largely on metabolic water formed internally.

Bird life in the southwestern scrub, or "hot," deserts includes quite a few predatory species — hawks, owls, Loggerhead Shrikes, and the Road-runner — which live largely on rodents, reptiles, and ground-dwelling arthropods, but erratically distributed seed-eaters take advantage of the sudden seed supply when occasional rains make any available.

Dixon (1959) found 17 species (not including wide-ranging raptors) on a 33-acre plot of scrub desert in western Texas. Austin (1970) recorded densities from as low as 6 pairs per 100 acres in the poorer mesquite areas in southern Nevada to 48.7 pairs in the richer mesic areas. Raitt and Maze (1968) found an average of 17.7 pairs per 100 acres in creosote bush communities in southern New Mexico, with lower densities in the xeric habitats and higher in the mesic. Quite a few species nest in cactus, either in cavities (Elf Owl, Gila Woodpecker; Fig. 13.15) or among the thorns (Cactus Wren). The more northern

FIG. 13.15. *The Gila Woodpecker is one of several species that utilize cactus for nest sites.* [Photo by Dale A. Zimmerman.]

sagebrush plains, commonly called "cold" deserts, are inhabited by the Sage Grouse, Sage Thrasher, and an interesting assortment of fringillids.

From this account of biomes it should be apparent that each can be broken down in successively smaller ecological units, or habitats, until perhaps a single kind of tree or level in trees can serve as a micro-habitat for a bird. This, in part, is the subject matter of the following chapter.

SELECTED REFERENCES

BARDEN, A. A., JR. 1941. Distribution of the families of birds. Auk, 58:543–557.

CLEMENTS, F. E., and V. E. SHELFORD. 1939. *Bio-ecology.* New York: Wiley.

DARLINGTON, P. J. 1957. *Zoogeography: the Geographical Distribution of Animals.* New York: Wiley.

DICE, L. R. 1943. *The Biotic Provinces of North America.* Ann Arbor, Mich.: Univ. of Mich. Press.

GRISCOM, L. 1945. *Modern Bird Study.* Cambridge, Mass: Harvard Univ. Press. Chapters VI–IX.

GRZIMEK, B., and M. GRZIMEK. 1961. *The Serengeti Shall Not Die.* New York: Dutton.

MATTHEW, W. D. 1939. Climate and evolution. Spec. Publ. N. Y. Acad. Sci., I. xii + 223 pages.

MAYR, E. 1946. History of the North American bird fauna. Wilson Bull., 58:3–41.

MERRIAM, C. H. 1894. Laws of temperature control of the geographical distribution of terrestrial animals and plants. Nat'l Geog. Mag., 6:229–238.

———. 1898. Life zones and crop zones of the United States. Bull. U. S. Bio. Sur., 10:1–79.

NEILL, W. T. 1969. *The Geography of Life.* New York and London: Columbia Univ. Press.

PITELKA, F. A. 1941. Distribution of birds in relation to major biotic communities. Amer. Mid. Nat., 25:113–137.

SERVENTY, D. L. 1960, Geographical distribution of birds, Chapter IV in A. J. Marshall's *Biology and Comparative Physiology of Birds,* New York: Academic.

WALLACE, A. R. 1876. *The Geographical Distribution of Animals.* 2 vols. New York: Harper.

14

Ecology

REVISED BY HAROLD D. MAHAN

FROM anatomical, physiological, and behavioral standpoints birds are intimately tied to their environments. The interactions between birds and their natural surroundings—the physical and biological aspects of their environment—constitute their *ecology*. Although every aspect of the bird's environment is important to its life style, considered here are only those that should be of interest to the beginning student: habitats and niches, succession, bird populations, and ecological adaptations.

Habitats and Niches

To field ornithologists it is almost inconceivable to think of a bird without calling to mind the type of *habitat* (or kind of place) where it lives as well as the *niche* it occupies (or role it performs) in its habitat. As with all organisms, the habitats of birds contain both living and nonliving components. The nonliving or *abiotic* component consists of sunlight, water, and soil. These, in turn, make possible the growth of green plants, which birds use either directly or indirectly for food and cover or as a source of nesting materials.

Birds may be categorized by the feeding level that they occupy in relation to their utilization of green plants for food. For example, the herbivorous Cedar Waxwing is a first feeding ("trophic") level organism since it eats plants directly; warblers are mostly insect-eaters, feeding at the second trophic level; hawks and owls feed at the third trophic level, since they eat mice that have eaten plant-eating insects. Each feeding level basically, then, is a function of the plants that are available in a certain habitat either directly or indirectly for food.

Not only are plants important to birds as food items at different trophic levels, but also because they produce a certain overall "quality" in every habitat, even including those that are mainly aquatic. The growth habit of plants (their size, density, and species composition) found in a particular habitat helps determine that habitat's avifauna. It is the overall nature of a mature beech-maple woods, for instance, that attracts Red-eyed Vireos. A woods of this type will not contain this species at a very early age in its development. This may be due to a lack of suitable branches for nest placement; it may be due to insufficient moisture and/or to other physical factors that inhibit the production of an adequate supply of insects. Generally, though, it is due to less than a minimal level of *all* such ecological factors taken collectively. Every habitat has, then, a certain overall quality. This quality can be understood better if we examine each aspect of it separately.

The principal qualitative characteristics of any habitat are stratification, sociability, vitality, and periodicity. *Stratification* refers to the arrangement of organisms, especially plants, in a particular habitat. It is most apparent when we look at the layered effect in a mature woods: below the ground are soil-inhabiting organisms, mostly invertebrate animals; near the ground are low herbaceous plants, such as fungi ferns, mosses and low growing seed-bearing plants; next is a shrub layer; and above this is the canopy of the larger trees. Different bird species occupy different strata in such woods. Allen (1961) divides birds of the woodlands, primarily on the basis of nest sites, into (1) birds of the forest floor, (2) birds of the undergrowth, (3) birds of the lower branches, (4) birds of the tree trunks and hollow branches, and (5) birds of the higher branches and treetops. Such stratification is even more pronounced in tropical rain forests (Fig. 14.1).

A similar horizontal spacing is seen in aquatic habitats: ducks and coots build nests on or just above the water's surface, Least Bitterns and Red-winged Blackbirds utilize the cattails and reeds (sometimes several feet above the water's surface) for nest attachment, and, when dead trees are available in such habitats, their upper branches may be occupied by Great Blue Herons, Ospreys, and even Bald Eagles. On a larger scale different birds also characterize different altitudinal levels in mountains. Through natural selection processes, bird species have evolved patterns of spacing themselves in a stratified fashion that has led to greater utilization of the existing space and available feeding opportunities.

At various times in their lives, like other organisms, birds are spaced fairly uniformly as individuals, in clumps, or in colonies. This ecological phenomenon is termed *sociability*. It is similar to stratification from the standpoint that it provides for a greater use of the entire habitat but on a different plane. Like stratification, bird sociability is quite variable. The nearness of nesting Herring and Ring-billed Gulls, where the nearest neighbor may be only inches away, allows for greater protection from predators than if their nests were widely spaced; on the other hand, such species generally spread out more when they feed; hence they more effectively utilize the feeding space by covering a greater area.

F IG. 14.1 *A tropical cloud forest in Chiapas, Mexico, showing the luxurious develop-ment and great diversity of stratified plant life, which in turn supports a great variety of bird life, from strictly terrestrial forms to those seen only in the treetops.* [Photo by Dale A. Zimmerman.]

In addition to spacing, the quality and composition of every habitat are determined by the numbers of organisms that are present at dif-ferent times of the year, by the reproductive potential of all the orga-nisms present, and by seasonal events that have an effect on these orga-nisms. The *vitality* of the habitat refers to the overall biotic potential it has for reproducing itself; the seasonal changes are referred to, collec-tively, as the *periodicity* of the habitat. Both help determine the nature of a bird's habitat. It is important to realize, that within each habitat are numerous *microhabitats*, each of which may harbor a colony, a single family, or perhaps only an individual bird.

Because the factors that determine habitat types are so variable, it is not surprising that habitats vary considerably from one location to another. Even the most monotonous and uniform landscapes, such as deserts and grasslands, may consist of ecologically diversified habitats. American deserts vary from the virtually birdless Bonneville Salt Flats in Utah to southwestern scrub areas that have an interesting diversity, though not a great density, of birds. The jack-pine "plains" of central and northern Michigan, famous for Michigan's unique species, the Kirtland's Warbler (Fig. 14.2), contain a low number of birds, both individuals and species, owing in part to the low number and the homogeneity of plant species found in such a relatively dry habitat. The veld in South Africa, a grassland that may have scattered shrubs and trees, is divisible into grass veld, bush veld, and tree (acacia) veld, each with its characteristic bird life. More heterogeneous areas also may be described based on major physiographic features that are present. Batts (1958) divides nesting habitats on a southern Michigan farm into (1) open water, (2) marsh, (3) woods, (4) woodslope, (5) fencerow, and (6) fields (alpha and beta).

Where two habitats come together, the ecological phenomenon called the "edge effect" occurs. Such areas contain plants and animals of both overlapping communities, and the greater diversity that results from this intermingling draws more organisms. (It is a well-known field technique to visit such areas if your purpose is to see a large variety of birds in a small area. Thus, for example, you can see more birds at the edge of the woods than you can within it.)

F IG. 14.2. *The Kirtland's Warbler, a species nesting only in Michigan, utilizes a relatively dry and homogenous plant community of jack-pines.* [Photo by William Coates.]

Like bird habitats, the niches that birds occupy within them are quite variable. Some birds act as scavengers (vultures), some as predators (hawks), some as herbivores (parrots). Certain species (Common Crow, Starling) are very adaptable at niche-filling and, as a result, occupy a variety of habitats; others such as the Kirtland's Warbler (pages 362–363 and 389) and the northern race of the Everglade Kite (page 361), are restricted to more limited niches. Still others (Marsh Hawk, kingfishers) may not be abundant in any one place because the niche they fill is a limited one, but they utilize a habitat that is widespread throughout the world—that is, there are many fish-eating birds in aquatic places, but only the kingfishers, in most parts of the world, fish small streams. Such niche-filling reduces competition, and also provides for more complete utilization of all ecological units.

Closely related species, occupying similar habitats, usually have slightly different requirements. Salt (1957) noted that Fox Sparrows, Lincoln's Sparrows, and Song Sparrows living in the same marsh in Wyoming occupied different niches and had slightly different food habits; the Lincoln's Sparrows fed on 40 percent plant food and 60 percent animal, while the Song Sparrow reversed these percentages and the Fox Sparrow was intermediate. Davis (1957) observed different foraging behavior (depth of scratching, food selection, and habitat preference) in Brown and Spotted Towhees in California. Often two or more similar species in a similar habitat are somewhat segregated ecologically because they tend to fill slightly different niches, but the absence of one or more of them allows the others to overflow the unoccupied space. Woodpeckers are also excellent examples of this *competitive exclusion* principle: the males and females on the same territory forage on different parts of a tree and select different food (Kilham, 1970).

Succession

One feature of all natural habitats is an orderly growth pattern, termed *succession*. Such growth, whether in terrestrial or aquatic habitats, involves progressive changes in the numbers and kinds of species present and an accompanying accumulation of organic matter.

There are two types of succession—primary and secondary. Primary succession begins at a relatively inorganic, primitive plant level and, as more complex plants and animals replace more simple ones, culminates in the *climax community*, which is the ultimate community of organisms possible under the existing climatic conditions. *Secondary succession* takes place if primary succession is interrupted or if a climax community already has been established and is then disrupted. Whether primary or secondary succession occurs in an area, the result is the same—the establishment of the ultimate mature community that can maintain itself indefinitely as long as the climatic conditions do not change drastically.

In each successional sequence (a *sere*) there are a number of fairly well-marked stages (*seral stages*) that lead to the climax community, and in each stage there are changes in bird life. Field, Grasshopper, and Vesper Sparrows are common in terrestrial communities during the early herbaceous plant stages (the *seral field community*). They nest either on the ground or only a few inches from it. As the low herbaceous vegetation dies, it produces more organic matter in the soil and the soil becomes more moist. This leads to the establishment of taller, more complicated field plants (the *seral old field*), and at this stage the Field Sparrow may become more numerous, the Grasshopper Sparrow less so. Possibly new species (e.g., meadowlarks, Bobolinks) may enter the area. With further increase in organic matter and moisture, the seral field stage is replaced by a *seral shrub* stage. In this shrub stage the Grasshopper and Vesper Sparrows usually disappear, the meadowlark and Bobolinks may also disappear or decrease considerably, and towhees and Song Sparrows may move into the area. There still may be a few Field Sparrows but, along with Song and Grasshopper Sparrows, and towhees, they disappear from the area at the end of this growth period, as the seral shrub community is replaced by mixed species of trees. During the early tree stage there still will be numerous shrubs as an understory. Birds typical of this stage are the Ruffed Grouse, Gray Catbirds, Brown Thrashers, and perhaps American Robins, all using the trees invading the area. Ultimately, of course, the shrub community and its avifauna will be vastly reduced as the climax-community trees create increasing amounts of shade. In this final stage, usually dominated by a few (often only two) dominant tree species, the community and its bird life (Veeries and Wood Thrushes, for example, in the eastern beech-maple climax forests) become somewhat stabilized.

A similar community progression occurs in aquatic habitats but with aquatic-adapted organisms. A typical progression in such habitats might begin with aquatic birds such as grebes or Black Terns that build floating nests. As aquatic vegetation develops and cattails and reeds become established, Red-winged Blackbirds, Long-billed Marsh Wrens, American and Least Bitterns will begin to nest in the area and replace the former species that required more open water. As cattails give way to woody aquatic vegetation (e.g., willows), these birds are replaced by species that nest in this vegetation, such as Yellow Warblers and Common Yellowthroats. Eventually, as more and more vegetation creates more organic material and dryer conditions, the tree stages and their accompanying avifauna (e.g., Red-eyed Vireos, Wood Thrushes) become the ultimate climax community.

Bird Populations

As previously mentioned, habitats vary greatly in diversity and density of bird life. In general, deserts and grasslands are low both in number of species and numbers of individuals (birds per acre); mixed woodlots are high compared to uniform stands; marshes and swamplands, especially if harboring such semicolonial forms as blackbirds and rails, have high concentrations. Kendeigh (1948) calculated the density of breeding birds in several biotic communities in northern Michigan and found 56 pairs per 100 acres in grasslands, 112 pairs in pine-aspen, 146 pairs in cedar-balsam, and 155 pairs in beech-maple-pine, thus disclosing a rising trend from the more open areas to the more diversified woodlots. Dixon (1959) found desert habitats in western Texas varying from only 15 males (pairs) per 100 acres to 51 in desert scrub, but of course lower densities, presumably approaching zero, could be found in completely barren deserts. At the other extreme, much higher densities have been recorded in spruce-fir forests in Ontario (Kendeigh, 1947) and Maine (Hensley and Cope, 1951) during budworm infestations (385 pairs per 100 acres in the Maine sample). The spruce-fir forest populations were augmented by a concentration of breeding warblers with small territories.

Still greater densities, or selection for desired species, can be achieved in managed areas. Since diversity is to some extent a reflection of the amount of interspersion of different vegetation types, it is a common practice in management to create as much edge and mixture as possible by planting and cutting, or other management measures, either to improve conditions for selected species (waterfowl, gamebirds, or songbirds) or merely to create a more favorable environment for all wildlife. A "farm" in Maryland is said to have a nesting population of 59 pairs per acre, or many times the number found on ordinary farmland, but nearly half of these were box-nesting Purple Martins not confined to their nesting area. Stine (1959) listed 41 species nesting in an 18-acre woods on Bird Haven Sanctuary in Illinois over a 10-year period, most of them present every year or nearly every year. Nineteen additional species occurred on adjacent farmland in the sanctuary. The total density of bird life (pairs per acre) was not given, but the great diversity of nesting species (60) attracted to the sanctuary seems phenomenal. Sprunt (1960) has recorded an extraordinary total of 156 species at his one-acre home outside Charleston, South Carolina, but these include visitors to the grounds over a long period of time.

Some amazing aggregations of nesting birds have been recorded

FIG. 14.3. *The several million Guanay Cormorants inhabiting this island in the Chincha group off the coast of Peru constitute one of the greatest concentrations of bird life in the world.* [Courtesy of Amer. Mus. Nat. Hist.]

among colonial species on islands (Fig. 14.3), rocky ledges, and in marshes. A few of these, perhaps the world's greatest concentrations of breeding birds, are listed in Table 14.1.

Bird populations are ascertained by inventories and censuses. One important aspect of a census is to evaluate the status of various species for management purposes, either for harvesting safe numbers of a game species, or to try to preserve vanishing forms. Hence, we often know by actual count the approximate numbers of critical species— 50–60 wild Whooping Cranes, 40–60 California Condors, 200 pairs of Kirtland's Warblers. Censuses also enable ornithologists to estimate populations over larger areas. Peterson (1948) calculated that there are 5–6 billion breeding birds (3 billion pairs) in the United States.

Another important aspect of such inventories is to evaluate population changes from year to year. Bird numbers fluctuate markedly, sometimes for unknown reasons, sometimes from obvious causes. Many species have declined or become extinct because of man's activities, but some birds have increased. McAtee (1951), in a study of bird records in

TABLE 14.1. SOME LARGE COLONIES OF BREEDING BIRDS

Species	Density	Authority
Adelie Penguin	215,000 in 55-acre rookery	Eklund, 1959
Slender-billed Shearwater	150 million, Bass Strait, Australia	Peterson, 1948
Gannet	7,500 pairs on Bonaventure Island	Fisher, 1954
Guanay Cormorant	22–25 million on Chincha Islands, Peru	Ullman, 1963
Flamingos	3–4 million,* Lake Nakuru, Kenya	Brown and Root, 1971
Sooty Terns	14,520 pairs/acre, Dry Tortugas	Peterson, 1948
Common Murre	750,000 on 17-acre island, Oregon	Peterson, 1948
Dovekie	5 million in Scoresby Bay, Greenland	Snyder, 1960
Auklets (3 species)	1,527,000 on St. Lawrence Island, Alaska	Bédard, 1969
Tricolored Black-birds	5,000–10,000 nests/acre, California	Emlen, 1941

* Postbreeding.

the District of Columbia over an 80-year period (1861–1942) listed 37 species that had increased, 50 that had decreased, and 44 that had suffered little or no change. Stine (1959), comparing her 1945–1955 records at the Bird Haven Sanctuary in Illinois with those of Ridgway in 1907–1910, found that 52 of Ridgway's species were still present but that 21 were gone, owing apparently to changes in habitat (see also page 14 for a more recent and broader comparison).

Some of the modern hazards to birds are discussed in Chapter 16, but it should be noted here that population trends seem to have been drastically downward in recent years. *Audubon Field Notes*, a journal that digests and summarizes the field records of a large corps of observers all over the United States, portrayed a dismal picture of the situation in the late 1950's. Rookery birds declined sharply in the southeast, apparently because of a succession of dry breeding seasons, but recovered somewhat after a rainy season. Duck declines, often due to drought and drainage in the prairie marsh lands, are reflected in the ever-changing hunting regulations. By 1958 songbird populations (except for such hardy and adaptable species as Starlings, House Sparrows, Common Grackles, and Red-winged Blackbirds) were reported to have plunged to an all-time low. This was attributed, by most observers, to the unprecedentedly severe southern winter of 1957–58 followed, in the northeast, by a cold rainy breeding season, and to the unremitting use of insecticides in so many large-scale "eradication" programs (see pages 456–60). However, some of the "disaster" species seem to be increasing again, indicating that such declines may be only temporary and that short-term fluctuations are not a dependable criterion for long-term trends.

Ecological Adaptations

The anatomical, physiological, and behavioral adaptations that birds have developed in their long evolutionary history result from having been exposed to numerous environmental changes in the past. The long legs of the Great Blue Heron, the dim-light-adapted eyes of owls, the ability to unlock metabolic water during digestion in desert-inhabiting Roadrunners, and even the habit of building a nest covered with lichens in the Ruby-throated Hummingbird are ecological reflections of each species' relationship to its environment.

ANATOMICAL ADAPTATIONS

A Great Blue Heron while fishing is a study in coordinated effort. In stalking prey it lifts one long foot at a time, hardly causing a ripple on the surface of water as it moves to a new location. It barely twists the trunk of its body, its head moves only slightly as its eyes refocus with each step. Finally it pauses and remains motionless, often for several minutes. Suddenly it plunges its bill into the water and a large fish is speared. The bird's neck is quickly drawn upward, the bill opens rapidly, and the fish is swallowed. There is little motion associated with any of the heron's fishing activities, except for stabbing its prey. All parts of its anatomy are coordinated for the task of locating prey, capturing it, and eating it with minimal environmental disturbance. Were this not so, there would shortly be one less heron to initiate a new generation.

Anatomical adaptations have permitted birds to invade almost every habitat on the earth's surface. The development of a salt gland (see page 161), for example, has enabled pelagic species to utilize salt water and live for long periods on the surface of the largest (but driest from a hypertonic standpoint) habitat of all—the oceans. Procellariiform birds, for example, could not survive at sea without corresponding ability to stay aloft (see page 88) for long periods of time. Nor would this be possible without the development of keen eyes and adjacent arterio-venous blood vessels in the feet and legs of those species that stand in cold water (Irving, 1960b). These and other adaptations in swimming birds (see page 101) have enabled them to take advantage of a seemingly hostile habitat.

Similarly, the internal and external structures of terrestrial species reflect environmental adaptations. As pointed out in Chapter 5, the Common Redpoll in Alaska has evolved a croplike diverticulum that permits greater food storage for the longer Arctic nights, and, like other northern forms, has a dense plumage and thick layers of insulating fat that reduce heat loss. Ptarmigans have feathered toes that serve the same function. Unlike penguins in the northern parts of their range, the Adelie Penguin in the Antarctic has nostrils covered with feathers that permit inhaling colder air. Even the color of plumages (e.g., white in ptarmigans during winter) and body size (e.g., larger in subspecies of more northern Fox Sparrows) are reflections of evolutionary adjustments to environmental pressures.

PHYSIOLOGICAL ADAPTATIONS

Paralleling anatomical adaptations have been physiological adjustments that enable birds to succeed in a variety of habitats. Although every aspect of a bird's physiology reflects this to some extent, there are several exceptional adaptations that illustrate how physiology and environment are closely interrelated.

The physiology of temperature regulation in birds is one example. In young altricial species (e.g., all passeriforms) temperature regulation is partly developed at hatching. After being exposed to a controlled 10°C environment for 1 hour, a one-day-old Eastern Phoebe has a body temperature of only 1 or 2° above the 10°C ambient temperature. At this body temperature its heart rate, blood flow, and breathing rate decrease until they are barely perceptible. After 10 days of age, however, the young birds maintain a fairly constant body temperature (comparable to adults), even when exposed for one hour or more to low ambient temperatures. Correspondingly, other physiological mechanisms in young birds of this age function at levels that approximate those in adults under similar environmental conditions. The inability of very young altricial birds to maintain constant high body temperatures in low ambient temperature has adaptive value because it slows metabolic activities, hence conserves energy. Interestingly, many adult birds (e.g., goatsuckers, swifts; see page 159) undergo periods of torpidity during periods when insects are less available. Indirectly, at least, the distribution of birds over extensive areas is affected by their ability to regulate their internal temperatures.

In some species, however, distribution is more directly related to the bird's physiological requirements. It is well known, for example, that certain galliform birds are physiologically dependent on specific soil types; the Ring-necked Pheasant apparently requires calcium-enriched soils, and the Gray Partridge limestone soils.

Another physiological adaptation involves the relationship between daylight periods and metabolic activities. The day-night light cycle (*circadian rhythm*) of Chaffinches can be altered in captivity simply by exposing them to different periods of light (Aschoff, 1965). In fact, increasing the *intensity* of light causes a shortening of the diurnal metabolically active period. Also, there is an accumulative effect on a bird's physiology when photoperiods are lengthened. During migration, for

instance, increasing day length appears to have an accumulative physiological effect on gonadal development that increases the pace of migration (see pages 337–39).

BEHAVIORAL ADAPTATIONS

Success of birds in any habitat is attributable not only to proper anatomical and physiological adaptations; of equal importance are those
patterns of behavior that have evolved to enable birds to find food,
acquire a mate, reproduce, and avoid predators or avoid being destroyed by the harsh conditions that sometimes occur in their world.

Although an amazing variety of feeding techniques have been
evolved (see Chapter 5), sometimes behavior intended for securing
food becomes nonadaptive. Black Vultures are gregarious feeders,
seldom feeding alone. Group feeding has its advantages, but not when
too many vultures choose to feed on the same carcass since so much
time is spent in fighting that little time is left for feeding. A similar situation occurs in other birds, even nectar-feeders. Recently Emlen (1973)
made brief observations of the feeding behavior of various species that
"dominated" at agave blossom feeding sites on Grand Bahama Island
and concluded that their overt aggression reduced their own feeding
time when energy was devoted to territorial defense instead of feeding.

Activities associated with avoiding predators can also be nonadaptive.
Bobwhites huddle so that the head of each member of the covey points
outward from the covey. Huddling in Bobwhites is adaptive behavior
since it serves to warn the whole covey'of predators. In a recent study,
however, Case (1973) found that huddling was metabolically advan-
tageous at 5°C but was "energetically disadvantageous at higher temperatures." This suggests that huddling may prevent excessive predation on the one hand; on the other hand this behavior may be overly
expensive metabolically on warm nights.

Numerous examples of adaptive behavior exist for activities associated with the breeding behavior of birds — those involving territorial
defense and other courtship displays, nest-site selection and nest-
building activities, as well as feeding and other activities associated with
raising young. Possibly, however, the most striking example of adaptation in the breeding biology of birds has to do with care-soliciting
behavior. Young birds have evolved a multitude of adaptations that
stimulate their parents (and even nest mates and other adults) to
provide them with care. Such adaptations, like most discussed in this
chapter, are not only behavioral but also anatomical (e.g., brightly
marked gapes help direct feeding motions of adults) and physiological
(e.g., body temperatures of young cause them to increase movements
while being brooded, which affects brooding response in adults).

In studying the ecology of birds, it is necessary to consider not only the places where they live and their role in such places, but also their internal and external structures, their biochemistry, and their coordinated activities.

SELECTED REFERENCES

ALLEN, A. A. 1961. *The Book of Bird Life,* 2nd ed. New York: Van Nostrand. Chapters III–VI.

BENTON, A. H., and W. E. WERNER. 1966. *Field Biology and Ecology.* New York: McGraw-Hill.

EVERHARDT, L. L. 1969. *Wildlife Management Techniques.* Washington, D.C.: The Wildlife Society.

FISHER, J. 1954. *A History of Birds.* Boston: Houghton. Chapters VIII–XI.

GIBB, J. A. 1961. Bird populations. Chapter XXIV in A. J. Marshall's *Biology and Comparative Physiology of Birds.* New York: Academic.

GRANT, P. R. 1968. Bill size, body size, and the ecological adaptation of bird species to competitive situations on islands. Syst. Zool., 17:319–333.

HOWELL, T. R. 1971. An ecological study of the birds of the lowland pine savanna and adjacent rain forest in northeastern Nicaragua. Living Bird, 10:185–242.

KENDEIGH, S. C. 1961. *Animal Ecology.* Englewood Cliffs, N.J.: Prentice-Hall.

KING, A. J., and D. S. FARNER. 1961. Energy metabolism, thermoregulation, and body temperature. Chapter XII in A. J. Marshall's *Biology and Comparative Physiology of Birds.* New York: Academic.

LACK, D. 1966. *Population Studies of Birds.* London: Oxford Univ. Press.

———. 1971. *Ecological Isolation in Birds.* Cambridge, Mass.: Harvard Univ. Press.

PETERSON, R. T. 1948. *Birds over America.* New York: Dodd. Pp. 56–70.

SMITH, R. L. 1966. *Ecology and Field Biology.* New York: Harper.

WALTERS, J. H. 1967. Avian populations in a recently disturbed old field succession. Bird-Banding, 38:17–37.

15

Birds in Our Lives

I N 1966 the U. S. Fish and Wildlife Service published *Birds in Our Lives*, a large (8½″ × 11″) volume with more than 500 pages of text contributed by 61 different authors, who described the many ways in which birds affect our lives and how we affect them. This chapter, without pretending to emulate that classic book, merely describes some of the uses we make of birds and how they affect our lives. Some of these relationships have been pointed out in other parts of this text, especially in Chapter 1, but some of them merit further treatment here.

Birds as Food

Undoubtedly man's first interest in birds, in prehistoric times, was for food. Birds were well established when man appeared on earth, and the many flightless forms, far more than exist today, could be captured without modern weapons. Earliest history records the use of birds as food; the ancient laws of Moses even discriminated between "clean" and "unclean" fowl. Domestic fowl, now the most abundant bird in the world, were in use in India before 2000 B.C. and have since spread to all parts of the world (Coltherd, 1966).

In many instances birds aided immeasurably in man's conquest and settlement of new regions: penguins in the Antarctic are credited with saving a Dutch expedition from starvation en route to the East Indies in 1599, the Cahow or Bermuda Petrel averted a famine in Bermuda in 1614 (Murphy and Mowbray, 1951), and the dodos and perhaps elephantbirds were exterminated primarily by navigators stopping off at Mauritius and Madagascar to replenish supplies. Similarly, the early settlement of Patagonia probably would not have been possible without rheas and their eggs, and Eskimos probably could not have survived in Greenland without the Dovekie (Chapman, 1943; May, 1963). Archeological investigations of burial sites of prehistoric Indians in the Ohio River valley testify to the abundant use of birds, particularly of the wild Turkey and Passenger Pigeon, for food (Black, 1953). Bird bones identified at four Indian sites at the western end of Lake Erie include 15 bones of Trumpeter Swans, now exterminated from the midwest, 8 of turkeys, and 85 of ducks, the birds presumably having been secured with traps, snares, and bows and arrows (Mayfield, 1972).

The role of game in the early settlement of this country, briefly touched on in Chapter 1, has been well documented. Figures cited for the numbers of wild game on city markets in market hunting days are literally appalling. Chapman (1943), for instance, writes of an 1864 shipment of 20 tons of prairie chickens, of 14,850,000 Passenger Pigeons shipped from a Michigan nest site in 1861, and of 5,719,214 game birds on the New Orleans market in 1909. Figures cited in the works of Forbush (1907), Hartley (1922), Henderson (1927), and others are equally staggering. Up to about 1913, millions of songbirds, particularly American Robins, Eastern Meadowlarks, and Bobolinks in the southern states, were also included in the legal kill.

Today birds and their eggs are still used — are indeed often virtually indispensable — for food by primitive people in many lands, but in most civilized countries birds can no longer support indiscriminate hunting. Hence the harvest is now restricted to game species during special seasons, and the birds are taken primarily for sport rather than food. Nevertheless, wild game as meat on the table is no small item. The legal take of waterfowl in the United States in a relatively short open season runs to millions of birds annually; the pheasant harvest in South Dakota, though highly variable, usually reaches several millions each fall; and a few northern states, such as New York, Michigan, and Minnesota, take half a million or more grouse per year, periodically low years notwithstanding. The harvest of Mourning Doves, in states that permit dove shooting, runs to more than 20 million per year. Woodcock (nearly 200,000 annually in Michigan), rails, Bobwhite, and several species of southwestern quail also are prominent in the hunter's bag. Introduced game birds, such as Gray or Hungarian Partridges and the Chukar, are established in some states, but as in the case of the native Turkey, the legal harvest is comparatively small. Thus wild birds serve as a supplement for many people to the more extensive supply of poultry and other fowl.

The history of the use of birds for food in most European countries does not closely parallel the trend in the United States. For 2,000 years the game birds, carefully protected by seasonal regulations, were more or less reserved for the people of higher rank, whereas songbirds and waterfowl were the property of the common people. After the French Revolution, in a wave of freedom for the individual, France even abolished all game laws for a time. Italy has long been noted for its systematic methods of trapping and netting songbirds, both for food and for pets. In 1967, in large part because of the lifelong efforts of Professor Ghigi, Italy's great conservationist, Italy passed a law prohib-

iting spring shooting and netting of birds, but the law was rescinded
in 1970, in spite of the protests of other European countries and the
International Council for Bird Preservation. Japan, the source of mist
nets for United States banders, also is noted for the art of netting birds,
for scientific purposes, to protect crops, and for food.

Eggs of wild birds are no longer harvested for food in this country,
but some European countries, notably Holland and Germany, permit
collecting of the first eggs of the Lapwing (*Vanellus vanellus*), though
they allow the birds to retain their later clutches (Fig. 15.1). Icelanders
also take wild duck eggs early in the season, but carefully protect
subsequent nestings. Indians in the high Andes have been harvesting
flamingo eggs, somewhat ruthlessly, for the market for centuries, but
inaccessibility of the nesting sites (14,800 feet) helps protect the birds.

Harvesting of eggs of colonial birds on islands is more practical. On
Isla Raza in the Gulf of Mexico natives collected the eggs of gulls and
terns for many years. In the early sixties one raiding expedition with 22

F I G. 15.1. *This lapwing, an attractive Eurasian plover with spectacular aerial displays
in the breeding season, is used for food in some countries and its eggs are sold in the
market.* [Courtesy of Georg Hoffman, Bremen, Germany.]

boats took 400,000 eggs in less than a week (Walker, 1967). Then conservationists persuaded the Mexican government to declare the island a
sanctuary, and it is now a tourist attraction; apparently some of the
displaced eggers serve as guides. Natives on Tristan da Cunha in the
south Atlantic visit neighboring islands in the nesting season and harvest eggs and young of the abundant shearwaters. More famous are the
huge colonies of Slender-billed Shearwaters (*Puffinus tenuirostris*) in Bass
Strait off the coast of Australia. The colonies have been estimated to
number 150 million birds. Under strict regulations Australians are permitted to harvest the first setting of eggs and later the young, which are
sold as "Tasmanian squab" and said to be delicious and not a bit fishy.

Plumages

In the past birds have served man with a variety of economic products.
Among minor items might be mentioned the beaks of Ivory-billed
Woodpeckers, which were in considerable demand by American Indians for decorative wear. They made crowns and belts of the bills and
northern Indians bought them from southern Indians, giving
deerskins in trade (May, 1963). Bird bones were sometimes used for
tools and at one time the leg bones of kiwis were in demand for pipe
stems among New Zealanders. Kiwi pie was also considered a special
delicacy and was once served to Queen Elizabeth on one of her trips to
New Zealand.

However, the plumages of birds have had the most widespread use.
Feathers have been used from time immemorial for arrow making, for
headgear, and for ornamental wear. Such practices still prevail among
primitive people, especially in New Guinea, but the commercial use of
wild bird plumages is now illegal in this country.

Down from various species of waterfowl—ducks, geese, and
swans—has been used since early times for clothing and bedding, but
the highest quality of down comes from eider ducks. In Arctic regions
a number of legal eider down industries harvest feathers on a large
scale. The down is collected from nests, usually twice each season, while
the ducks continue incubating minus most of their original nest material. The extreme lightness and superior quality of eider down may
be inferred from the fact that it requires the feathers from 35 to 40
nests, each profusely supplied with down, to make a pound. Though
eider down has often been wastefully exploited in the past, in Iceland
and Norway the natives carefully guard the supply and have practically
domesticated their eider ducks.

In South America the skins of rheas are still used for rug making
and their feathers for dusters. Chapman (1943) reports 60 tons of rhea

feathers once found in a warehouse in Buenos Aires, baled and ready for shipment to New York. Another Argentine warehouse had hundreds of thousands of skins of the Black-necked Swan to be used solely for powder puffs. The wings and tails of Andean Condors were also in great demand. One Argentine hunter admitted shipping 16,000 to Paris, at $20 per bird, but when the price dropped to $10 he refused to participate further "in the destruction of such a noble bird for such a low price."

The history of the use of special plumes and plumages for millinery purposes and the near extermination of many of our most beautiful birds (mentioned briefly in Chapter 16) is too long a story to relate here. In special demand were the nuptial plumes (aigrettes) of egrets (see Fig. 3.6) and herons, which at one time commanded a price of $32 an ounce, but dried skins of many smaller species, such as hummingbirds and even terns, were also used. In 1885 Chapman (1943) counted 173 birds of 40 species on women's hats on Fourteenth Street, then the shopping district in New York City. Figures on the annual kill here and in South America (the source of hummingbirds for the world market) are appalling; Hartley (1922) reported 55 million birds taken in one season in the United States, and England imported 35 million annually.

In the Old World, birds of paradise from New Guinea were in great demand in the hat shops of Paris, Amsterdam, and London. This traffic in feathers began in Magellan's time (1522) and lasted until 1924, when legal restrictions made ghost towns of New Guinea ports that lived on the plume trade (Mayr, 1945). Japanese poachers nearly exterminated the Short-tailed Albatross, which once nested by the millions in the Bonin Islands; the birds were plucked of their coveted feathers and their bodies used for fertilizers. The most elaborate feather work in the world was perfected in Hawaii, where leis, head dresses, capes, and robes were made. Cloaks worn by the nobility came chiefly from Hawaiian Oos (an extinct meliphagid) and Mamos (an extinct drepanid), the former supplying the more abundant tufts of yellow feathers, but often interspersed with the brighter yellow feathers of the Mamo (Berger, 1972). A robe in the Bishop Museum in Honolulu is composed of the yellow shoulder feathers of 80,000 Hawaiian Oos (Peterson, 1960).

Perhaps the most valuable plume producer at the present time is the Ostrich, which fortunately, unlike egrets and many other ornamental types, lends itself fairly readily to domestication. Ostrich farms have been a flourishing industry in South Africa in the past, once bringing in a revenue of about $10,000,000 annually (Reese, 1942). A single pair of tame birds was worth nearly $1,000. Feathers, particularly the numerous ornamental wing and tail plumes, are plucked or clipped once or twice (usually twice) annually. In the millinery trade the feathers are usually dyed a bright shade of red or blue, though black is also a popular color.

Now Ostrich farms in Africa serve also as tourist attractions; the Ostriches are harnessed for "horseback rides" for visiting children. They are also in considerable demand for zoos, and there is some market for the large but not especially palatable eggs.

Other Economic Products

Birds are sometimes used for oil, though this is hardly practical for commercial purposes. In northern South America the young of the Oilbird (*Steatornis caripensis*), which are fed on rich fruit of palm trees until they become helpless globs of fat, are collected by natives who melt out the oil into earthen pots and use it in cooking. On St. Kilda, an island off the Scottish coast, the oil regurgitated by fulmars during nesting is collected and used by natives for medicinal purposes, externally and internally, particularly for rheumatism. Another strange use is the lighting of the oily bodies of small petrels to serve as torches. Oil rendered from penguins is sometimes used for candles in the Antarctic.

The greatest single economic product (other than food) from birds is guano. In certain dry climates the unleached excrement of large colonies of fish-eating birds builds up deposits that furnish an extremely valuable fertilizer, rich in phosphates and nitrogen. Islands in the Pacific and Indian oceans, and the coasts of Africa, Australia, and Mexico are important guano-producing regions; the guano comes from various species of penguins, boobies, gannets, cormorants, and terns. The most productive beds, however, are on the islands off the coast of Peru (Fig. 15.2), among colonies of the Guanay Cormorant (*Phalacrocorax bougainvillii*). Like most other natural resources, the guano deposits were wastefully exploited in the past; at one time revenue from them defrayed the entire cost of the Peruvian government. Fortunately, the country finally became aware of the permanent value of this resource and, with the help of American advisers, set up a commendable long-range plan of sustained harvest. The Guanay Cormorant has been called the most valuable bird in the world.

Currently, however, the situation is changing drastically. The demand for anchovies for fish meal, mainly to supply the poultry industry in the United States, has produced a competitive industry, and the cormorants are being destroyed because they feed chiefly on anchovies. Perhaps it boils down to a question of values, that of fish meal as opposed to guano. To some extent artificial fertilizers have reduced the

FIG. 15.2. *Photo of guano cutting at the Chincha Islands taken about 1860, showing thickness of old beds. The present-day guano harvest, though not comparable to early days, is an extremely valuable resource.* [Courtesy of Amer. Mus. Nat. Hist.]

demand for guano, but organic farming, especially in southern France, places a greater value on natural fertilizers. Moreover, chickens fed on fish meal do not taste as good as grain-fed birds and represent a less efficient food chain than direct human consumption of fish.

Food Habits in Relation to Man

The economic value of birds in suppression of destructive insects and other pests has long been realized. Economic ornithologists in the old Bureau of Biological Survey (now Fish and Wildlife Service), formerly in the Department of Agriculture, were employed to evaluate the usefulness of birds in relation to agriculture, and many Farmers Bulletins on the detailed food habits of the birds resulted (see Selected References for Chapter 5). At that time it was a common practice to try to place a monetary value on the services of birds and to predict how long mankind could survive without them; but now we know that the work of the birds was overrated in many of these estimates and that the chief controls of insects are among the insects themselves (parasitic and predaceous forms) and weather. Now, also, modern insecticides, apparently unsuccessfully in many cases, have taken over many of the pest-control functions formerly performed by birds.

The high value formerly placed on the services of birds has been followed by a debunking trend — an inclination to question whether or not birds play any significant role in the control of insects or rodents, but there are also indications that the pendulum may swing the other way again. W. L. McAtee, after a long lifetime of study on the food habits of birds, concluded that they play an important part in the constant suppression, rather than outright control, of insect pests, that they sometimes curb potential outbreaks, and on occasion may even avert disaster.

Perhaps the most widely publicized example of the role of birds in averting crop disaster was the "cricket plague" in Utah in 1848, when hordes of the pests threatened to destroy the Mormons' fields of grain. Seven California Gulls appeared, then more and more until they "darkened the sky," cleared the fields of crickets, and saved the settlers from probable starvation. In grateful appreciation of the services of the birds and of divine providence, the Mormons erected the famous monument to the gulls in Salt Lake City. A somewhat similar case is that of the Rocky Mountain locust plague, which swept over the Great Plains in the summers of 1873–1876. Many, but not all, fields of grain were supposedly saved by birds; in Nebraska more than 200 species were found to be feeding in part on the locusts. However, great damage was done before the locust was brought under control, and it is a moot question whether it died out from natural causes (it is now believed by some to be extinct) or from man's persistent war against it.

Many such examples, mostly less spectacular, might be cited. McAtee (1920) listed 70 instances of local extermination of insects and other pests by birds. A more recent example is the help of woodpeckers in combating an outbreak of bark beetles in Colorado spruce forests in

1951. A large buildup of woodpeckers that year, perhaps due to the local availability of beetles, effectively supplemented roadblocks and insecticides in preventing the spread of the pest. In some areas 75 percent of the beetle population was taken by birds (100 percent in some trees), and some specimens collected had 90 percent bark beetles in their stomachs (Olson, 1953).

The insect-eating propensity of birds and the thoroughness with which they cover all habitats were described in Chapter 5, but there are many ways in which other birds, particularly the granivorous and frugivorous forms, come into conflict with man. Transient sparrows, especially White-throated and White-crowned Sparrows, congregate on newly planted lawns and consume grass seed. Resident House Sparrows are a perennial threat to such enterprises. Debudding of fruit trees by grosbeaks and finches may be severe enough at times to be an economic loss. But it is chiefly the depredations of "blackbirds" in grainfields that cause economic loss — Bobolinks and Red-wings in rice fields in the southern states, other icterids and crows and Starlings on other grains. Greatest losses are in the Great Lakes states from damage to ripening kernels of corn by migrating blackbirds, and later to mature grains before harvest. The loss to farmers in Indiana is estimated at more than $2,000,000 annually; it is also high in Pennsylvania, Ohio, Michigan, and other midwestern states (Fig. 15.3R), but less in southern grainfields if the hordes of blackbirds arrive after most of the grain has been harvested (Meanley, 1971; Stone et al., 1972). Monocultural practices in agriculture tend to aggravate problems with grain-feeding birds by concentrating the birds in favorable feeding areas in fall and winter (Buchheister, 1960).

Control methods for preventing damage in grainfields run the gamut of shooting, trapping, scaring devices (Fig. 15.3L), repellents, poisons, and gametocides. The U. S. Fish and Wildlife Service, state experiment stations, and Japanese and European farmers have been working on these problems for a long time, and are continually working out new techniques, but many of the problems are still unresolved. Dykstra (1960), Boudreau (1968), and Meanley (1971) outline some of the recent methods of control.

Depredations of birds in orchards and commercial berry fields are also often severe (see page 137). The loss in cherry orchards, for instance, particularly in Michigan and Wisconsin, and more recently from Starlings in Oregon, is very high. Sometimes small stands are almost completely harvested by birds; in large orchards the proportion of damage is smaller, but the total monetary loss greater. Usually the whole fruit is consumed, but additional damage is done by pecking into the cherries and ruining them for the market. Similarly, private and commercial berry patches, grape vineyards, olive groves in California,

FIG. 15.3. LEFT *An acetylene exploder, set to explode with a loud report at timed intervals.* RIGHT *Corn damage by Red-wings in Michigan — a somewhat exaggerated example, as only a few spots in a large cornfield would normally suffer damage like this.* [Photos by D. W. Hayne; courtesy of Mich. Agr. Exp. Sta.]

and even plum, pear, peach, and apple orchards suffer varying degrees of damage. Cultivated blueberries, often in settings naturally attractive to birds, may suffer considerable damage (Fig. 15.4). In small operations in Michigan, Hayne and Cardinell (1949) thought that the loss was not great enough to warrant much expenditure of funds for control, and that much of the fruit eaten was from the ground, which might reduce insect infestation from blueberry maggots, but blueberry production has expanded greatly in recent years with a consequent increase in damage.

In the past there has been much persecution of predatory and fish-eating birds for their alleged attacks on game birds, poultry, and sport and commercial fisheries, but it is now generally conceded that these birds may have a beneficial and desirable culling effect on prey populations. Poultry can be protected by enclosures, and rearing pools at hatcheries can be screened for protection. The last diehards seem to be ranchers and sheepherders in the western states (who may be motivated more by desire for more subsidies and grazing privileges than by real enmity toward predatory birds). The salmon industries in

FIG. 15.4. TOP *A blueberry field, in a setting attractive to birds, inevitably suffers some damage from birds.* BOTTOM *Blueberries injured by birds; the berry on the right was probably fed on by insects also.* [Photos by D. W. Hayne; courtesy of Mich. Agr. Exp. Sta.]

Alaska are flourishing in spite of an abundance of eagles that prey on
spent salmon *after* spawning.

The food habits of ducks have been studied in great detail, primarily
for management purposes, but since they live primarily on aquatic vegetation and/or invertebrates, they do not usually pose an economic
problem (Fig. 15.5). Exceptions are ducks wintering on agricultural
lands in the southwest, where they damage winter grains and early
garden produce. In California one report estimated the damage at
$1,500,000, but gave the meat value of birds harvested as $4,000,000,
with another $500,000 spent by the hunters. In other, often unappre-
ciated ways, grazing waterfowl may be quite useful. They keep down
excessive vegetative growth and fertilize waters for plankton (fish food)
production (Sokolowski, 1960).

Miscellaneous Damage by Birds

There are numerous other ways in which birds come into conflict with
man. A few of these special cases are enumerated below.

The nest-robbing tendency of certain birds has always been a matter
of deep concern to bird lovers, perhaps needlessly, for it is merely one
of Nature's methods of balancing populations. Notorious in this respect
are the corvids (crows, jays, and magpies), grackles, gulls and, out at
sea, Skuas and jaegers. Most nest robbers, plunder more or less incidentally, or to supplement their regular diet, but Skuas and jaegers
are confirmed marauders, not only robbing nests but also purloining
the catches of other sea birds. The House Wren, though not a nest
robber, has the malicious habit of visiting the nests of other birds and
puncturing the eggs so that they do not hatch.

The various interspecific relationships among birds may be mutually
beneficial in some cases, or operate to the disadvantage of some in
other cases. The relationship between avian brood parasites and their
hosts takes a heavy toll of the host species (page 204). Competition
of aggressive introduced species with native species for nesting sites
also often seriously affects the reproductive efforts of the losers.
Starlings frequently dislodge flickers from nesting holes, and House
Sparrows take over the nest sites of bluebirds and other hole-nesting
species, sometimes actually killing nestling occupants and building their
own nest on top of the deceased.

Roosts of birds, especially of Starlings, blackbirds, and pigeons, often
become a public nuisance. Starlings in particular congregate in immense flocks in late summer and fall; the chatter of the noisy birds may
become almost deafening about city parks and suburbs, and accumulations of their droppings may kill valuable shrubbery or arouse public
indignation in other ways. Soil fertility may be increased by moderate

FIG. 15.5. *A formation of Lesser Scaup on the Indian River in Florida. The Lesser Scaup is quite omnivorous and becomes a problem in agricultural lands less commonly than the more vegetarian ducks.* [Photo by Allan D. Cruickshank.]

accumulations of droppings, but at dense roosts over a period of time undesirable ammonia compounds tend to build up.

Some of the dispersal methods mentioned for grain-eating birds have been used with varying degrees of success against Starlings, but aluminum owls have not lived up to the reputation accorded them by newspapers. At the State Educational Building in Albany, New York, where the birds have long been a public nuisance, the alcoves were electrically charged in a way that prevented Starlings from entering them, at least temporarily, but the cost of the project was prohibitive for general application. Some communities have been successful in dispersing Starlings by playing recordings of their alarm calls over a loudspeaker in the streets, but in most cases success has been only temporary. Use of this method in blueberry fields in Michigan was ineffective after a few trials.

Woodpeckers sometimes do damage to trees or even buildings in

Fɪɢ. 15.6 *The Pileated Woodpecker* (ᴏᴘᴘᴏsɪᴛᴇ) *can drill deep, rectangular holes into decaying trees in its energetic quest for carpenter ants and wood-boring larvae.* [Courtesy Cleveland Mus. Nat. Hist.] ᴀʙᴏᴠᴇ *A red pine in northern Michigan on which woodpeckers have been at work.* [Photo by Lawrence A. Ryel; courtesy of *Jack-Pine Warbler.*]

their quests for wood borers or for drumming sites. The loud drumming of flickers on wood or metal surfaces about dwellings can be an annoyance, especially in the early morning hours. Pileated Woodpeckers have been known to drill holes completely through the logs of Canadian cabins, and often tunnel into trees (Fig. 15.6), usually already in a state of decay, in their persistent quest for deeply imbedded grubs or carpenter ants. Yellow-bellied Sapsuckers sometimes attack buildings, for no apparent reason, gouging out holes in shingles or siding. Their worst offense, however, is in their regular habit of drilling little wellholes into the bark of trees for sap. Such practices disfigure and devitalize fruit and ornamental trees and greatly reduce the value of forest trees for lumber. The loss to lumber industries in this way is extensive in northern forests. McAtee (1926) reported sapsucker injury in 174 species of trees, in 90 of which it was serious enough to spoil the appearance or workability of the wood. Damage to fence posts and utility poles by Golden-fronted, Ladder-backed, and Pileated

Woodpeckers in the south and southwest is quite extensive (Dennis, 1964, 1967a). Often they drill completely through a pole or cross bar, necessitating replacements. Creosoting the poles does not protect them; treated poles may be attacked more frequently than untreated poles (Rumsey, 1970), but nest failures result from the toxic effects of the creosote — eggs fail to hatch and nestlings die.

An extraordinary case of birds making a nuisance of themselves is the entanglement of divers, especially Oldsquaws, in fishermens' nets. Oldsquaws dive to great depths (page 106) for deep-water crustaceans and molluscs, which constitute their principal food. In several Great Lakes fishing centers and in the Finger Lakes region of New York, they get caught in the nets, tearing the mesh and restricting the catch of fish; in some cases more than 1,000 birds have been brought up in a single haul of the nets. Unfortunately the dead birds, virtually tons of them, are of little use except as fertilizer. Bartonek (1965) in a study of mortality to diving birds from commercial fishing on Lake Winnipegosis, Manitoba, found many ducks, especially Redheads, and loons and grebes included in the catch.

An increasingly pressing modern problem is the frequent collision of birds with aircraft. Formerly serious collisions were largely with flocks of migrating birds, such as geese, ducks, and cranes, at high altitudes, rarely resulting in loss of life, but more recently there have been disastrous collisions with smaller birds congregating on runways at airports and interfering with the take-off and landing of planes. In 1960 more than 60 airline passengers were killed as a result of a bird strike (Graber, 1972). Recommendations made at several conferences for averting this danger at inland airports have been tried with varying degrees of success: (1) eliminating city garbage dumps, which attract large numbers of gulls around runways, (2) making airports unattractive to birds by clearing away vegetation, (3) use of repelling colors on runways, (4) use of warning signals, and (5) use of trained falcons as scaring devices. The seriousness of the situation was indicated by the calling of an international conference (Cahalane, 1969) in 1969 at Kingston, Ontario, in which 130 people from 19 different countries participated. Forty-eight papers were presented. Use of radar to detect flights, altitudes, and concentrations of flying birds has been helpful but not foolproof; a single bird can down a plane (Graber, 1972).

Similar problems have developed at air bases on Pacific Islands. On Midway Atoll large numbers of "gooney birds" (albatrosses) used to congregate in and over the runways, causing frequent collisions and damage to the planes. The Navy's original plan to destroy the albatrosses, which constituted 35 percent of the world population of Laysan Albatrosses and 16 percent of the Black-footed species, met with great public protest. Subsequent studies by biologists of the Fish and Wildlife

Service indicated that levelling the dunes paralleling the runways, and thus eliminating favorable soaring conditions, might be a more practical solution than killing the birds. Collisions decreased 80 percent after the dunes were levelled, whereas killing 30,000 birds and breaking up their nests had only aggravated the problem by increasing the number of "unemployed" (nonnesting) birds over the airstrips (Rice, 1959). Later studies (Fisher, 1966b), however, recommended restoring the dunes and vegetation along the runways because the birds crossing them flew at higher levels (above the level of grounded planes) than when the land was flat. Paving 50 percent of Sand Island reduced the number of birds around the airstrips but caused a loss of ancestral nesting grounds. Moving eggs and birds met with little or no success. In the nine years of grappling with the problem up to 1966 little progress had been made, but it was felt the whole matter would eventually be settled when the island was no longer needed for a refueling station. Changing of flight schedules, such as landing and taking off at night, would avoid most bird strikes.

The Hand of Man

Since early times man has tried, often unsuccessfully, to "improve" on nature by "managing" wildlife. Among these management methods are such time-honored practices as predator control, including "anti-vermin" campaigns by trapping, shooting, and poisoning; payment of bounties for "pest" animals destroyed; and the introduction of exotic species, either deliberately, for pest control or restocking purposes, or accidentally. Sometimes such enterprises are eminently successful; more often they seem to backfire, as indicated in the following sections.

PREDATOR CONTROL

It is probably already clear that Nature's system of predator-prey regulation is fundamentally sound, and that short-sighted attempts by man to improve things commonly result in upsets. Usually the elimination of predators, whether mammalian, avian, or insect, is followed by an increase in destructive herbivorous forms, or results in an upsurge of some hitherto unsuspected pest, which may necessitate still further control efforts.

An example of the complex interrelationship between predator and prey was once described by Fisher (1908). A New York marsh with a well-balanced population of ducks, skunks, and snapping turtles was heavily trapped when the market price of fur went up, so that snapping turtles, formerly held in check by the skunks eating their eggs, increased rapidly, and ducks, as a result of excessive turtles, all but disappeared. Then the market price of fur declined so that trapping was no longer profitable, skunks increased, snapping turtles decreased, and ducks were restored to normal numbers.

A more recent example of the apparent consequences of interference with basic predator-prey relationships has been reported in the outbreak of mice in the fertile Klamath basin of northern California and southern Oregon. Intensive predator control, particularly of coyotes, was followed by a buildup of mice that reached a peak in the spring of 1958. To combat the plague, poisoned bait (1080 and zinc phosphide) was widely distributed in an area used by 500,000 waterfowl in the spring. More than 3,000 grazing geese were said to have been poisoned, so driving parties attempted to keep the geese off the treated fields. Here it seems conceivable that the whole chain of costly events—cost of the original predator control, economic loss to crops from the mouse plague, another poisoning campaign to combat the mice, loss of valuable waterfowl resources, and man-hours involved in flushing geese from the fields—might have been averted by a policy of noninterference with the original predator-prey relationship.

Another example of mismanagement of predators is the dilemma in Israel, where a barrage of poisons, ordered by the Israeli Ministry of Agriculture to protect increasingly intensive cultivation, has been released in the environment (Mendelssohn, 1971). At first thallium sulfate was widely used to control rodents in grainfields. Secondary poisoning, however, virtually eliminated both resident and wintering birds of prey. Then a ban on thallium sulfate, because so many humans were poisoned, was followed by an uncontrollable outbreak of rodents, sparrows, bulbuls, and blackbirds because there were no predators. Another error was the attempt to control woodpeckers with poisoned pecan nuts, because the birds were alleged to damage plastic irrigation pipes. The result was the freeing of a destructive fruit borer (*Zeuzera pyrina*) previously controlled by the woodpeckers.

Government poisoning operations, primarily directed against mammalian predators and herbivores but usually involving widespread secondary poisoning in birds, have long been a controversial issue. To try to resolve these conflicts a committee of wildlife experts, the Leopold committee, was appointed in the early 1960's. Their recommendations included a sharp curtailment in such operations, many of which were deemed unnecessary, and the elimination of most poisons. However, such deeply entrenched programs die hard, and bitter opposition to the recommendations has developed among woolgrowers, ranchers, and politicians in the western states.

There are many instances in modern management, however, where man may, with good reasons, want to favor certain species at the expense of others. It is obvious that on game farms and about fish hatcheries, or sometimes in other special projects, species that interfere with the outlined program of the area need to be controlled. But where control measures are necessary they should be in competent hands, and executed only after careful study.

BOUNTIES

Payment of bounties to encourage people to participate in control campaigns has repeatedly proved to be ineffective. Bounties are costly, often are grossly mishandled, and usually do not produce the sort of control desired. The Goshawk bounty in some eastern states is an example of misapplication. One of the most difficult of the hawks to identify, at least of the immatures in fall and winter, thousands of other hawks have been shot in states where the Goshawk is uncommon, or almost unknown. Young (1952) reported 9,000 other hawks shot in Virginia, where there were only 10 authentic Goshawk records for the state.

A glaring example of an inefficiently administered bounty was that on the House Sparrow in Michigan (Barrows, 1912). In 1898 about $50,000 was paid out of county funds in one- and two-cent bounties on House Sparrows. But in the dispensation of funds, graft and fraud arose to an alarming degree; a great deal of property damage and even fatal accidents resulted from careless use of firearms; many native songbirds were needlessly destroyed; and, after the smoke of the barrage had subsided, the House Sparrows seemed to have suffered little if any reduction in numbers.

Fortunately, bounties on birds in this country are largely a thing of the past, but they still persist in many states on some mammals, despite the opposition of most game biologists.

INTRODUCED SPECIES

Introduction of exotics into a new environment often poses severe management or control problems, usually because an introduced species is relieved of contact with its natural enemies and competitors. Perhaps the worst phase of this is the accidental introduction of foreign insect pests, which necessitates severe importation restrictions, annoying quarantines, and subsequent control measures if the pest secures a foothold. Among vertebrates one need only cite problems created by rabbits in Australia, goats on various islands, and rats everywhere. Introduced House Sparrows and Starlings are still a problem in this country, after a century of control measures.

Introduction of exotics on islands often creates havoc among the native birds; in the Hawaiian islands exotics have almost completely replaced the native land birds. In this country escaped cage birds can become a problem. In most cases such escapees die out harmlessly, but sometimes they become established. In Florida some 12 species have

FIG. 15.7. *Monk Parakeet* (Myiopsitta monachus). *Possibly the latest cage bird "escapee," this medium-sized parrot poses a threat to endemic species in areas where it has become established. It now occurs from southern Florida to New York and westward to Ohio and Michigan.* [Photo by Bruce Frumker; courtesy of Cleveland Mus. Nat. Hist.]

"gone wild," including the Monk Parakeet (*Myiopsitta monachus*), which now threatens to "take over" other states (Fig. 15.7). The figures on the importation of cage birds into this country, some of them illegally, are staggering. In 1970 more than 937,000 birds—745 species belonging to 108 families from 23 different countries—were imported (Banks and Clapp, 1973). Three of the five leading families were the psittacids, sturnids, and ploceids, all of which have contributed ominously to our pest species in the past.

Another aspect of introductions is the intentional stocking of game habitats with new species. This has been eminently successful with the Ring-necked Pheasant, but less so, or only on a more local scale, with Gray (Hungarian) Partridges, Chukars, Migratory Quail (*Coturnix*), Capercaillie (almost a total failure), and other species. Now a special committee carefully screens potential foreign gamebird introductions and recommends trial liberations of only a small part of those actually screened (Bump, 1963).

Diseases

Birds as carriers of diseases of man or other animals are a problem at times, but ordinarily there is not much cross-infection between birds and man, since infective organisms are rarely common to both (Worth, 1949; Herman, 1966). Pigeons are considered a menace to public health in some communities, chiefly because of *ornithosis*, a transmissible form of virus pneumonia, or air-borne respiratory infection. For this reason control campaigns have been carried out in some cities. *Psittacosis*, or parrot fever, which is similar, or perhaps identical, to ornithosis (Worth et al., 1957), is a serious and sometimes fatal disease to humans, particularly among parakeet handlers. Epidemics in 1929 and 1930 caused restrictions on importations and a decline in the disease. Then following World War II came a resurgence of the disease owing to popularity of parakeets (Boyd, 1958). Tests on 33 bird banders, who handle wild birds in large numbers, disclosed only one case of ornithosis, but railway express agents who handle parakeets were found to be 71 percent infected (Worth et al., 1957).

Several types of *encephalitis*—eastern, western, St. Louis, and equine—occur in man and horses. There have been occasional epidemics and many deaths, but the role of birds in transmitting the disease, in spite of recent studies sponsored by state and federal public health agencies, is not well understood. Encephalitis is a virus disease transmitted by blood-sucking arthropods (hence arboviruses), such as ticks and mosquitoes. Birds, especially poultry and pheasants, and perhaps wild birds at large roosts, can be active carriers for short periods if infected by mosquitoes, but usually they build up antibodies quickly and become immune. Large-scale banding and netting projects to capture and take blood samples of wild birds, both in this country and the tropics, have revealed few cases of active carriers; hence the role of birds in transmitting the virus to humans via mosquitoes in probably a minor one. Even the exact mosquito vector is not well known (Anderson, 1961; Herman, 1962, 1966).

Histoplasmosis, a pulmonary disease caused by inhaling fungus spores from bird droppings, is less serious because it is rarely fatal, but it is of importance here because of the involvement of birds in spreading the disease. Persons working in poultry litter, in old buildings housing bats or starlings, and under roosts of birds can contract the disease by inhaling dust from the disturbed debris. The obvious remedy is precaution, such as wearing a mask when working in poultry yards or under roosts of birds. Most persons exposed to the disease (95 percent) soon acquire immunity (Jackson, 1973).

Salmonella, a form of bacterial food poisoning, can be obtained from improperly handled poultry and egg products (Herman, 1966) and could, theoretically, be obtained from wild game birds, but birds are rarely the cause of *Salmonella* poisoning.

On the other hand, birds sometimes benefit man by consuming disease-carrying organisms, such as malaria and yellow fever mosquitoes. In 1880 Pasteur, experimenting with chickens, stumbled onto a method of making vaccines that has saved millions of lives (Herman, 1966). Avian malaria also helped solve problems in human malaria, and chicks are used in making vaccines (Boyd, 1958). The ancient Egyptians were aware that the Sacred Ibis helped keep down the incidence of *schistosomiasis* but did not know it was by feeding on the secondary hosts (snails) of the schistosome parasite. Unfortunately, subsequent persecution and near extermination in some areas of a bird once held sacred, as well as impounding water for irrigation and building dams, which create favorable conditions for the parasite, have caused a resurgence of the disease.

Thus birds enter in many ways into relation with man. The relationships are primarily beneficial and enjoyable, but, as indicated above, there are some exceptions.

SELECTED REFERENCES

BENTON, A. H., and L. E. DICKINSON. 1966. Wires, poles and birds. Pages 390–395 in A. Stefferud (ed.), *Birds in Our Lives.* Washington, D. C.: Fish and Wildl. Ser.

CHAPMAN, F. M. 1943. Birds and Man. Guide Leaflet Series. Amer. Mus. Nat. Hist, 115:1–52.

DRURY, W. H. JR. 1966. Birds at airports. Pages 384–389 in A. Stefferud (ed.), *Birds in Our Lives.* Washington, D.C.: Fish and Wildl. Ser.

FORBUSH, E. H. 1907. *Useful Birds and Their Protection.* Boston: Mass. State Board of Agr.

HARTLEY, G. I. 1922. *The Importance of Bird Life.* New York: Macmillan.

HENDERSON, J. 1927. *The Practical Value of Birds.* New York: Macmillan.

HERMAN, C. M. 1966. Birds and our health. Pages 284–291 in A. Stefferud (ed.), *Birds in Our Lives.* Washington, D.C.: Fish and Wildl. Ser.

REESE, A. M. 1942. *Outlines of Economic Zoology.* Philadelphia: Blakiston.

STEFFERUD, A. 1966. *Birds in Our Lives.* Washington, D.C.: Fish and Wildl. Ser.

16

Conservation:
To Save or Not to Save

As INTIMATED at frequent intervals throughout this text, the future of birds is seriously threatened by our modern way of life—the demands for more living space for a rapidly expanding population and the highly developed but destructive technology employed to meet these needs. The following pages review some of our inglorious past history in exterminating bird life, the status of some endangered species, current efforts to try to save what is left, and our hopes and fears for the future.

Extinct Birds

The passing of a species, however minor, is regrettable. Often it means the end of a phylogenetic line that has been millions of years in the making. The California Condor and Whooping Crane, for instance, are highly specialized end products of long lines of development that go far back into the Tertiary; within a few decades we may witness the extinction of two magnificent types that have taken 50 million years to evolve.

Commonly man assumes some responsibility, or one faction blames another, for exterminating a species, but in many cases it is a natural process with mankind perhaps putting the finishing touches to an inevitable event. Thousands of ill-adapted or outmoded species of birds died out before man's appearance on earth. Among the New World vultures, for example, there are seven North American fossil forms, compared to three living species, one of which is nearly extinct. Even today we see relic species, such as the kiwis (see Fig. 7.2) and rail-like *Notornis* (Fig. 16.1) in New Zealand, making a probably futile bid for survival, in spite of the protection now accorded them.

The hazards of island life are strikingly borne out by the fact that more than 90 percent of the birds that have become extinct in the last 200 years have been island species, many of them flightless and unable

F IG. 16.1. Notornis, *a quaint, rail-like, flightless bird of New Zealand, rediscovered after 50 years of supposed extinction. Originally discovered in 1849 by a Maori native who captured a specimen with a dog, Notornis was rediscovered in 1948; at first only footprints were seen; then nesting birds were found in a remote corner of South Island.* [Courtesy of Amer. Mus. Nat. Hist.]

to spread elsewhere when their island sanctuary was invaded by man. Rodriquez, the last stand of the dodo-like Solitaire, has lost 10 of its 12 endemic land birds since the island was settled by man, and another is very rare; only one of the original 12 survives in good numbers on uncleared portions of the island (Gill, 1967). Hawaii has lost 23 of its original 68 endemic species, and 29 more are on the endangered species list; the remaining 16 are on a "blue list" (Arbib, 1972a).

Greenway (1958), in his comprehensive *Extinct and Vanishing Birds of the World,* lists 87 species and subspecies of birds that are definitely known to be extinct, 19 others that are probably extinct, 18 that are known only from osseus remains (not including prehistoric forms), 2 for which the collecting locality is indefinite, and 27 hypothetical birds (known only from pictures, descriptions, or unconfirmed reports)—a pathetic total of 153 forms of bird life that have probably disappeared from the earth, largely because of man's activities, within the last 300 years. Fisher (1964) gives the known total since 1680 as 78 species and 49 wellmarked subspecies, a total of 127 forms. More disturbing than the total already gone, however, is the alarmingly increasing rate of extinction. More birds have been lost in the past 200 years than in all of the preceding 1800 years. Mathematically, with the current rate of extinction, it is only a matter of time before all bird life will be gone, although it is conceivable that some "nuisance species" may outlive man.

Unfortunately, North America, perhaps because of its rapid growth and expansion in the past two centuries, has the worst record of any comparable land mass for exterminating its animal life, although some of the developing countries may soon outdo us by emulating our ecological blunders. The history of the extinction of several North American birds has been well documented; other less known birds, chiefly on peripheral islands (Guadalupe alone has lost four or five endemic forms), as well as a few questionable birds described by early naturalists but not since rediscovered, are also gone. But every student should be familiar with some of the facts associated with the passing of the following forms.

GREAT AUK (*Pinguinis impennis*)

This strange bird (Fig. 16.2), the largest of the alcids and the only flightless one in recent times, was apparently the first American species to become extinct. Though Funk Island, Newfoundland, was its main North American breeding site, the birds visited the North Atlantic Coast in large numbers, and were killed by sailors, fishermen, and sealers who used the birds and their eggs for food or fish bait, their feathers for bedding, and their carcasses for oil. The birds showed little fear of man and could not outrun him, so they were easily clubbed to death; in fact, they were often herded into corrals or even driven on board ships where they were easily dispatched. The main stock was gone soon after 1800, but some persisted up to 1844 when two were captured on Eldey Island off the coast of Iceland on June 3. All that

F I G. 16.2. *The Great Auk was probably the first of the North American birds to be exterminated by man. The little-known Pallas' Cormorant* (Phalacrocorax perspicillatus) *in the north Pacific disappeared at about the same time.* [Sketch by Vito Cangemi from Fuertes' drawing on cover of the *Auk*.]

remain today are about 80 preserved specimens, 20 or more skeletons, and about 75 eggs. Aside from unrestricted killing, known factors that hastened the extinction of this species were the single egg, which hindered recovery from heavy losses, and the bird's comparative helplessness because of lack of flight.

L A B R A D O R   D U C K (*Camptorhynchus labradorium*)

Little is known of this eiderlike duck (Fig. 16.3), which formerly visited the New England and North Atlantic states in winter from a probable summer home in Labrador or further north. Perhaps it was never abundant. Perhaps it was persistently hunted by natives for its flesh, down, or eggs, but there seem to be no records of heavy kills by white men. Some maintain that it was a stupid bird, lacking the wariness of other wild ducks. The last known member of the species was either killed on Long Island in 1875 (Greenway), or in New York

FIG. 16.3. *Labrador Duck habitat group in American Museum.* [Courtesy of Amer. Mus. Nat. Hist.]

on December 12, 1878 (A.O.U.). About 50 specimens have been saved in various museums.

PASSENGER PIGEON (*Ectopistes migratorius*)

The most regrettable, as well as the most dramatic, extermination of any North American bird was that of the Passenger Pigeon, whose history has been fully documented by Schorger (1955). So abundant was it in the 19th century that counts of single flocks were set at more than a billion birds, exceeding any single species known today and outnumbering tenfold the whole continental waterfowl population. But the pigeons dwindled rapidly with relentless persecution and within a few decades had completely disappeared. The last two specimens were taken in Michigan and Wisconsin in 1898 and 1899. The species survived in captivity somewhat longer; a bird in the Cincinnati Zoological Gardens died on September 1, 1914, after having lived in captivity for 29 years.

The full explanation of the extinction of the Passenger Pigeon may never be known, but it is mainly a pathetic example of incredible slaughter. The birds nested in dense colonies, millions at a single site, with trees so heavily laden with birds and their nests that the branches often broke under the weight. Such habits made the birds exceedingly vulnerable; they were shot, clubbed to death, captured in huge nets, and burned out of trees; wagonloads of birds were hauled away and shipped in barrels to the New York and Chicago markets. One Indian hunter at Northport, Michigan, shooting from a ridge as the birds flew by in layers, would get 1,000–1,200 birds before breakfast, then quit while other hunters continued shooting throughout the day (Wilson, 1969). In 1861, 14,850,000 birds were shipped from the famous Petosky (Michigan) nest site, a "pigeonry" said to be 44 miles long and 3–10 miles wide. There was an even larger nesting colony in Wisconsin in 1871, estimated to be 100 miles in length. The great nesting colonies rapidly shrank in size and number, and the persecuted birds were pursued by professional market hunters from one nesting site to another, even up into the wilds of Canada; thus in their later years the pigeons were largely unsuccessful at nesting. Legislation was enacted in some states in an attempt to save the last birds, but it was too little and too late. Michigan passed a law in 1897 to prohibit further killing, and in 1905 transferred the bird from the game to the nongame list—to protect a bird that was probably already extinct.

Various reasons, other than obvious overkilling, are given to account for the bird's rapid decline: the lack of suitable nesting sites due to deforestation, inadequate mast supply (Wilson's flock of 2 billion birds would have required 17 million bushels of food daily), the breakup of organized flocks that eliminated the social stimulus perhaps necessary for successful breeding, the usually single egg per clutch, storms during migration, and disease. But whatever the reasons we can only reminisce with regret. The bird is commemorated only in books and museums, and by a monument (Fig. 16.4) at Wyalusing State Park in Wisconsin, unveiled on May 11, 1947, with touching tributes to a vanquished bird by Aldo Leopold, A. W. Schorger, Walter Scott, and others. (See *Silent Wings: A Memorial to the Passenger Pigeon* by Scott et al., 1947.)

HEATH HEN (*Tympanuchus cupido cupido*)

This race of the Greater Prairie Chicken, in contrast to the preceding species, died out with man aware that it was going and making determined but futile attempts to save it. Probably the Heath Hen (Fig. 16.5) was never widely distributed and perhaps not abundant, but it was known to occur from New Hampshire to Chesapeake Bay and was an important game bird in the 19th century. But after 1835 it was restricted to its last stronghold on Martha's Vineyard, where a nucleus of some 200 birds increased to about 2,000 by 1916 but dwindled rapidly thereafter. A disastrous fire in 1916, disease in 1920, a heavy influx of Goshawks one winter, and an unbalanced sex ratio (2 females and 11 males left in 1927) gradually reduced the population to a single

FIG. 16.4. *Passenger Pigeon Memorial Monument at Wyalusing State Park, Wisconsin and close-up of plaque.* [Photos by F. R. Poe.]

F IG. 16.5. *Heath Hen at nest. Note close resemblance to the Greater Prairie Chicken* (*Fig. 9.7*). [Courtesy of Amer. Mus. Nat. Hist.]

much-pampered individual, which was last seen on March 11, 1932. Attempts to cross the lone survivor with its closest relative, the Greater Prairie Chicken, were unsuccessful (Gross, 1928; A.O.U. Checklist, 1957).

C AROLINA (L OUISIANA) P ARAKEET
(*Conuropsis carolinensis*)

This parakeet, consisting of a southeastern (*carolinensis*) and a midwestern (*ludovicianus*) form, was widespread during the 19th century, ranging north to Wisconsin, Ohio, and central New York and south to Texas and to Florida. But several factors combined to eliminate it from its entire range. The birds were very destructive in fruit orchards and grainfields; hence they were destroyed in large numbers by farmers. They were also shot as game, collected for their plumage, and trapped for cage birds. The fact that they were tame and unsuspicious made them easy prey. McKinley (1964, 1965) documents known records for *ludovicianus,* citing the paucity of satisfactory breeding records even in states where it was common. Apparently the last specimen was taken in Kansas in 1904. In the southeast the last specimen of *carolinensis* was taken in Florida in 1901, but a flock was sighted there in 1920 and there were reports of its existence in the southeast as late as 1938 (Greenway, 1958).

Endangered Species

Currently, endangered species are receiving much attention, and some financial support, at state, national, and international levels. At the meeting of the International Council for Bird Preservation held in England in 1966, Colonel Vincent distributed a world list of 318 rare and endangered species and subspecies (see Appendix III), but not all of these would be in the "endangered" category. The Federal Register of the U.S. Department of Interior (Jan. 4, 1974) lists 129 endangered species and subspecies of birds for the world. Vincent painted a gloomy picture for the future, saying that additional birds had to be put on his list each year, whereas a species is seldom removed because of recovery. He attributed this dilemma to the appalling rate of habitat destruction to provide more food and living space for an ever-increasing human population. A *Red Data Book,* revised from time to time to keep it up to date, has been prepared by the International Union for the Conservation of Nature.

At the national level the Department of Interior, in January, 1974, has published a list of 53 endangered "native" birds: 24 on the continent, 3 in Puerto Rico, and 26 in Hawaii. Arbib (1972b) has published a "blue list" (the "amber list" of the U. S. Fish and Wildlife Service) of 42 additional continental forms not currently considered endangered though suffering population decline or range shrinkage. Wallace (1969) published a similar analysis for Michigan birds.

Obviously space forbids a discussion of the hundreds of endangered birds throughout the world, but the status of some of the better known American birds is characterized below. (See Appendix III for more complete lists.)

ESKIMO CURLEW (*Numenius borealis*)

The status of this bird (Fig. 16.6) is uncertain. Since it breeds on the tundra in the far north and winters in southern South America, it is known in this country only as a bird of passage and can be easily overlooked or confused with the somewhat similar Hudsonian and Eurasian Curlews. Apparently the last continental specimens were taken in Maine in 1929 and in Labrador in 1932 (A.O.U.), but,

F I G. 16.6. *Eskimo Curlew. Formerly thought to be extinct, this species occasionally appears along the Atlantic and Gulf coasts.* [Photograph of mounted specimen by Walter P. Nickell.]

dramatically, a specimen was shot (illegally) on the island of Barbados in the West Indies in the fall of 1963. Fortunately, it was retrieved from the hunter and its identification verified. Additional sight records, from the wintering grounds in Argentina as well as along migration routes, have persisted over the years. Reports of birds seen in Texas in 1959 and 1960 are summarized by Emanuel (1961).

In some respects the passing of the curlew parallels that of the Passenger Pigeon. They traveled in large flocks and were not very wary; hence market hunters easily collected the birds congregating along the Atlantic flyway in the fall and in the Mississippi valley during the spring flight. On the western plains, hunters followed the flocks with wagons to haul away the kill of the day. The birds were abundant up to about 1875, after which a rapid decline took place.

IVORY-BILLED WOODPECKER (*Campephilus principalis*)

The status of the Ivory-billed Woodpecker (Fig: 16.7) in the United States is also uncertain and very critical. When the small but well-studied population in the Singer Wilderness tract in Louisiana disappeared with the cutting of the timber in the mid-1940's, the species was thought to be gone, but reports of its occurrence in South Carolina (where one bird responded to a tape recording), Florida, and Louisiana, and particularly in the Big Thicket area of eastern Texas, have persisted over the years. Although a certain amount of mystery (uncertainty and hearsay) appears to cloud these records, there seems little doubt about the reliability of some of them. The secrecy of the

FIG. 16.7. *Ivory-billed Woodpecker. Photo of the lithograph after a painting by J. J. Audubon. There is still some uncertainty about the existence of this species in the United States, but a closely related form still survives in Cuba. Two other similar species inhabit Mexico and South America, respectively.* [Photo by Vito Cangemi.]

birds and their great mobility in seeking new feeding areas hinders followup observations.

Studies of the Lousiana birds by Tanner (1942) disclosed two serious limiting factors: (1) the birds lived on a restricted diet of wood-boring larvae that infest dead and dying trees of a certain age and thus required large tracts of mature forest for living space; and (2) though the birds mated and had nests, they were unsuccessful in rearing young. Apparently these huge woodpeckers were evolved to fit a particular ecological niche that is now practically nonexistent; if gone, they provide an example, perhaps the only one in North America, of an avian species lost through deforestation.

WHOOPING CRANE (*Grus americana*)

The most publicized of North America's rare birds is the Whooping Crane (Fig. 16.8), one of our largest and most magnificent birds. Though a monographic study by R. P. Allen (1952) indicates that the species was never as abundant as popularly supposed, it used to be a fairly common migrant over the western plains. Now careful counts of the survivors on their wintering grounds and much publicity keep the public well informed on the year-to-year status of this spectacular bird. In 1972 there were 51 known wild birds, a loss of 6 from the 1971 population; the 1973 figures showed a slight further decline, but the number represents an encouraging increase over the 21 wild birds in 1953. In recent years the Fish and Wildlife Service has been taking one egg from the two-egg nests of wild birds, which rarely raise both young, and rearing the young at Patuxent Research Center.

Fortunately for management measures, the cranes winter on the Aransas Wildlife Refuge in Texas, where they are carefully protected. Their northern nesting grounds, in Wood Buffalo Park in northwestern Canada, also accord them some security. The lumbering industry's demand for pulpwood, and minings for strategic minerals near the park, pose some hazards for the birds, but the Canadian government has been quite cooperative in trying to assure the cranes' privacy. On the wintering grounds there is urgent need for more space, as the aggressive territorial adults force the young off the refuge. More space is not available, however, because an air force bombing range and other land uses prevent expansion of the refuge. In view of these other demands for the land they occupy, R. P. Allen (1960), a leading authority on the cranes, asks the pertinent question: "Do we want to save the Whooping Crane?"

CALIFORNIA CONDOR (*Gymnogyps californianus*)

Great interest attends the current status of this condor, a carrion-feeding vulture and one of the largest and most majestic of flying birds,

F IG. 16.8. *Two Whooping Cranes (one injured), photographed on their wintering grounds in the Arkansas Wildlife Refuge in Texas in 1949.* [Photo by L. H. Walkinshaw.]

now largely confined to a refuge in southern California. Probably never abundant, though once much more widespread, it has been victimized in the past by gunners who prized specimens because of their large size, egg collectors who coveted the rare and valuable condor egg, and ranchers who may have unwittingly poisoned many of the birds with bait intended for coyotes. The actual effect of such poison bait on the birds is not well known, however, and poisoned ground squirrels may be an important source of food for the condors. Also many ranches in California have been converted into fruit farms, which do not provide carrion for the condors.

Discouraging features in the management of the condors are that they are easily disturbed by sight-seeing tourists, photographers, and bird lovers and that their reproductive rate is very low. The clutch consists of a single egg and the adults usually nest *every other year* because their slowly maturing young are dependent on them for so long. Only eight young are known to have hatched successfully from 1966 to 1971, an average of less than two a year (Wilbur et al., 1972). The total condor population is estimated to be about 40 birds, but censusing the birds is very difficult and considerable disagreement exists between the "optimists" and the "pessimists."

On the more encouraging side, however, is the fact that the condors have not declined much in recent years, that mortality is low (one known loss since 1966). and that a large tract of land has been set aside as sanctuary for the birds, albeit the sanctuary is continually threatened by demands for "developments" and exploitation. A committee, with representation from five cooperating agencies, has the condors under constant study. Opinion differs sharply as to whether or not attempts should be made to perpetuate a few of the survivors in the San Diego Zoo. A detailed monograph on the species by Koford (1953) presents a fairly complete picture of the condor situation up to 1952, and, apparently, there have not been any really significant population changes in the subsequent 20 years.

EVERGLADE KITE (*Rostrhamus sociabilis plumbeus*)

This hawk (Fig. 16.9) is now confined to the Everglade region of Florida (other races occur in the West Indies and Central and South America), and in 1972 was believed to number less than 100 individuals. Though it used to be shot by alleged sportsmen, particularly visiting tourists, probably the chief reason for its precarious status is its peculiarly limited diet of snails (*Pomacea paludosa*). Drainage projects for real estate and other developments, which eliminate the snail populations, threaten the survival of the kite. Evidence that diet may not be the sole factor, however, is the fact that the range of the snail is more extensive than that of the kite.

EAGLES AND OSPREY

Eagles throughout most of the world are in a precarious position. The magnificent Monkey-eating Eagle (*Pithecophaga jefferyi*) in the Philippines now survives in small numbers on only a few of the islands; deforestation and the demand for zoos and trophies have caused its

FIG. 16.9. *An Everglade Kite at its nest in the Florida Everglades.* [Photo by Alan D. Cruickshank from Natl. Audubon Soc.]

downfall (Gonzales, 1971). The Spanish Imperial Eagle (*Aquila heliaca*) is nearly gone from its former strongholds in the western Mediterranean; simple persecution seems to be the reason. The Harpy Eagle of South America, largest and most powerful of the eagles, is on the endangered species list. The Golden Eagle is considered endangered throughout its European range. Wanton killing in continental Europe and the use of dieldrin in sheep dips in Scotland have been its undoing.

But it is our national emblem, the Bald Eagle, that merits our greatest concern. Although the northern race (*alascanus*) is still abundant in Alaska, the more southern race *(leucocephalus)* is in serious trouble—the existence of the few thousand, often nonbreeding, scattered survivors in the contiguous states does not augur well for the future of our national emblem. Broley (1958) called attention to the sharp decline in his study areas in Florida, from approximately 150 eaglets produced annually in 100 nests up to 1947 to only 8 eaglets in 1957. He attributed the loss to reproductive failure caused by persistent pesticides in fish eaten by the eagles, but his claims (typically) were discredited. Now the phenomenon of reproductive failure from persistent pesticides is well known (see pages 458–459). Illegal killing and loss of nest trees also contribute to the Bald Eagle's problems.

The Golden Eagle in the Rockies fares better because it lives primarily on less contaminated mammalian prey, but it is subject to much illegal persecution from sheepherders and ranchers, and from airplane pilots who seem to enjoy (and often get paid for) shooting down the eagles from planes. The deliberate slaughter of hundreds of eagles in Wyoming and Colorado in the 1960's and the release of the offenders with only minimal fines is a disgrace to our legislative processes.

The declining status of the Osprey in this country and in Europe closely parallels the decline of the Bald Eagle, and for the same reason—its dependence on contaminated fish. Peterson (1965) described the most dramatic (30 percent per year) decline in the long-standing Connecticut River valley populations and Ames and Mersereau (1964) found persistent pesticides in the eggs to be the probable cause for reproductive failure. Ospreys have declined severely in the northeastern states but less so in the Chesapeake Bay and Great Lakes areas, where they are declining slowly or holding their own. Fortunately for continued survival, the Osprey is widespread throughout the world, inhabiting all continents and many oceanic islands.

PEREGRINE FALCON (*Falco peregrinus*)

Perhaps the most critical of the North American birds of prey is the Peregrine Falcon. Formerly widespread, though not abundant, it is now considered extinct as a breeding bird everywhere east of the Mississippi from the Gulf of St. Lawrence to the Gulf of Mexico. Up until the mid-1940's, 8–10 pairs of the falcons had eyries along the Hudson Palisades and 16 or more birds wintered in New York City, subsisting on pigeons and Starlings (Herbert and Herbert, 1965). They even nested on buildings in New York City (one pair raised four broods on skyscrapers), Montreal, and Hartford. But a postwar crusade against the "cruel" hawks by homing pigeon addicts, which was supported by Humane Societies and newspapers, soon eliminated the falcons from their strongholds. (The falcons "select" homing pigeons, which attempt to escape by flying, whereas other pigeons gain some immunity by not flying.)

Up until the mid-1960's Peregrine Falcons were reproducing well in the far northwest where pesticides were rarely used. Now a collapse in this population seems imminent because the birds winter along the Pacific Coast, where they subsist largely on sea birds containing DDT and other hydrocarbons. The consequent calcium deficiency (page 458) in the falcons causes them to eat, break, or desert their eggs and/or young (Cade et al., 1967, 1968). Even in the more favorable Rocky Mountain areas, where the falcons are living on less contaminated prey, Enderson (1965) has shown that many formerly occupied nest sites are abandoned. He attributed this largely to human interference, but subsequent evidence tends to incriminate pesticides.

Concern for the future welfare of the Peregrine Falcon was strongly evidenced by the calling of an International Conference in Madison, Wisconsin, in 1965, where "experts" from European and North American countries discussed all possible angles affecting the plight of the noble birds [see *Peregrine Falcon Populations*, J. J. Hickey (ed.), 1969]. As in the case of the California Condor and Whooping Crane, some controversy exists as to the feasibility of raising falcons in captivity, either for falconry purposes or for restocking wild birds. Breeding falcons in captivity poses difficult but probably not insurmountable problems; the raptor breeding bird program at Cornell University has already achieved dramatic initial successes.

OTHER RAPTORS

Nearly all raptorial birds have shown varying degrees of decline throughout all or parts of their range, presumably because they are at the end of food chains (pages 458–459) and suffer most from environmental pollution, and because they are tempting targets in spite of protective legislation. The bird-eating accipiters, especially the Cooper's and Sharp-shinned Hawks, have become increasingly scarce over much of their range. The Red-shouldered Hawk, in contrast to the ostensibly similar Red-tailed Hawk, has decreased in all states except California and West Virginia, with the severest declines in New England and the Midwest (Brown, 1971). In 1942, the Craigheads (1956) found 22 pairs of Red-shoulders and 2 pairs of Red-tails on a 37-square-mile tract in southeastern Michigan. By the end of their study in 1948 the Red-shouldered Hawks had decreased to 17 pairs and the Red-tailed had increased to 5 pairs. Now the replacement of the Red-shoulders by Red-tails is nearly complete; the latter are more tolerant of habitat changes.

The Marsh Hawk has shown similar declines in parts of New York, Ohio, and Wisconsin. In Wisconsin, Hamerstrom (1969) reported that the nesting population of Marsh Hawks in her study area dropped from 25 nests in 1963 to 3 nests in 1967, with no recovery when voles (mice) started to increase in 1966. Arbib's Blue List (Appendix II) includes other hawks.

FISH-EATING BIRDS

Among the many fish-eating birds that have suffered setbacks, the Brown Pelican has received the most publicity. In Louisiana, the "Pelican State," where thousands of pelicans used to nest, no nesting occurred from 1961 to 1970; in 1971 nesting birds were reestablished by imports from Florida (Schreiber and Risebrough, 1972). Formerly thriving colonies with thousands of nests have completely disappeared from some islands off the coast of California, and on other islands populations have been much reduced (Schreiber and De Long, 1969). Again egg-shell breakage and consequent reproductive failures have been shown to be the cause. Some recovery has been noted since the ban on DDT in California (Jehl, 1973). Local declines, also usually attributed to pesticides, have been reported for cormorants, herons, and egrets, but the severe, almost catastrophic decline of Black-crowned Night Herons in New York, Ohio, Indiana, and Michigan began before the heavy use of pesticides. A 500-nest colony in Michigan suddenly disappeared, without any apparent reason.

GALLINACEOUS AND OTHER NONPASSERINE BIRDS

Several upland game birds are on endangered species lists. The Attwater's Prairie Chicken population, originally a probable million or more birds in coastal Texas, was reduced to 8,700 in 1937 and to a "few hundred" in 1960. It appears headed for extinction unless a costly land acquisition program, already started, can save it (Cottam, 1962). Loss of the original blue-stem coastal prairie to the plough, overgrazing, and "clean farming" may have spelled its doom. Likewise, the wider-spread Greater Prairie Chicken has experienced rapid shrinkage in both range and numbers in Ohio, Illinois, Indiana, and Michigan (Madsen, 1969). In Michigan it increased with the clearing of the land between 1910 and 1930 and was widespread in both the Lower and Upper Peninsulas, but by 1956 all the colonies in southern Michigan and most of the birds in the Upper Peninsula were gone (Ammann, 1957). In 1972, only one sizable colony and scattered individuals remain. Loss of habitat, changing farm practices, competition and hybridization with the more aggressive Sharp-tailed Grouse, and possibly genetic factors—reminiscent of the Heath Hen—appear to be responsible for the decline. Other galliform birds on endangered species lists include the Masked Bobwhite in the southwest and the Horned Guan, a preglacial relict, in Mexico and Guatemala (Andrle, 1967). (See Appendix III for other nonpasserine endangered species.)

PASSERINE BIRDS

Passerine bird populations are difficult to estimate, especially if scattered over extensive areas, but some are so rare and/or so restricted in distribution that fairly accurate counts or estimates can be made. Among these are two unique parulids. The rarest, and the hardest to find, is the Bachman's Warbler, one of the "most wanted" of American birds by bird listers. There have been few records since 1950: a maximum of six reported in 1960, one in 1966 and none since (Stevenson, 1972). If not already gone, it is on the verge of extinction. Various reasons have been suggested for its demise, but perhaps the best one is offered by Stevenson: its "time has come".

In a somewhat different category is the Kirtland's Warbler. It is usually conspicuous and easy to find because of its loud, explosive, persistent, and easily identified song during its short breeding period in the few counties of north-central Michigan where it nests. But thorough censuses over the entire breeding range in 1971 and 1972 disclosed only about 200 pairs, fewer than 500 birds, a 60 percent decline from censuses taken in 1951 and 1961 (Mayfield, 1972a). Management measures to try to save the bird, carried out under the supervision of a special committee composed of several state and federal agencies, include cutting, planting, and burning to maintain areas with the right developmental stages of pines. High heat, best produced by fire, is necessary to unseal the resin bond binding the scales on the cones and to release the seeds for regeneration. The worst enemy of the Kirtland's Warbler is the Brown-headed Cowbird, a relatively recent and abundant invader in the warbler nesting areas. The cowbird takes a heavy toll, often 100 percent of the parasitized nests (Mayfield, 1960, 1972a; Walkinshaw, 1972).

Other passerine birds on the U. S. endangered species list include the Dusky Seaside Sparrow and the Cape Sable Sparrow (now considered conspecific), both restricted to brackish marshes on the coasts of Florida; the Ipswich Sparrow on Sable Island, Nova Scotia, now considered a local geographic race of the widespread and abundant Savannah Sparrow; and the Pribilov race of the Winter Wren in the Aleutians.

HOPE FOR THE FUTURE

Lest the impression be left that declining species are doomed, attention should be called to species that have responded to protection. The Trumpeter Swan (Fig. 16.10) became alarmingly scarce in this country in the 1930's, but with nearly complete protection of nonmigrating birds on western refuges the population increased from a low of 35 birds in 1931 to 642 in 1954 (Banko, 1960) and to nearly a thousand in

FIG. 16.10. *This Trumpeter Swan family, on a lake in Yellowstone National Park, is evidence of the usefulness of complete protection for endangered birds. Trumpeters have increased on western refuges from 35 birds in 1931 to nearly a thousand in 1972.* [Photo by Edward M. Brigham, Jr.]

the Rocky Mountain states in 1972. Alaskan Trumpeters are more than double this figure. Transplantation experiments in which swans were stocked on other refuges have been quite successful. Even more encouraging was the dramatic comeback of egrets and southern herons after their near extermination at the turn of the century (page 403); with protection from shooting and preservation of their southern breeding haunts, they have increased and spread over much of the United States. In contrast to the rare Everglade Kite in Florida, White-tailed Kites in California have virtually exploded and are also doing well in the Gulf Coast region.

Some game birds have responded well to modern management, while others have not. The native Turkey, exterminated from much of the range it occupied in colonial days, has been restocked in several states, which now permit limited open seasons. The American Woodcock has withstood reasonably heavy hunting, yet the more widely distributed Common Snipe has not. The Sharp-tailed Grouse has expanded its range, with transplanted birds taking well in new habitats.

Various species of waterfowl are periodically put on and then taken off the protected list, as the fluctuations in numbers seem to dictate. In the last decade the Ross' Goose has shown a spectacular increase (Dzubin, 1965) that now permits an open season on them. The open season on Whistling Swans in the west, however, has been severely criticized; it might jeopardize the survival of two species—the Trumpeter Swan and Whooping Crane—that we have been trying to save.

Legislation

One more or less effective method of trying to conserve bird life is by protective legislation. Perhaps because of their great aesthetic and economic value, birds have enjoyed more efforts on the part of man to try to help them than any other group of animals. Seasonal protection of game birds was initiated in Massachusetts in 1818, and in New Jersey in 1820. Other states soon established similar regulations for game and more or less complete protection for songbirds. The following paragraphs describe some of the major legislative actions that have been undertaken in behalf of birds.

THE TARIFF ACT OF 1913

In about 1875 a sudden boom in the popularity of bird plumages in the millinery industry threatened the security of some of America's most beautiful birds (see page 403). In special demand were the breeding plumes of egrets and herons, which could be secured only in the nesting season, a fact that encouraged exceedingly destructive methods of obtaining them. As early as 1877 Florida passed an antiplumage law to prevent the appalling waste of bird life, but legislation was ineffective as long as feathers were in such demand—one Audubon Society warden was killed trying to enforce the law.

Eventually, however, feathers for hats went out of fashion, partly because of an aroused public sentiment, partly because of the decline of the birds most in demand. The whole matter culminated in a provision or clause of the Tariff Act of 1913 that prohibited the importation into this country of wild bird plumages from any part of the world. Feathers used for millinery purposes must come from birds raised in captivity. Unfortunately this law was amended in 1922 to permit importation of feathers for the manufacture of artificial flies, resulting in substantial illegal diversions to the millinery industries in New York. Supposedly this loophole in the law was plugged in the early 1950's, but heavy illegal traffic in jungle fowls for fly tying persists (Lane and Hartgen, 1971). Even heavier smuggling of cage birds for pets takes place (see page 418).

MIGRATORY BIRD TREATIES, 1913–1936

The various punitive and often ineffective efforts of the states to protect migratory birds finally culminated in federal action. A Migratory Bird Treaty, passed in 1913–14 for the United States (Weeks–McLean Migratory Bird Law) was ratified in a convention between the United States and Great Britain in 1916, and went into full effect as the Migratory Bird Treaty Act in 1918. In 1936 a convention with Mexico extended this treaty to include that country in a remarkable demonstration of international cooperation for the protection of the bird life of a whole continent. Still later, in 1972, amendments to the treaty were ratified by the United States and Mexico to extend protection to 63 families of birds common to both countries, including 6 families of birds of prey and many families of fish-eating birds not covered in the original treaty. In 1972, also, a treaty with Japan accorded protection to 189 species of birds that fly between the United States and Japan, mainly from Alaska.

Hence, *all native birds* in this country, theoretically at least, are protected by law, but special permits can be obtained to deal with cases of nuisance birds causing damage to domestic animals or crops, or creating other annoyances. Predatory birds raiding poultry or domestic animals, Bobolinks in southern rice fields, blackbirds in grainfields, sapsuckers defacing buildings, and shrikes interfering with banding traps are examples.

SUPPLEMENTARY STATE REGULATIONS

The various states have regulations of their own, primarily for the nonmigratory game birds (grouse, pheasants, and quail) not under the jurisdiction of federal law, but often for other species as well. Some 30 or more central and southern states have an open season on Mourning Doves, in contrast to about 18 northern states that do not. Sharp controversies exist about the merits of an open season on doves in the closed states. Waterfowl regulations are primarily a federal responsibility, but state officials make recommendations for opening and closing dates in their own states. Heretofore, predatory birds came under state laws, which varied from no protection in some states to complete or nearly complete protection in others, but now all predatory birds come under the new federal regulations. Even crows are protected by the 1972 amendments, but some states have open seasons on crows, thus treating them as game birds.

OTHER FEDERAL LAWS

Most states have had laws to protect the Bald Eagle, but the federal government formally supplemented the various state actions by a special act in 1940 that prohibited killing this species in the 48 contiguous states though not in Alaska. The regulation extending much-needed protection for the Bald Eagle in Alaska was passed in 1952, in spite of the objections of the salmon industries. The Golden Eagle, not formerly protected in most western states, now comes under the 1972 amendment according protection to all predatory birds.

The Migratory Bird Conservation Act (Norbeck–Andersen Sanctuary Bill), enacted in 1929 and amended in 1935, was another milestone in conservation, as it provided for badly needed refuges at a critical time of nation-wide droughts. The Migratory Bird Hunting Stamp Act (1934) provided funds for the land acquisition program visualized in the Conservation Act. Currently, in the early 1970's, the federal refuge system is in dire straits for lack of adequate financing.

Parks, Sanctuaries, and Refuges

The preservation and maintenance of natural areas as nearly as possible in their natural state and the effort to improve them for wildlife have probably done more for birds than any other measure. Birds must have a suitable place to live. Now many millions of acres in federal, state, municipal, or private holdings, where wildlife is held more or less inviolate, are available for birds. It is not possible to deal at length with these vast sanctuary areas, but a few remarks about them are pertinent here.

NATIONAL PARKS, MONUMENTS, AND FORESTS

This category includes not only the better known national parks, but also national monuments, seashores, forests, and historic sites. These areas furnish badly needed nesting grounds for critical species, such as the Everglade Kite in the Everglades National Park and the Trumpeter Swan in Yellowstone; the parks also feature valuable instructional programs for visitors. Some of the more popular or conveniently located parks attract millions of visitors each year (Fig. 16.11); the total for all units in 1972 was close to 100 million. Bird watchers, particularly easterners in the west, have found the national parks meccas of bird life; there are posted lists and museum exhibits, as well as trained park naturalists conducting nature hikes, to help visitors in their identification of new forms.

Unfortunately, the national parks are currently beset with many problems: inadequate financing, more visitors than can be handled properly, destructive vandalism, pressure from state game departments to open the parks to hunting, overgrazing by protected herbivores because of the absence of large predators, demands for more facilities and modern conveniences, and the removal of the wildlife the

Fɪɢ. 16.11. *Acadia National Park, Maine. Accessibility for summer visitors and diversified terrain — seacoast, mountains, woods, and fields — account for an abundance of both human and avian visitors. More than 250 species of birds have been recorded in the park.* [Photo by R. H. Manville.]

parks are supposed to protect if they pose a threat to visitors (bears) or carry contagious diseases (bison, prairie dogs).

The national forests, by contrast, are less developed. Since they usually lack modern facilities, they are wilder and have fewer people problems. Lumbering in the national forests, and sometimes in the national parks, is a moot question: whether to harvest mature timber and protect diseased trees by spraying (with its attendant harmful consequences), or to leave them in their natural state.

FEDERAL WILDLIFE REFUGES AND STATE GAME AREAS

The development of federal wildlife refuges, initiated in a small way with the establishment of Pelican Island in 1903, really got under way in 1929 with the passage of the Migratory Bird Conservation Act (page 443). In the 1930's, largely by means of funds obtained through the Migratory Bird Hunting Stamp Act, a vast program of land acquisition was started. Now more than 300 wildlife refuges totaling nearly 30 million acres, primarily for waterfowl but often for other species as well, are well distributed over the continent.

Some of the more northern refuges (Figs. 16.12 and 16.13) are strategically located for breeding grounds. Other refuges are along migration routes, or scattered over Atlantic, Gulf Coast, or southwestern wintering grounds. A few are strategically located for jeopardized species: the Aransas Refuge in Texas as the wintering grounds of the Whooping Crane, the Okefenokee for the Limpkin, and the Florida Keys for the rare and localized white phase of the Great Blue Heron. Such refuges have also become important research centers for the study of the many current problems relating to waterfowl management. Like the national parks, the refuges often suffer from inadequate financing. In the early 1970's, for instance, refuge personnel had to be cut back drastically and the operation of some areas discontinued.

Whereas waterfowl are primarily a federal responsibility, nonmigratory game birds come under the jurisdiction of the individual states. State game areas, usually operated as public shooting grounds, and experiment stations, set up primarily for research, serve as refuges where

FIG. 16.12. *J. Clark Salyer National Wildlife Refuge in North Dakota. The marshes created by impounding waters here produce more than 100,000 waterfowl annually.* [Photo by C. J. Henry.]

FIG. 16.13. *Seney National Wildlife Refuge in Michigan is noted particularly for its wild goose propagation and management program.* [Photo by C. J. Henry.]

the birds find partial protection and good habitat. Such areas are also important as research centers for the study of game problems and for the training of wildlife personnel. State game areas own and operate areas for upland game and for migratory waterfowl as well as for nongame species.

OTHER SANCTUARIES

Vast sanctuary holdings are widely scattered over the continent. They are primarily dedicated to the preservation of birds and/or other wildlife but often conduct well organized educational programs for visitors. Prominent among these are the 40 sanctuaries maintained by the National Audubon Society. Many of these are strategically located, often deliberately selected, to protect endangered species. A good example is Corkscrew Swamp Sanctuary in southern Florida, a 6,000-acre tract that preserves the country's largest remaining stand of virgin bald cypress and its associated fauna and flora—alligators and orchids as well as the rare Limpkin and declining Wood Stork.

Most state ornithological organizations, as well as more localized bird clubs, maintain sanctuaries. The Massachusetts Audubon Society, for example, has 32 sanctuaries with a permanent staff on some of them. These are open the year round to the public and are visited by thousands of people annually. Nearly all known modern devices for attracting birds are also utilized.

State parks sometimes serve as sanctuaries where essential habitats are preserved for species that might otherwise be eliminated by commercial developments. However, the current trend in many state parks is toward increased camping and recreational facilities at the expense of wildlife. Vandalism has become rampant in many state parks.

Quite unique among other types of sanctuaries are (1) Bird City at Avery Island, Louisiana, where E. A. McIlhenny built up a large rookery of water birds by constructing a dam to impound water and introducing eight Snowy Egrets, which attracted other kinds of birds; (2) the heron sanctuary at Stone Harbor, New Jersey, which has attracted to it nearly all the ardeid species that occur in North America; (3) Hawk Mountain Sanctuary in Pennsylvania (page 340); and (4) the system of refuges in Michigan for the rare Kirtland's Warbler (page 439).

Less heralded, but in toto very effective, are the numerous sanctuaries maintained by private citizens, sometimes small backyards, sometimes large estates, where studied efforts are made to attract as many birds as possible to the home grounds.

Diseases and Parasites

Avian diseases communicable to man were discussed in Chapter 15, but another important aspect of this topic is the effect on the birds themselves. Avian populations are sometimes regulated by diseases, which are responsible in part for periodic die-offs. Sometimes losses are severe enough to have economic significance. Among ducks, *botulism, fowl cholera,* and *lead poisoning* destroy thousands of birds annually. Botulism, or western duck sickness (Fig. 16.14), is a bacterial disease caused by *Clostridium botulinum,* which thrives on decaying organic matter at times of low water, particularly in alkaline marshes. The best remedy known to date is control of fluctuating water levels, including periodic flooding to eliminate the shallow water zone that favors the development of *Clostridium.* Inoculations against the disease have been used successfully, but of course it is time consuming and costly to trap and inoculate the ducks. During the 1952 outbreak at Bear River Refuge in Utah, 6,233 sick ducks were treated and released, but 20,000 ducks had already died (Tinker, 1958).

The mysterious die-off of loons and grebes in Lake Michigan in the mid-1960's when thousands of dead and dying birds were washed up on the beaches, was eventually determined to be caused by type-C botulism. Rising lake levels in the late 1960's seem to have eliminated the disease. Gulls shared to a less extent in the die-off. White (1963) ob-

FIG. 16.14. *Ducks, dead or dying from botulism on alkali flats at Bear River Refuge in Utah. Many sick birds are picked up and given antitoxin treatments at the refuge's duck hospital.* [Photo by W. F. Kubichek, Bureau of Sport Fisheries and Wildlife.]

served 27 cases of type-C botulism in Peregrine Falcons and suggested that they might get it from drinking and bathing in infected waters as well as from sick prey. Fowl cholera is another bacterial disease affecting waterfowl. In the winter of 1950, 4,400 birds died of it on the Muleshoe Refuge in Texas (Petrides and Bryant, 1951).

Lead poisoning is a result of the ingestion of bullets, which ducks accidentally pick up while feeding in areas heavily shot over during the hunting season (Fig. 16.15). Sometimes a few pellets in the digestive tract are sufficient to cause death, and yet many pellets, up to 76 or more, have been removed in postmortem examinations of single birds. In spite of considerable research, as with nonlead alloys for bullets and with stomach pumps for removing ingested lead, a workable solution to this problem has not been put into practice. A leafy diet of pondweeds *(Potamogeton)*, as opposed to a corn or grain diet, seems to lessen the incidence of the disease: hence artificial feeding may sometimes have harmful consequences. Bellrose (1959), in the late 1950's, concluded that the losses, which varied from 1 to 10 percent of the population in different species, were not high enough to warrant drastic measures to try to curb them, but suggested that the situation might get worse if hunting pressure increased in congested areas. It is now much worse,

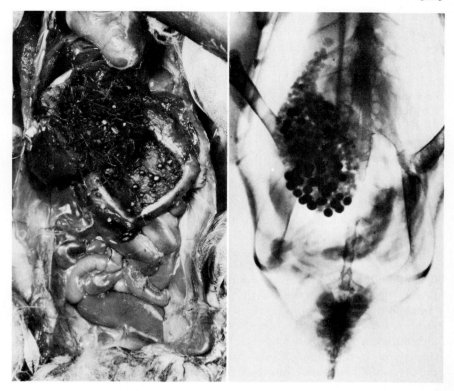

FIG. 16.15. LEFT *Lead poisoning in Mallard, showing shot in opened gizzard and rectum distended with greenish material.* RIGHT *X-ray of duck that died from lead poisoning.* [Photos by Mich. Dept. Nat. Res.]

according to a recent Audubon report. Losses are greater than those estimated by Bellrose, and there are additional hazards in decreased reproduction, anemia, and decreased alertness. The Department of the Interior aims for a ban on lead shot by 1975, substituting steel bullets (of slightly lower ballistic quality and higher cost) for lead, but ammunition manufacturers are reluctant to convert to new techniques. It is now estimated that 2–3 million waterfowl are lost each year by lead poisoning.

Trichomoniasis, a protozoan disease caused by the flagellate *Trichomonas gallinae,* has been known in doves, pigeons, turkeys, and other birds for some time. It was largely responsible for the die-off of Mourning Doves in the southeastern states in 1950 (Haugen, 1952). The disease results in the formation of a yellowish cheesy mass ("canker") in the posterior part of the mouth or crop, which sometimes blocks off the passage of food or air, causing starvation or asphyxiation. Hawks can become infected by feeding on diseased birds. A similar infection called "frounce," caused by the flagellate *Capillaria*, has been reported in Gyrfalcons in Greenland, probably obtained from feeding on ptarmigans (Trainer et al., 1968).

Ulcerative enteritis (quail disease), a disease of uncertain origin causing lesions in the lower digestive tract, is a common occurrence in gallinaceous birds in captivity but also occurs in wild grouse, pheasants, and quail (Buss et al., 1958). Strict sanitation seems to be the best preventive in captives. Game birds often have to be raised on wire, so that droppings, the source of further infection, will fall through the wires.

Other endoparasites often have serious and sometimes lethal effects on birds, although ordinarily their purpose is to find a safe home and not destroy themselves by destroying their host. Endoparasites of waterfowl and game birds include *Leucocytozoon* in the blood of ducks and grouse, gizzard worms in geese, and acanthocephalid worms in the intestinal tract of eider ducks (see Herman, 1955, for specific references). Cornwell and Cowen (1963) found Canvasbacks in the Detroit River heavily infested with helminth parasites, five times as many as in "normal birds," yet some birds on their breeding grounds in Manitoba were as heavily infested as the Detroit birds. One duckling had 40,464 helminths in its digestive tract and showed no obvious effects, but indirect or long-range effects are not known. Ducks feeding on muskgrass *(Chara)* were mostly free of intestinal parasites.

Ectoparasites sometimes inflict serious harm to their hosts, although in some cases birds merely harbor the parasite without physical injury to themselves. Heavy infestations (Fig. 16.16) may cause anemia (blood-sucking types) or other run-down conditions. Ectoparasites produce lacerations in the skin that become portals of entry for bacteria. They also decrease egg production in poultry, transmit avian diseases, and cause the death of nestlings (Boyd, 1951). Bird lice (Mallophaga) in particular have been studied extensively by Hopkins and Clay (1952) and Malcomson (1960). Malcomson lists 800 species of Mallophaga from 500 birds in North America (including some nonnative zoo birds). He reports that some 2,600 species are known and that the total may yet exceed 3,000. Ectoparasites of birds may be divided into six general types (Boyd, 1951):

1. Mallophaga (bird lice)—obligatory parasites, chewing, feather-eating types; spend complete life cycle on host.
2. Fleas—adults on birds, larvae nonparasitic.
3. Hippoboscid flies—obligatory parasites, common on many species of birds, especially nestlings.

Fɪɢ. 16.16. *Head of Gray Partridge heavily infested with louse eggs.* [Photo by Mich. Dept. Nat. Res.]

4. Protocalliphorid flies—larvae, hatched from eggs laid in nests, attack nestlings at night, sometimes causing death.
5. Ticks—blood-sucking types, often about the head; cause anemia and weakened condition; often vectors of disease.
6. Mites (Acarina)—blood-sucking and dermatophagous (skin and itch mites), including chiggers; cause scaly-leg in icterids (Herman et al., 1962).

Herman (1955, 1966) gives a good review of diseases and parasites in birds, including ecto- and endoparasites, and bacterial and virus diseases.

Man-made Mortality: The Modern Dilemma

Chapter 11 mentioned some of the factors, natural and man-made, responsible for high mortality to birds, but did not discuss the latter. In minor ways birds and man have been in competition or conflict since man first appeared on earth, but in the last few decades, with sharply increasing demands on the land and its resources, this conflict has become greatly accentuated. The following briefly outlines some of the most pressing problems that birds, and sometimes man, face in the struggle for survival.

LAND USE

One of the most critical present-day problems that threatens the security not only of birds but also of man himself is the ever-increasing demand for living space and the needs that go with it. More and more wild land is being taken over for highway construction, airports, industrial expansion, housing and suburban developments, and food and fiber production. The advanced technology required to fulfill these needs introduces new problems — pollution of the air, and soil, and water on which we depend for life.

One of the many examples that might be cited to illustrate the controversy over land use is what happened at the heron sanctuary in Stone Harbor, New Jersey. Formerly low-value brushland set aside as sanctuary for the herons, the price for its 31 acres sky-rocketed to $500,000 for real estate developments. Through the intervention of several influential conservationists the herons won the first round in the conflict over the land, but as its value, and the need for it, increase, the herons may have to go. Thus one of the most unique bird sanctuaries in North America may become yet another cluster of houses.

Similar conflicts arise from the need for greater food production. There is an old saying in Turkey that the farmer sows a handful of grain for himself, another for the birds, and another for the gods, but in this country the farmer must reap most of the crop or he doesn't stay in business. Formerly birds and agriculture were somewhat complementary, one aiding the other, but this is no longer always true. Monocultural practices create an imbalance, favoring a buildup of grain-feeding birds, which become economic pests, at the expense of the insectivorous species whose food supply is depleted or poisoned. Other land-use developments that imperil birds have already been mentioned (see Endangered Species).

A continually recurring threat to wildlife is the need for more space for military operations. One example is the demand for a bombing range near Lake Mattamuskeet National Wildlife Refuge in North Carolina, where nearly 200,000 waterfowl winter annually. New construction threatens other wildlife areas. Murchison (1968) describes the past and proposed whittling away of the 35,000-acre Wheeler National Wildlife Refuge in Alabama — 10,000 acres already lost to the Redstone Arsenal, 1,500 acres requested for a public recreation area, 60 acres wanted for a hospital for the mentally retarded, and the threat of a highway through the refuge. The battle over the jetport in the Florida Everglades and the proposed extension of the Kennedy International Airport into Jamaica Bay Wildlife Refuge are other examples.

DRAINAGE

Drainage is a special type of land usage, conducted for varying purposes, but it is in a special category as far as waterfowl management is concerned. The Departments of the Interior and Agriculture have long been engaged in a struggle over the use of marsh lands, with government-subsidized projects draining millions of acres of the most productive duck-breeding grounds in North America—jeopardizing a dwindling multimillion-dollar resource to produce more grain. Yet even the most ardent duck hunter may have to concede that wheat has greater food value than ducks. Animal proteins, whether in ducks on the wing or beef on the hoof, are uneconomical. Wetlands—bogs, swamps, marshes, and estuaries—are considered "cheap" lands to be exploited, after being drained and filled, but they are extremely important for maintaining water balance and for the production of wildlife and other resources (Niering, 1968; Wayburn, 1972; Mitchell, 1973).

OIL POLLUTION

With the ever-increasing demand for oil and its products and the advanced technology involved in its extraction and transport, oil pollution is becoming an increasingly aggravating, world-wide problem, still largely unresolved despite international meetings, resolutions, and agreements. And the situation threatens to get worse. The trans-Alaskan Pipeline and its environmental implications pose serious problems. Then there is the prospect of extraction, by exceedingly costly and destructive methods, of deep reserves underlying several western states—which cannot be extracted by present methods without literally tearing down the Rocky Mountains (Soucie, 1972).

Oil spills and their consequences are too numerous to chronicle here. Aldrich (1970) describes some of the worst disasters since 1937, some of them involving losses of more than 100,000 birds. But perhaps the greatest disaster, or at least the most publicized, was that of the super-tanker Torrey Canyon (Fisher and Charlton, 1967; Bourne, 1970), which went aground off the English Coast in 1967, spilling 119,328 tons of crude oil and scattering an estimated 20,000–30,000 oiled birds, mostly guillemots, along the coasts of Britain and France. Some 6,000 birds were "rescued," but attempts at their rehabilitation, as in other similar rescue operations, were largely unsuccessful. Detergents used to clean feathers render the oil gland inoperative (Nero, 1964) and destroy the waterproof structure of the feathers. Only a small fraction of the Torrey Canyon birds were saved, and there is some doubt about the survival of those released. Some improvements have been made recently in the technique of cleaning feathers and in rehabilitation of oil-soaked birds (Stanton, 1970).

Other recent major oil spills, mostly well publicized, include one at Santa Barbara, California, in 1969, when a "black tide" bathed the coast; another on the coast of Texas in 1970, when wells lacking adequate safety devices spewed 1,000 barrels a day over the coastal waters (Chevron Oil Company was fined $1,000,000 for its negligence); and another in San Francisco Bay in 1971, when two tankers collided in dense fog and spilled 840,000 gallons of fuel oil over 55 miles of beaches. Of the 7,000 oiled birds recovered and treated at 30 impromptu cleaning stations only a small number recovered.

The immediate kill from oil is only part of the problem. Oil on and in the water persists, often unobserved, for a long time, spreading out to sea and settling where it produces long-lasting effects on bottom organisms (Blumer et al., 1971). Birds not killed immediately often suffer from sublethal effects that may eventually produce death (Holcomb, 1969).

Though the chief concern of oil pollution has been in coastal waters, contamination in inland lakes and rivers is sometimes severe. Peller (1963) vividly describes the attempts to salvage 10,000 waterfowl and other victims when a broken oil pipe and a soybean oil storage tank spilled millions of gallons of oil along the upper Mississippi in Minnesota. Anderson and Warner (1969) put 3,333 of the salvaged ducks to good use by a detailed morphological analysis and measurements of the dead birds. The Detroit River losses (Fig. 16.17), which run to thousands of ducks annually, are largely from oil pollution, although other industrial wastes are also involved.

A less publicized hazard comes from pools of oil that inevitably collect in oil fields. These seem to attract birds for drinking and bathing purposes. The problem is particularily acute in the San Joaquin Valley in California, where each year "at least 150,000 waterfowl, shorebirds, raptors, and songbirds perish in these modern-day La Brea tar pits" (Audubon, 75(3):114–115, 1973). Oil companies have been lax in cleaning up these oil sumps. Similar sumps occur in other oil-producing states, including the North Slope of Alaska, where oil fields are already trapping birds.

CEILOMETERS AND TELEVISION TOWERS

Collisions of birds with man-made structures (tall buildings, monuments, lighthouses) and moving vehicles (aircraft and automobiles) have constituted a hazard to birds for a long time, but now ceilometers and high television towers pose a more serious threat. Birds are attracted to encircling beams of light from ceilometers and fly in and out

FIG. 16.17. *A Canvasback* (LEFT) *that died from oil pollution in the Detroit River and a normal Canvasback* (RIGHT) *free from oil. An estimated 10,000 to 12,000 ducks died in the Detroit River and western Lake Erie from oil pollution in the winter of 1959–60.* [Photo by Mich. Dept. Nat. Res.]

of the beams until exhausted. The actual kill varies greatly. At the Albany airport in New York, Bartlett (1952) watched hundreds of birds flying in and out of the light, but no dead or exhausted birds were found. Later, during a heavy flight, some 300 dead or injured birds were retrieved (Bartlett, 1956). The heaviest losses from ceilometers have been reported at Warner Robins Air Force Base in Georgia (Johnston and Haines, 1957), where 50,000 birds of 53 species were killed on October 8, 1954.

High television towers are a hazard on cloudy nights with a low ceiling. In ordinary weather nocturnal migrants fly higher than the highest towers; even under cloudy skies there have been few reports of kills at towers under 800 feet, but high mortality has been recorded at some of the higher structures. The most notorious TV tower is at Eau Claire, Wisconsin where an estimated 20,000 birds have been killed in one night (Kemper, 1958). This 1,000-foot tower continues to take a heavy toll: 30,000 estimated killed in two nights in September 1963—and there is a proposal to build a 1,700-foot tower at the same site (Kemper, 1964). Somewhat lower, but still significant, mortality has been reported by Tordoff and Mengel (1956) at Topeka, Kansas, and by Brewer and Ellis (1958) in east-central Illinois. Warblers, vireos, and thrushes usually predominate in television tower kills.

On the brighter side, many valuable scientific data have been derived from studies of TV tower kills — weights, measurements, age and sex ratios, fat extractions, and migration data. The most comprehensive of these studies have been by H. L. Stoddard and cooperators at a tower in north Florida, where 29,451 specimens of 170 species have been handled in an 11-year period (Stoddard and Norris, 1967).

A similar type of destruction has been recorded for deadly antennas of a radio communication system on Midway Island, which threatened to destroy one sixth of the world's population of Laysan's Albatrosses and took a toll of 30–40 Sooty Terns *per day* while the birds were on the island (Fisher, 1966c). Fortunately, the antennas were demolished in 1967, not to save the birds but because the antennas were outmoded as a means of communication (Fisher, 1970).

PICTURE WINDOWS

Windows have always presented a minor obstacle to flying birds, but the hazard increased with the vogue for large picture windows. A mass mortality has rarely if ever been recorded; it is the demise of a few birds at each of thousands, perhaps millions, of picture windows, though not all windows experience kills. Two types of window problems occur: (1) male birds fighting their image in a window until they become exhausted — more often an annoyance to the homeowner than a fatality to the bird, and (2) birds flying against the windows and stunning or killing themselves. No very effective solution to these problems has been devised. Sometimes pulling the shades (which defeats the purpose of a picture window) helps. More effective is a silhouette or model of a large predatory bird placed in front of the window, or a row of two or more whirl-a-gigs.

PESTICIDES[1]

The use of chemical poisons in the control of insects and other pests has long been a controversial issue, especially since the use of thallium-treated grain for ground squirrel control in California in the late 1920's, which resulted in the death of thousands of birds from feeding directly on the poisoned grain or, in the case of predatory birds, from secondary poisoning. But pesticide-wildlife problems multiplied rapidly

[1] See also Endangered Species, pages 429–439, and Appendix III.

with increased use of DDT and other hydrocarbons after World War II. Control programs, some of them highly indiscriminate and often unnecessary, expanded so rapidly that by the late 1950's more than a billion pounds of formulated pesticides were being applied annually to about 100 million acres of land in this country.

A few preliminary studies in the 1940's seemed to indicate that some of the less toxic chemicals, when applied to small acreages in approved ways, could be used without noticeable effects on birds; but as control programs expanded, and heavier dosages or more lethal chemicals were used, reports of bird mortality increased. Bird lovers in particular protested the use of such lethal materials, but their claims of birds losses were generally discredited. The heaviest losses were in the fire-ant "eradication" campaign in the southeastern states (Allen, 1958; Lay, 1959), in Dutch elm disease "control" in the midwest (Barker, 1958; Hickey and Hunt, 1960; Wallace et al., 1961), and in Japanese beetle programs in Illinois and Michigan (Scott et al., 1959).

And of course pesticides do not necessarily remain where applied. They are distributed by air, wind, water, and mobile organisms to all parts of the biosphere. DDT, for instance, is found on vegetation, in caribou, and in Eskimos, in the Arctic; it is found in seals and penguins in the Antarctic, in all the major rivers in North America, and in the atmosphere over cities where it comes down in rain. Fish and Wildlife Service personnel have suggested that probably no living thing on earth, plant or animal, is entirely free of DDT.

Some early warnings about excessive use of pesticides were sounded, but generally ignored. The on-rolling juggernaut, deeply entrenched and well financed, could not be stopped. Most research under way in the Fish and Wildlife Service was terminated by budget cuts, and people who were securing incriminating data were threatened, dismissed, or reassigned.

Then in 1963 came Rachel Carson's *Silent Spring*, which stirred government agencies into action and jolted the agri-chemical industries. This was followed by President Kennedy's Science Advisory Committee's *Use of Pesticides* (1963), which in general vindicated Rachel Carson and recommended phasing out DDT and other persistent hydrocarbons as rapidly as possible. It wasn't done. Twice the prestigious National Academy of Sciences–National Research Council was commissioned to review the whole matter, but the investigating committees producing the first reports (Parts I, II, and III, 1962–1963) were dominated by agri-chemical interests and failed completely to come to grips with the problems involved. A later, equally lamentable effort (1969 report) was well characterized by *Audubon* magazine (The Audubon View, 1969): "The Academy of Sciences lays another thin-shelled egg." Hence, for nearly a decade following *Silent Spring* and the recommendations of the President's Science Advisory Committee, little progress was made in phasing out the persistent hydrocarbons.

But data on the effects of pesticides on animal organisms continued to accumulate, at first largely by independent investigators, then (mainly after *Silent Spring*) by a reorganized and revitalized Fish and Wildlife Service and other government agencies. Early reports of dead and dying birds were followed by detailed censuses, analyses of tissues, feeding experiments to determine lethal levels, and finally more sophisticated studies of sublethal effects on behavior, reproduction, and physiology. One of the dramatic, totally unexpected findings was that sublethal levels of DDT caused egg-shell thinning and breakage by interfering with the enzyme-hormone balance and creating a calcium deficiency in egg-laying birds (Hickey and Anderson, 1968; Peakall, 1970; Ratcliffe, 1970). Hence, the dilemma in nesting eagles, Ospreys, Peregrine Falcons, and Brown Pelicans (see Endangered Species, especially pages 434–438). And reproductive failures do not stop with thin and breaking egg shells; Hays and Risebrough (1971), 1972) found young terns of Long Island hatching with serious birth, defects—gross morphological malformations. Bernard (1963) had determined much earlier that DDT accumulates in the ovaries of exposed robins and is transferred to the eggs and developing embryos and if they hatch, to the young birds (Fig. 16.18).

Another surprising finding was the buildup or biological magnification in food chains. Earthworms can accumulate up to 10 or more times the level of DDT found in the soils in which they live, and dying robins have much higher levels than the earthworms. Bottom sediments in Green Bay, Wisconsin, which receive runoffs from adjacent cherry orchards, had only 0.014 ppm of DDT and its metabolites, but amphipods in the mud averaged 0.41 ppm (a 30-fold buildup), fish of three species had much higher levels, and the body fat of 12 nesting Herring Gulls contained from 1,041 to 2,441 ppm (Hickey et al., 1966).

Although much of the past research on pesticides has been on DDT and other persistent hydrocarbons (aldrin, dieldrin, heptachlor, etc.) and more recently on "safer" substitutes, much recent attention has also been directed toward mercury contamination in fish, waterfowl, game birds, in our food supplies, and on the widespread occurrence of PCB's (polychlorinated biphenyls). As yet only limited testing for mercury compounds in birds has been done in this country (the problem has been more severe in Sweden and Japan), and PCB's are still something of a mystery. Indeed, it now seems possible that the PCB's found in so many unexpected and illogical places may actually be derivatives of DDT.

In summary, some of the current major concerns arising from the

FIG. 16.18. *A part of the dead and dying American Robins picked up on and around the Michigan State University campus in the spring of 1961 after the spraying of the elms. Nearly all of the 200 or more robins analyzed from the same area had DDT; more than 90 percent of them had levels in the brain comparable to levels recovered from experimentally poisoned birds.* [Photo by Robert L. Fleming, Jr.]

past and present use of persistent pesticides can be categorized as follows:

1. World-wide dissemination in all conceivable environments.
2. Biological magnification or buildup in food chains (robins, oysters, Lake Michigan ecosystems).
3. Effects on reproduction (egg-shell thinning, birth defects).
4. Possible association with diseases (cancer, leukemia, etc.).
5. Altered behavior (slowdown of reactions and alertness, impairment of physiological functions).

Many committees, organizations, and government agencies are still trying to evaluate and resolve conflicting pest control–wildlife relationships. At the meeting of the International Council for Bird Preservation in Cambridge, England, in 1966, delegates from 28 nations discussed problems relating to the use of pesticides in their countries; several criticized the United States for shipping to their countries pesticides no longer legal, or for uses no longer legal, in our country.

And at long last, after more than 20 years of controversy, it appears that DDT and other persistent hydrocarbons are on the way out, and that the newer pesticides, hopefully, will be safer for birds and man.

SELECTED   REFERENCES

MONOGRAPHS ON EXTINCT AND ENDANGERED SPECIES

ALLEN, R. P. 1952. The Whooping Crane. Research Report (Nat. Aud. Soc., New York), 3:1–246.

BANKO, W. E. 1960. The Trumpeter Swan. No. Amer. Fauna, 63:1–214.

GREENWAY, J. C., JR. 1958. *Extinct and Vanishing Birds of the World*. Spec. Publ. No. 13. New York: Amer. Comm. for Intern. Wildl. Prot.

GROSS, A. O. 1928. The Heath Hen. Boston Soc. Nat. Hist, 6:491–588.

HICKEY, J. J. (ed.). 1969. *Peregrine Falcon Populations: Their Biology and Decline*. Madison: Univ. Wisc. Press.

KOFORD, C. B. 1953. The California Condor. Research Report (Nat. Aud. Soc., New York), 4:1–154.

MAYFIELD, H. 1960. *The Kirtland's Warbler*. Bloomfield Hills, Mich.: Bull. 40, Cranbrook Inst. Sci.

MCNULTY, F. 1966. *The Whooping Crane: The Bird That Defies Extinction*. New York: E. P. Dutton.

SCHORGER, A. W. 1955. *The Passenger Pigeon; Its Natural History and Extinction*. Madison, Wisc.: Univ. Wisc. Press.

TANNER, J. T. 1942. The Ivory-billed Woodpecker, Research Report (Nat. Aud. Soc., New York), 4:1–111.

DISEASES AND PARASITES

BELLROSE, F. C. 1959. Lead poisoning as a mortality factor in waterfowl populations. Bull. Ill. Nat. Hist. Sur., Vol. 27, Art. 3:233–288.

BOYD, E. M. 1951. The external parasites of birds: a review. Wilson Bull., 63:363–369.

HERMAN, C. M. 1955. Diseases of birds. Chapter 13 in A. Wolfson (ed.), *Recent Studies in Avian Biology*. Urbana, Ill.: Univ. of Ill. Press.

———. 1966. Birds and our health. Pages 284–289 in Stefferud (ed.) *Birds in Our Lives*. Washington, D. C.: Fish and Wildlife Service.

LAND USES, POLLUTION, AND PESTICIDES

ALDRICH, J. W. 1970. Review of the problem of birds contaminated by oil and their rehabilitation. Resource Publ. 87, Washington, D. C.

BERNARD, R. F. 1963. Studies on the effects of DDT on birds. Mich. State Univ. Mus. Publ., Biol. Series, 2:161–191.

CARSON, R. 1962. *Silent Spring*. Boston: Houghton Mifflin.

GRAHAM, F., JR. 1970. *Since Silent Spring*. Greenwich, Conn.: Fawcett Publ.

MURPHY, R. 1969. *A Heritage Restored: America's Wildlife Refuges*. New York: E. P. Dutton.

RUDD, R. L. 1964. *Pesticides and the Living Landscape*. Madison, Wisc: Univ. Wisc. Press.

UDALL, S. L. 1965. *The Quiet Crisis.* New York: Holt, Rinehart and Winston.

WALLACE, G. J., W. P. NICKELL, and R. F. BERNARD. 1961. Bird mortality in the Dutch elm disease program in Michigan. Bloomfield Hills: Cranbrook Inst. Sci. Bull., 41.

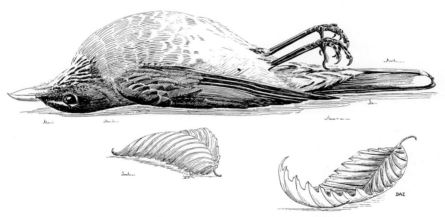

[Drawing by Dale A. Zimmerman; courtesy Cranbrook Inst. Sci.]

APPENDICES

I

Ornithological Organizations and
Their Journals

I T I S not feasible here to enumerate the many ornithological organizations devoted to the study of birds, but those of greater ornithological importance need to be described, particularly from the standpoint of the journals they publish. Those that are national in scope, state organizations with *printed* journals, and the leading foreign journals are listed in this appendix. The three oldest, and most important since, until recently at least, they have published about three fourths of the world's original literature on birds, are (1) the *Journal für Ornithologie* (1853), (2) the *Ibis* (1859), and (3) the *Auk* (1884).

Baldwin and Oehlerts (1964) examined the status of the world ornithological literature for 1964. They listed 138 "basic" journals (those of major ornithological importance — mainly or exclusively ornithological), 76 "local" journals (those of local interest), and 244 others that publish "significant amounts of avian material," a total of 458 ornithologically important serials. However, if periodicals dealing with aviculture, poultry, and other "bird" magazines are included, the number of current periodicals runs to over 1,000. Biological Abstracts abstracted 13,176 "articles of avian interest" in the three-year period ending in 1964, an average of more than 4,000 per year. This should give the aspiring bird student some insight into the scope of current ornithological literature.

There is one international organization, known as the International Ornithological Congress, which meets at four-year intervals and publishes *Proceedings* of its meetings. The Congress is a gathering of representatives of various countries, particularly of Europeans and Americans. Fifteen such meetings have been held in the past, but only one of them in this country — at Cornell University in 1962.

National Organizations

The leading ornithological organizations in this country that are national in scope are listed and discussed briefly.

American Ornithologists' Union. Publishes the *Auk,* a quarterly (Fig. I.1).

The American Ornithologists' Union (A.O.U.) is the leading scientific ornithological organization in this country and currently the largest in the world. It is primarily North American in scope, including Canada and Mexico, but it has become increasingly involved in recent years in world-wide ornithology. Founded in 1883, largely from the stimulus of the earlier Nuttall Ornithological Club (1873), the Union has published more than 90 volumes of the *Auk,* a really pretentious and valuable shelf of books covering all aspects of technical ornithology. The A.O.U. has published five editions of the *Check-List of North American Birds* (fifth edition in 1957) and is working on a badly needed sixth edition. It is also sponsoring a several-volume *Handbook of North American Birds* and has published 15 *Ornithological Monographs.*

Bird-Banding Organizations. Publish *Bird-Banding* magazine, a quarterly (Fig. I.1).

The various bird-banding organizations in this country are divided on a regional basis into Northeastern, Eastern, Inland, and Western Bird-Banding Associations. A single journal, *Bird-Banding,* serves all of them, although some of the regional associations have their own special leaflets and newsletters (usually mimeographed) and hold regional meetings. Some states have their own local units (e.g., *Michigan Bird Banders*), separate from or more usually affiliated with the larger orga-

F I G. 1.1. *Covers of four national ornithological journals.* [Photo by Bruce Frumker; courtesy of Cleveland Mus. Nat. Hist.]

nizations. *Bird-Banding* magazine, started in 1929, publishes articles based primarily on banding studies, but is also known for its outstanding reviews of foreign literature.

Cooper Ornithological Society. Publishes the *Condor,* a quarterly (formerly bimonthly, Fig. I.1).

Though this organization, founded in 1898 and named for J. G. Cooper, author of the first comprehensive work on Californian birds, has members throughout North America, it serves primarily the Pacific Coast and Rocky Mountain states. The *Condor* is a thoroughly scientific journal, drawing its subject matter from both field and laboratory, from Alaska to South America. The Society also publishes *Pacific Coast Avifauna,* a series of outstanding monographs.

National Audubon Society. Publishes *Audubon Magazine* (formerly *Bird-Lore*), *American Birds* (formerly *Audubon Field Notes*), and *Research Reports* on rare and endangered species.

The National Audubon Society, incorporated in 1905 and named after John James Audubon (page 9), is a large, active, and influential organization, with its headquarters at 930 Fifth Avenue in New York City. Its membership (currently more than 275,000) is much greater than that of the more scientific organizations named above. Though originally founded for the protection and preservation of birds, it has since broadened its scope to include all phases of renewable natural resources.

In collaboration with the United States Fish and Wildlife Service, the Society publishes *American Birds*: six issues yearly with detailed summaries of bird records for the four seasons plus the Christmas counts and breeding-bird censuses. The Society also issues a great deal of educational material for teachers and youth leaders and keeps a vigilant eye on all developments that might affect the nation's natural resources. Audubon Wildlife Films, presented in 250 cities, reach an audience of millions. Still other activities of the Society have been indicated elsewhere in the text.

The Wilson Ornithological Society. Publishes the *Wilson Bulletin,* a quarterly (Fig. I.1).

Though founded in Massachusetts in 1888, the Society has centered most of its activities and interest in interior North America. Named in honor of Alexander Wilson, father of American ornithology, the Society early advocated field studies of birds as its goal, but the magazine later extended its coverage to include taxonomic and anatomical studies, especially if based on or supplementing field data. The Wilson

Ornithological Society Library, housed in the University of Michigan Museum of Zoology and named the Josselyn Van Tyne Memorial Library in honor of the man largely responsible for it, is an outstanding collection of books, periodicals, and reprints, acquired largely by donations, for the service of members.

American Birding Association, founded in 1969, publishes *Birding,* a quarterly.

An organization formed purely for the sport of birding and for publishing bird-listing records (see page 12).

Raptor Research Foundation, founded in 1967, publishes *Raptor Research,* news releases and has sponsored a two-volume bibliography on birds of prey.

An organization devoted to research on birds of prey, particularly on the care of raptors in captivity.

Other National Organizations. Several other organizations, not exclusively ornithological, merit mention for their substantial contributions to bird literature. The *Journal of Wildlife Management,* published by the Wildlife Society, includes all forms of wildlife, especially birds and mammals. Though major emphasis is on game species, papers cover many other aspects of wildlife, such as statistical studies, measurements of populations, and economic aspects of wildlife species. *Ecology* and *Ecological Monographs,* both published by the Ecological Society of America, are fruitful sources of bird literature in that field. *The American Midland Naturalist,* published by the University Press, Notre Dame, Indiana, sometimes has extensive papers on birds. *Natural History* magazine, published 10 months a year by the American Museum of Natural History, New York, features birds prominently in its popular coverage of topics. *National Geographic Magazine* frequently carries colorfully illustrated articles on birds from various parts of the world. More recently *National Wildlife* and *International Wildlife,* bimonthlies, published by the National Wildlife Federation, vie with *Audubon* for the world's most colorful wildlife magazines. On the more technical side, *Systematic Zoology.* published quarterly by the Society of Systematic Zoology since 1952, explores problems of systematics and taxonomy for both vertebrates and invertebrates.

There are of course many important institutional publications, such as those of museums, colleges, universities, and various scientific societies, that contribute substantially to bird literature. Journals in allied fields (morphology, physiology, genetics) often include articles that add materially to ornithological knowledge. Perhaps the ultimate in institutional publications is *The Living Bird,* a thoroughly scientific yet not highly technical annual publication of the Laboratory of Ornithology at Cornell University.

State Organizations

A survey of all of the state ornithological organizations is not practical here; some are set up on a restricted regional rather than a state-wide basis, and some locally active clubs, regional or state-wide, do not issue printed publications. The following survey, then, is limited largely to state-wide organizations that issue a printed magazine or journal that is mainly ornithological. Most of the journals of state societies are not very technical, but deal, often in considerable detail, with local distribution and field studies conducted within the state, and thus serve an important function.

Alabama. Alabama Birdlife, a quarterly published by the Alabama Audubon Society.

Arkansas. Arkansas Audubon Newsletter, a quarterly published by Arkansas Audubon Society.

Delaware. Cassinia, an annual issued by the Delaware Valley Ornithological Club, which includes southern New Jersey and eastern Pennsylvania (Philadelphia region).

District of Columbia. The Atlantic Naturalist (formerly called the *Wood Thrush*), published five times a year by the Audubon Naturalist Society of the central Atlantic States. Though founded in 1897, the Society has been issuing its magazine only since 1945.

Florida. The Florida Naturalist, a quarterly published by the Florida Audubon Society. Founded in 1900, the Society has been issuing its present journal since 1927.

Georgia. The *Oriole,* a quarterly published by the Georgia Ornithological Society, founded in 1936.

Illinois. The *Audubon Bulletin,* a quarterly published by the Illinois Audubon Society, which was founded in 1897 and is centered largely in the Chicago area.

Indiana. The *Indiana Audubon Quarterly* (formerly the *Indiana Audubon*

Society Yearbook), published by the Indiana Audubon Society, founded in 1898 and incorporated in 1939.

Iowa. Iowa Bird Life, a quarterly published by the Iowa Ornithologists' Union, founded in 1923. It has been issued under its present title since 1931, having been preceded by the *Bulletin* and by mimeographed newsletters.

Kansas. Kansas Ornithological Society Bulletin, a quarterly published by the Kansas Ornithological Society.

Kentucky. The Kentucky Warbler, a quarterly published by the Kentucky Ornithological Society, founded in 1923.

Maine. Maine Field Naturalist (formerly the *Bulletin*), a quarterly published by the Maine Audubon Society.

Maryland. Maryland Birdlife, a quarterly published by the Maryland Audubon Society.

Massachusetts. The Massachusetts Audubon Society, in cooperation with some adjacent states, publishes *Man & Nature,* now a yearbook, formerly a quarterly. Previously the Society published a monthly bulletin, which later became *Massachusetts Audubon,* a quarterly. *Records of New England Birds,* a monthly report of records of New England observers, is published separately. The Society is the oldest (founded in 1896) and largest of the state organizations. It has a salaried staff and maintains 32 sanctuaries, some of which have year-round directors.

Michigan. The Jack-Pine Warbler, a quarterly published by the Michigan Audubon Society. The magazine is named in honor of Michigan's unique bird, the Kirtland's Warbler (see Figs. 13.2 and 14.2). The Society, founded in 1904, is one of the largest and most active of the state societies. It maintains a library, a bookshop, and six sanctuaries, three of them with interpretative programs. The magazine, originating as a mimeographed newsletter in 1923, has been issued in substantially its present form since 1939. A *Newsletter* keeps members posted on items not pertinent to the more permanent journal. Spring, fall, and sometimes summer "campouts" are held yearly in addition to the regular Annual Meeting.

Minnesota. The *Loon* (formerly the *Flicker*), a quarterly published by the Minnesota Ornithologists' Union. Eight or more local Minnesota clubs are affiliated with the Union. The magazine includes notes from adjacent Ontario (the Canadian Lakehead) and North Dakota.

Missouri. The *Bluebird,* a quarterly published by the Missouri Audubon Society.

Nebraska. The *Nebraska Bird Review,* a quarterly published by the Nebraska Ornithologists' Union, founded in 1899.

New Jersey. New Jersey Nature News, a quarterly published by the New Jersey Audubon Society, one of the largest of the state societies.

New York. The *Kingbird,* issued four times a year by the Federation of New York State Bird Clubs, which is composed of the many bird clubs scattered throughout the state. Also, some of the local groups have their own literature: *Feathers,* published by the Schenectady Bird Club, and the *Goshawk* by the Genesee Ornithological Society. Active bird clubs are found at Buffalo (Buffalo Academy of Natural Science), Ithaca (Cornell University), Rochester, Albany, and of course in and around the New York City region.

North and South Carolina. The *Chat,* a quarterly published by the Carolina Bird Club. The Club was preceded by the North Carolina Bird Club (1937), which joined with several South Carolina natural-history groups in 1948 to form the broader organization.

Oregon. Audubon Warbler, published monthly (except in summer) by the Oregon Audubon Society. (The *Murrelet,* published by the Pacific Northwest Bird and Mammal Society, is a joint Oregon–Washington– Pacific Northwest enterprise.)

Rhode Island. The *Bulletin,* published (formerly) five times a year by the Audubon Society of Rhode Island, which was founded in 1897. The Kimball Bird Sanctuary is maintained by the Society.

South Dakota. South Dakota Bird Notes, issued quarterly by the South Dakota Ornithologists' Union.

Tennessee. The *Migrant,* a quarterly published by the Tennessee Ornithological Society.

Texas. Bulletin of the Texas Ornithological Society, a quarterly.

Virginia. The *Raven,* a quarterly published by the Virginia Society of Ornithology.

West Virginia. The *Redstart,* a quarterly published by the Brooks Bird Club (Wheeling, West Virginia).

Wisconsin. The *Passenger Pigeon,* a quarterly published by the Wisconsin Society for Ornithology.

Foreign Journals

The more progressive countries throughout the world have well-developed ornithological programs, particularly in northern and central Europe. Unfortunately, the foreign journals usually are not readily available in the smaller schools and libraries, although they are almost indispensable for doing advanced work with birds. Some of the better known or more accessible ones are listed below, alphabetically by countries.

Australia. The *Emu,* quarterly of the Australasian Ornithologists' Union, Melbourne; *Australian Natural History; The South Australian Ornithologist.*

Belgium. *Le Gerfaut* (The Falcon), Revue de Belge d'ornithologie.

Canada. *The Canadian Field-Naturalist,* a quarterly (Ottawa); *Canadian Audubon,* a bi-monthly (Toronto); *The Blue Jay,* a quarterly published by the Saskatchewan Natural History Society.

Denmark. *Dansk Ornithologisk Forenings Tidsskrift.*

England. The *Ibis; British Birds; Bird Study; Avicultural Magazine* (serves also for Avicultural Society of America); the *Oologists' Record* (worldwide oölogy); *Animal Behavior* (published jointly by England, Canada, and U. S.); *Journal of Animal Ecology.*

Finland. *Ornis Fennica; Suomen Riista,* Finnish Game Foundation, Helsinki. Articles in Finnish, Swedish and English.

France. *Alauda; L'Oiseau* et *La Revue Française d'Ornithologie; Oiseaux de France.*

Germany. *Journal für Ornithologie; Die Vogelwarte; Die Vogelwelt; Die Zeitschrift für Tierpsychologie.*

India. *Journal of the Bombay Natural History Society,* a quarterly devoted to natural history; *Pavo,* The Indian Journal.

Japan. *Tori,* Ornithological Society of Japan.

Netherlands. *Ardea* (some articles in English); *Limosa; Behavior* (an international journal).

New Zealand. *Notornis,* quarterly bulletin of the Ornithological Society of New Zealand.

Poland. The *Ring,* a quarterly journal of bird-banding data, published by the Polish Zoological Society.

Scotland, *Scottish Birds,* a quarterly published by the Scottish Ornithologists' Club.

South Africa. The *Bokmakierie,* South African Ornithological Society and Wikwatersprand Bird Club (popular magazine for bird watchers); the *Ostrich,* South African Ornithological Society.

Sweden. *Vår Fågelvärld.*

Switzerland. *Nos Oiseaux,* Bulletin de la Société Romande pour l'étude et protection des oiseaux; *Der Ornithologische Beobachter.*

II

Ornithological Collections

ORNITHOLOGICAL collections (Figs. II.1 and II.2) for study usually consist of *mounts* (birds set up on a pedestal in some lifelike pose), *skins* (specimens laid out flat), *alcoholics* (soft parts of birds preserved in alcohol or formalin), nests and eggs, and skeletons. Mounted birds are usually placed on exhibit, and in many museums and public institutions serve a widespread and useful educational purpose. Large metropolitan museums, such as the American Museum of Natural History in New York, the National Museum of Natural History in Washington, D.C., the Cleveland Museum of Natural History, the Carnegie Museum in Pittsburgh, Pennsylvania, the Chicago Museum of Natural History, the Museum of Natural History in Denver, and several museums in California, have elaborate displays of birds from all over the world, often placed in attractive settings simulating natural habitats. Less pretentious exhibits, sometimes merely of local avifauna, are available at most institutions; if these are accessible in the laboratory, they can be used for identification purposes.

Skins are more commonly used for specimen studies, since they take up less room and are easier to prepare. Classroom specimens need to be kept in protective containers, such as glass or plastic tubes (Fig. II.1), so that they can be handled by many persons without the otherwise inevitable destruction to the skins. Research collections used by fewer students for more advanced studies, are not thus protected, yet if properly handled they may last indefinitely. Collections intended primarily for detailed taxonomic work need to be large, with numerous examples of each species, in order to show variations of age, sex, and season, as well as geographical differences. About eight museums* in this country have world-wide collections numbering more than 100,000 specimens; the largest, at the American Museum of Natural History in New York City, now totals nearly a million birds (900,000 skins plus mounts in exhibits). Many smaller collections, however, perhaps not world-wide in scope, are adequate for many types of studies, and a laboratory or

---

* Museum of Comparative Zoology at Harvard, Smithsonian Collections in Washington, Academy of Natural Sciences at Philadelphia, Carnegie Museum at Pittsburgh, University of Michigan Museum of Zoology at Ann Arbor, Field Museum of Natural History at Chicago, University of California at Berkeley and Los Angeles.

FIG. II.1. *Collections for student use include mounted birds and skins, the latter often kept in tubes for protection in handling. Here a Cardinal (mount), a Pyrrhuloxia (skin) and a Painted Bunting (in tube) illustrate these devices.* [Photo by Robert L. Fleming, Jr.]

small museum with only a few hundred to a few thousand specimens of the right selection can be very useful for instructional purposes.

Mounting birds requires special techniques, largely beyond the scope of an ornithology course, but all advanced students need to know how to prepare a bird skin. This is a fairly simple and quick method of skinning out a bird by removing the body and substituting a proper-sized filler of cotton or other suitable material. The following condensed directions, with an instructor's guidance, should serve to get a beginning student started on the technique of making bird skins.

Directions for Preparing Bird Skins

1. Relax the bird, flexing the legs, wings, and head. Sponge off any dirt or blood stains on the feathers and dry with cornmeal. Use

F IG. II.2. ABOVE *Scientific skin collections are usually kept in cases, the specimens ar-*
ranged in phylogenetic sequence, with the names of larger categories at least on the out-
side of the door and names of smaller categories on the sliding trays. [Photo by Robert
D. Burns.] OPPOSITE *Museum workers identify specimens and record pertinent data*
preparatory to cataloguing and storing the skins. [From L. C. Pettit's *Introductory*
Zoology, copyright 1962 by the C. V. Mosby Company, St. Louis. Photo courtesy
of Mich. State Univ. Museum.]

cornmeal (or other absorbent) liberally through the whole skinning process.

2. Part the abdominal feathers and make an incision through the skin, being careful not to cut through the body wall into the viscera. Extend the cut from the posterior end of the sternum into the anus or around one side of it (Fig. II.3a).

3. Loosen the skin from the abdominal wall with fingers or scalpel until the knee joint is exposed. Cut the leg at this joint, cleaning the flesh from the tibia. Rub borax over the tibia, especially at the tarsal end, wrap it tightly with cotton until it simulates the size of

the original limb. Push back into natural position. Repeat with the other leg.

4. Make an incision across the base of the tail, being careful not to cut too close to the base of the tail feathers which might then be lost.

5. Invert the skin carefully, using fingers and scalpel, over the back and breast until it reaches the wings (Fig. II.3b).

6. Sever each wing at the shoulder joint (proximal end of humerus) and continue pushing the skin up over the head (Fig. II.3c). In most birds the skin will slip over the head easily, but in some large-headed birds (woodpeckers and ducks) it may be necessary to make a slit through the skin along the nape (or throat, if preferred) to free the head.

7. Pull the skin of the ear out of its socket with finger and thumb. Slip the skin over the eyes and remove them (Fig. II.3c).

8. Make an incision across the palate between the mandibles; extend the cut along the inner margin or each mandible and across the base of the skull. Then the body and most of the contents of the skull (brain) can be removed from the skin (Fig. II.4a).

9. Clean out any remaining portions of the brain and flesh from the skull, sprinkle the interior of the skull liberally with borax, and turn the skin back carefully over the skull.

10. Clean the wing bones (humeri) of flesh and remove as much as possible of the flesh from the radius and ulna, *without* loosing the secondaries from the ulna. In large birds it is better to remove flesh and tendons from the forearm by making an incision along the underside of the wing, cleaning the bones, then sewing the slit together with a few stitches.

11. The wings can be made to lie in a natural position in the final bird skin by (1) tying the two radii together in a parallel position about one-half inch apart (in medium-sized birds), (2) tying the humeri in a similar manner, or (3) bringing the skin between the wings closer together (taking up the slack) by a stitch through the skin.

12. Remove the oil gland, or at least its contents, from the base of the tail. Remove any fat, flesh, or blood stains still adhering to the skin or feathers. Skins with excess fat (e.g., ducks) may have to be degreased by immersing them in benzene or carbon tetrachloride

---

FIG. II.3. *Preparing a bird skin. After weighing and measuring the specimen, place it on its back and part the abdominal feathers. Make a medial incision in the skin* (a) *from the end of the sternum to the anus. Then separate the skin from the body* (b), *slipping it up over the skull, ears, and eyes* (c). [Photos by Bruce Frumker; courtesy Cleveland Mus. Nat. Hist.]

a

b

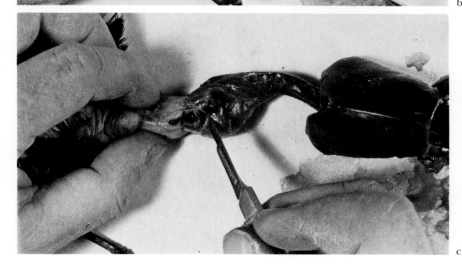

c

a

b

c

for several hours or more, then drying them with cornmeal and/or compressed air.

13. Put a small tight wad of cotton in each eye, inserting it up through the neck with forceps.

14. Prepare a pointed stick (medical applicators are ideal for small birds) and wrap it tightly with cotton to form a body roughly comparable to the size of the body removed (use excelsior or other material for large birds).

15. Insert the body into the skin, with the point of the stick pushed into the skull, or protruding into the mouth cavity.

16. Arrange the skin carefully around the cotton body and sew up the abdominal incision (Fig. II.4b). Close the bill by tying a thread around it; through the nostrils if necessary.

17. Tie the crossed legs together and tie on a *correctly prepared label — with locality, date, sex, weight, collector,* and any other pertinent data on the tag. Be sure to sex the specimen by internal examination of the gonads, recording the data (a sketch to size is useful) on the label (Fig. II.4c).

18. Wrap the completed specimen in strips of cotton, or pin it out on a pinning board, and allow it to dry for several days before putting it in storage.

Collecting birds needs only a brief explanation. Responsible persons who wish to collect protected birds for strictly scientific purposes can get special permits; a federal permit is required and usually one for each state in which specimens are to be taken. Collecting is usually done with shotguns of appropriate caliber, but traps and nets can also be employed. Bullets (bird or dust shot) need to be used judiciously, because dust shot will singe or burn the feathers at close range, and shot that is too large may damage the specimens. Many collections are augmented by specimens turned in by people who find dead birds. Our chief sources of dead birds at Michigan State University in recent years have been: (1) insecticide victims (but run heavily to robins), (2) television towers (warblers, vireos, thrushes), (3) road kills, and (4) picture windows. Also hunters often provide us with special needs, sometimes unwittingly through the confiscation of illegal kills turned over to us by conservation officers. However, there are strict laws about even the temporary possession of dead or injured birds without a permit, and

---

FIG. II.4. *Preparing a bird skin* (continued). *After removing the body* (a), *clean out the skull and wing bones; insert a cotton body wound tightly on a stick and sew up the incision* (b). *Tie the crossed legs together and attach a carefully prepared label* (c). [Photos by Bruce Frumker; courtesy Cleveland Mus. Nat. Hist.]

many unsuspecting students have run into difficulties with unsympathetic police and conservation officers.

Formerly many, if not most, ornithologists had a personal collection of birds; some private holdings in the past have rivaled large institutional collections, but perhaps fortunately for greater accessibility these have now largely gone over to responsible museums. The day of private egg collections, once the pride and joy of many oölogists, is also largely past (in this country though not in England) but most ornithological collections include birds' eggs, with or without accompanying nests (the latter are hard to exhibit in an attractive way). Less common in collections, but potentially more useful, are skeletal materials. In the laboratory a few representative skeletons are essential for study, but research collections need to be much more complete. In food-habit studies, for instance, one may need to identify a skull or fragment of bone, often a difficult if not impossible task without consulting a reference collection. Preparing skeletons is usually done with the aid of *dermestids,* the carrion-feeding larvae of certain beetles that live on flesh.

III

~~~~~

Endangered Species Lists

U. S. Endangered Species
(excluding Hawaii and Puerto Rico)

Species and subspecies listed as endangered by the Bureau of Sport
Fisheries and Wildlife, U. S. Department of Interior (Jan. 4, 1974).
Subspecies common names in parentheses.

Brown Pelican *(Pelecanus occidentalis)*
(Aleutian) Canada Goose *(Branta canadensis leucopareia)*
Mexican Duck *(Anas diazi)*
California Condor *(Gymnogyps californianus)*
(Florida) Everglade Kite *(Rostrhamus sociabilis plumbeus)*
(Southern) Bald Eagle *(Haliaeetus leucocephalus)*
(American) Peregrine Falcon *(Falco peregrinus anatum)*
(Arctic) Peregrine Falcon *(F. p. tundrius)*
(Attwater's) Prairie Chicken *(Tympanuchus cupido attwateri)*
(Masked) Bobwhite *(Colinus virginianus ridgwayi)*
(Mississippi) Sandhill Crane *(Grus canadensis pulla)*
Whooping Crane *(Grus americana)*
(California) Clapper Rail *(R. longirostris obsoletus)*
(Light-footed) Clapper Rail *(R. l. levipes)*
(Yuma) Clapper Rail *(R. l. yumanensis)*
Eskimo Curlew *(Numenius borealis)*
(California) Least Tern *(Sterna albifrons browni)*
(American) Ivory-billed Woodpecker *(Campephilus principalis principalis)*
Red-cockaded Woodpecker *(Dendrocopos borealis)*
Bachman's Warbler *(Vermivora bachmanii)*
Kirtland's Warbler *(Dendroica kirtlandii)*
Dusty Seaside Sparrow *(Ammospiza nigrescens)*
Cape Sable Sparrow *(A. mirabilis)*
Santa Barbara Sparrow *(Melospiza melodia graminea)*

The following four forms are listed in the *Red Data Book,* but are not on the U. S. List.

(Greater) Prairie Chicken *(Tympanuchus cupido pinnatus)*
(Florida) Sandhill Crane *(Grus canadensis pratensis)*
(Pribilov) Winter Wren *(Troglodytes troglodytes alascensis)*
(Ipswich) Savannah Sparrow *(Passerculus savannarum princeps)*

Arbib's Blue List for 1973

The following forms apparently are suffering population declines or range retractions in all or part of their range but are not yet considered endangered (Arbib, 1972). We would add the Eastern Bluebird and the Eastern Phoebe to this list.

Red-throated Loon *(Gavia stellata)*
Western Grebe *(Aechmophorus occidentalis)*
Fork-tailed Petrel *(Oceanodroma furcata)*
White Pelican *(Pelecanus erythrorhynchos)*
Double-crested Cormorant *(Phalacrocorax auritus)*
Black-crowned Night Heron *(Nycticorax nycticorax)*
Wood Ibis *(Mycteria americana)*
White-faced Ibis *(Plegadis chihi)*
White Ibis *(Eudocimus albus)*
Fulvous Tree Duck *(Dendrocygna bicolor)*
Sharp-shinned Hawk *(Accipiter striatus)*
Cooper's Hawk *(A. cooperi)*
Red-shouldered Hawk *(Buteo lineatus)*
Swainson's Hawk *(B. swainsoni)*
Ferruginous Hawk *(B. regalis)*
Harris' Hawk *(Parabuteo unicinctus)*
Marsh Hawk *(Circus cyaneus)*
Osprey *(Pandion haliaetus)*
Caracara *(Caracara cheriway)*
Prairie Falcon *(Falco mexicanus)*
Pigeon Hawk *(F. columbarius)*
Sparrow Hawk *(F. sparverius)*
Sage Grouse *(Centrocercus urophasianus)*
Limpkin *(Aramus guarauna)*
American Oystercatcher *(Haematopus palliatus)*
Piping Plover *(Charadrius melodus)*
Snowy Plover *(C. alexandrinus)*
Gull-billed Tern *(Gelochelidon nilotica)*
Least Tern *(Sterna albifrons)*

Royal Tern *(Thalasseus maximus)*
Ancient Murrelet *(Synthliborhamphus antiquum)*
Yellow-billed Cuckoo *(Coccyzus americanus)*
Barn Owl *(Tyto alba)*
Burrowing Owl *(Speotyto cunicularia)*
(Florida) Scrub Jay *(Aphelocoma coerulescens)*
Bewick's Wren *(Thryomanes bewickii)*
Loggerhead Shrike *(Lanius ludovicianus)*
Bell's Vireo *(Vireo bellii)*
Gray Vireo *(Vireo vicinior)*
Yellow Warbler *(Dendroica petechia)*
Yellowthroat *(Geothlypis trichas)*
Bachman's Sparrow *(Aimophila aestivalis)*

World Endangered Species
(excluding North America)

Colonel Jack Vincent's list of "rare and endangered birds," prepared for the I.C.B.P. meeting in 1966, contained 318 forms, including many of uncertain status. Fisher and Peterson's Red List *(The World of Birds,* pp. 268–271) included 143 species—those believed to have populations of less than 2,000. Many others were omitted for lack of adequate information. The Federal Register of the U. S. Department of Interior lists 129. The Red Data Book includes 277 but not all of these are "endangered." The following list, compiled mainly from these four sources, is a selection of those whose status seems particularly critical.

PODICIPEDIDAE (5 flightless species confined to single lakes).Titicaca Grebe *(Podiceps micropteras)*. Confined to L. Titicaca, Bolivia. Atitlan (Giant Pied-billed) Grebe *(Podilymbus gigas)*. About 100 pairs surviving on Lake Atitlan, Guatemala, but believed declining from indiscriminate shooting, devegetation of nesting sites for firewood, egg collecting for food, and predation on the young grebes by introduced bass (the bass pull the young grebes underwater), which the natives do not fish (Bowes, 1965)

DIOMEDEIDAE. Short-tailed Albatross *(Diomedea albatrus)*. Exterminated by Japanese poachers on the Bonin Islands in 1932 and believed extinct (Peterson, 1948), but rediscovered on Torishima, where a census in 1962 showed a population of 47 birds.

PROCELLARIIDAE. Cahow or Bermuda Petrel *(Pterodroma cahow)*. Believed exterminated by early settlers (pages 7–8) but rediscovered in 1949. Survives in small numbers in Bermuda where it is carefully protected. Its extinction now seems imminent because of reproductive failure from DDT-contaminated oceanic prey and nest-burrow take-overs by tropicbirds (Wurster and Wingate, 1968).

THRESKIORNITHIDAE. Nippon (Japanese) Ibis *(Nipponia nippon)*. Reported in 1965 to total only 12 birds in Japan, but there may be others along the mainland (Manchuria and Northern China).

ANATIDAE (Vincent lists 10 species).   Né Né or Hawaiian Goose *(Branta sandvicensis)*. Formerly nearly extinct but now being bred successfully in England and restocked in its native Hawaii. Some danger of "genetic swamping" by interbreeding of wild and hand-reared birds.

Laysan Teal *(Anas laysanensis)*. About 200 surviving in the Hawaiian Islands.

ACCIPITRIDAE (6–13 species, many others with low populations). Gundlach's Hawk *(Accipiter gundlachi)*. Confined to Cuba; apparently very rare; few data available.

Galápagos Hawk *(Buteo galapagoensis)*. About 200 left on the islands in 1962.

Hawaiian Hawk *(B. solitarius)*. A few hundred on Hawaii.

Monkey-eating Eagle *(Pithecophaga jefferyi)*. See page 434.

Spanish Imperial Eagle *(Aquila heliaca adalberti)*. See page 435.

Harpy Eagle *(Harpia harpyja)*. A recent addition to the Endangered Species List.

FALCONIDAE (all of the Falconidae — 58 species — are now included in a supplement to the Red Data Booklist).

Mauritian Kestrel *(Falco punctatus.)*. Possibly the world's rarest bird, on verge of extinction on island where the Dodo breathed its last; 6 or 7 birds survive (Audubon, 75:132, 1973).

MEGAPODIIDAE (3 species in Red Data Book).   Maleo *(Macrocephalon maleo)*. Celebes, very rare (page 281).

PHASIANIDAE (nearly all ornamental pheasants are in jeopardy because of demand for zoos; two seem especially critical).

Imperial Pheasant *(Lophura imperialis)*. Known only from very restricted areas in Annam and Laos; good stock in captivity.

Mikado Pheasant *(Syrmaticus mikado)*. Restricted to high mountains in Formosa; rare but many in captivity.

GRUIDAE. . Hooded Crane *(Grus monacha)*. A few hundred winter in Japan; others in southeast Asia; breed in Siberia (Walkinshaw, 1973).

Manchurian Crane *(Grus japonensis)*. Probably a few hundred left in Manchuria and Japan.

Siberian White Crane *(Grus leucogeranus)*. Breed only in Arctic Siberia; 60–100 wintering in India, others scattered over southern Asia (Walkinshaw, 1973).

RALLIDAE (about 10 mostly insular species are in danger; some flightless or nearly so and cannot escape introduced predators).

Zapata Rail *(Cyanolimnas cerverai)*. "Confined to within a mile of the high ground in the wooded part of the Zapata Swamp, Cuba" (Fisher and Peterson 1964?).

Takahe *(Notornis mantelli)*. About 300 survive on South Island, New Zealand, where carefully protected. May have been rescued from extinction (see Fig. 16.1 and legend).

Horned Coot *(Fulica cornuta)*. High Andes (see page 260).

RHYNOCHETIDAE. Kagu *(Rhynochetos jubatus)*. Rare and little known crepuscular form in forests of Caledonia. Unique type, the world's only rhynochetid.

LARIDAE. Audouin's Gull *(Larus audounii)*. In Mediterranean area, rare and local, about 160 breeding birds in the main colony on Corsica.

COLUMBIDAE (9–13 species in varying degrees of danger).

Giant Imperial Pigeon *(Ducula goliath)*. In mountain forests of Caledonia, very restricted.

Tooth-billed Pigeon *(Didunculus strigirostris)*. On two Samoan Islands, where it seems to be recovering from a low population.

PSITTACIDAE (more than 20 species of parrots and parakeets are rare or endangered).

STRIGIDAE (5–9 species critical). Seychelles Owl *(Otus insularis)*. Recently rediscovered on one island; last previous record in 1906.

Forest Spotted Owlet *(Athene blewitti)*. Resident of forests in central India; no record since 1872!

CAPRIMULGIDAE. Least Pauraque *(Siphonorhis americanus)*. Jamaican race probably extinct; Hispaniolan race barely survives.

Puerto Rican Whip-poor-will *(Caprimulgus noctitherus)*. Thought extinct (only one previous specimen known) until rediscovered in 1961. Detailed studies from 1969–1971 (Kepler and Kepler, 1972) disclosed possibly 500 breeding pairs in three restricted areas in Puerto Rico.

TROCHILIDAE (status of many not well known; some known only from 1 or 2 specimens, others only from dubious trade skins; some "rare" specimens may be hybrids).

CAPITONIDAE. Kinabulu Barbet *(Megalaima pulcherrima)*. Confined to the Kinabulu mountains in Borneo.

PICIDAE. Fernandina's Flicker *(Nesoceleus fernandinae)*. Confined to palm groves in two provinces of Cuba.

EURYLAIMIDAE. Grauer's Green Broadbill *(Pseudocalyptomena graueri)*. Known only from three places in mountains of Tanzania.

FURNARIIDAE. Sclater's Spinetail *(Asthenes sclateri)*. Known only from Argentina; perhaps extinct.

FORMICARIIDAE. Red-rumped Ant Thrush *(Myrmotherula erythronotos)*. Known only from one place in Brazil; not collected recently.

ACANTHISITTIDAE.   Bush Wren *(Xenicus longipes)*. Three races persist on three islands in New Zealand; all rare. Well protected by law but threatened by introduced predators.

ATRICHORNITHIDAE.   Rufous and Noisy Scrubbirds *(Atrichornis rufescens* and *A. clamosus)*. Australia. Both species quite rare; the latter thought extinct until 1961.

ALAUDIDAE.   Raza Island Lark *(Calandrella razae)*. Confined to a three square-mile Cape Verde island.

CORVIDAE.   Hawaiian Crow *(Corvus tropicus)*. Slopes of Haleaka Mountain; 25–30 left?

TIMALIIDAE.   Spiny Babbler *(Turdoides nipalensis)*. None seen between 1844 and 1948, when rediscovered in Nepal.

PYCNONOTIDAE.   Olivaceous Bulbul *(Hypsipetes borbonicus)*. Two races, Réunion and Mauritius, both rare and local.

TROGLODYTIDAE.   Zapata Wren *(Ferminia cerverai)*. Restricted to five square miles of Zapata swamp, Cuba.

TURDIDAE (about 15 species in danger, including two Tristan da Cunha forms.

Seychelles Magpie Robin *(Copsychus seychellarum)*. Formerly on five islands, now on only one. About 20 birds in 1959.

Cebu Black Shama *(Copsychus niger cebuensis)*. Possibly extinct, along with 6 other species exterminated on Cebu by deforestation.

SYLVIIDAE.   Long-legged Warbler *(Trichocichla rufa)*. From Fiji Islands, not seen in present century; may be extinct.

MUSCICAPIDAE.   Chatham Island Robin *(Petroica traversi)*. Confined to a one-acre island, where 40–70 birds existed in 1937–38.

CALLAEIDAE (three species formerly occurred in New Zealand; the Huia is extinct; the Saddlebacks *(Creadion carunculatus)* and Wattled Crows *(Callaeus cinerea)* survive in small numbers).

MELIPHAGIDAE (one species each in Hawaii, New Zealand, and Australia, all barely surviving; four other Hawaiian species extinct).

ZOSTEROPIDAE.   Truk and Ponape Great White-eyes *(Rukia ruki* and *R. sanfordi)*. One on each island; both very rare.

VIREONIDAE.   Slender-billed Vireo *(Vireo gracilirostris)*. Confined to a three-square-mile island in the Atlantic.

DREPANIDIDAE.   Eight species of Hawaiian honeycreepers, very restricted on various islands, some barely surviving; 8 other species extinct.

PARULIDAE.   Semper's Warbler *(Leucopeza semperi)*. Confined to mountain forest of St. Lucia, West Indies, where one seen in 1947, one heard in 1962.

ICTERIDAE.   Arment's Cowbird *(Tangavius armenti)*. Found in Colombia, few specimens known.

Slender-billed Grackle *(Cassidix palustris)*. Lived in marshes near Mexico City; may be extinct.

FRINGILLIDAE. Puerto Rican Bullfinch *(Loxigilla portoricensis)*. One race extinct, one surviving, in Puerto Rico.

Sao Miguel Bullfinch *(Pyrrhula murina)*. Formerly thought extinct, but apparently a few individuals survive on Sao Miguel. Persecuted by natives because it is destructive to fruit trees.

The Federal Register lists other birds but without data as to how critical their status is.

LITERATURE CITED FOR ENDANGERED SPECIES

ARBIB, R. 1972. The blue list for 1973. American Birds, 26:932–933.

BERGER, A. J. Hawaiian birds. 1972. Wilson Bull., 84:212–222.

BOWES, A. LaB. 1965. An ecological investigation of the Giant Pied-billed Grebe, *Podilymbus gigas* Griscom. Bull. Brit. Ornith. Club, 85:14–19.

Federal Register. Fish and Wildlife Service, U. S. Department of Interior, Vol. 29, No. 3, Part III (Jan., 1974).

FISHER. J., and R. T. PETERSON. No date given. *The World of Birds.* New York: Doubleday. The Red List, pp. 268–271.

KEPLER, C. B., and A. K. KEPLER. 1972. The distribution and ecology of the Puerto Rican Whip-poor-will, an endangered species. Living Bird, 11:207–240.

PETERSON, R. T. 1948. *Birds Over America.* New York: Dodd, Mead.

VINCENT, J. 1966. List of rare and endangered species, and threatened species of birds. Reports prepared for Intern. Council for Bird Preservation (I.C.B.P.), Cambridge, England.

WALKINSHAW, L. H. 1973. *Cranes of the World.* New York: Winchester Press.

WURSTER, C. F., JR., and D. B. WINGATE. 1968. DDT residues and declining reproduction in the Bermuda Petrel. Science, 159:979–981.

Literature Cited

(See also chapter-end references, which usually are not repeated here.)

ABLE, K. P. 1970. A radar study of the altitude of nocturnal passerine migration. Bird-Banding, 41:282–290.

AGEE, C. P. 1957. The fall shuffle in central Missouri Bob-whites. Jour. Wildl. Mngt., 21:329–335.

ALDERTON, C. C. 1961. The breeding cycle of the Yellow-bellied Seedeater in Panama. Condor, 63:390–398.

———. 1963. The breeding behavior of the Blue-black Grassquit. Condor, 65:154–162.

ALDRICH, J. W., and K. P. BAER. 1970. Status and speciation in the Mexican Duck *(Anas diazi)*. Wilson Bull., 82:63–73.

ALLEN, F. H. 1948. The quails of the Sinai Peninsula—another interpretation. Auk, 65:451–452.

———. 1951. Audubon on territory. Wilson Bull., 63:206.

ALLEN, R. H., JR. 1958. Wildlife losses in the southern fire ant program. Pass. Pigeon, 20:144–147.

ALLEN, R. P. 1960. Do we want to save the Whooping Crane? Aud. Mag., 62:122–125, 134–135.

ALTMANN, S. A. 1956. Avian mobbing behavior and predator recognition. Condor, 58:241–253.

AMADON, D. 1962. A new genus and species of Philippine bird. Condor, 64:3–5.

———. 1966a. Another suggestion for stabilizing nomenclature. Syst. Zool., 15:54–58.

———. 1966b. Avian plumages and molts. Condor, 68:263–278.

AMES, P. L. 1967. Overlapping nestlings by a pair of Barn Owls. Wilson Bull., 79:451–452.

——— and G. S. MERSEREAU. 1964. Some factors in the decline of the Osprey in Connecticut. Auk, 81:173–185.

AMMANN, G. A. 1937. Number of contour feathers of *Cygnus* and *Xanthocephalus*. Auk, 54:201–202.

———. 1957. *The Prairie Grouse of Michigan*. Mich. Dept. Cons., Lansing.

ANDERSON, K. S. 1961. Eastern encephalitis in Massachusetts. Mass. Aud., 46:86–89.

ANDERSON, B. W., and D. W. WARNER. 1969. A morphological analysis of a large sample of Lesser Scaup and Ring-necked Ducks. Bird-Banding, 40:85–94.

ANDRLE, R. F. 1967. The Horned Guan in Mexico and Guatemala. Condor, 69:93–109.

ANNAN, O. 1963. Experiments on photoperiodic regulation of the testis cycle in two species of the thrush genus *Hylocichla*. Auk, 82:166–174.

ARBIB, R. S., JR. 1967. Considering the Christmas count. Aud. Field Notes, 21:39–42.

———. 1972a. The Hawaiian candidates for the blue list. Amer. Birds, 26:703–704.

———. 1972b. The blue list for 1973. Amer. Birds, 26:932–933.

ASCHOFF, J. 1965. Circadian rhythms in man. Science, 148:1427–1432.

ASHMOLE, N. P. 1968. Breeding and molt in the White Tern *(Gygis alba)* on Christmas Island, Pacific Ocean, Condor, 70:35–55.

———. and H. TOVAR S. 1968. Prolonged parental care in Royal Terns and other birds. Auk, 85:90–100.

AUSTIN, G. T. 1970. Breeding birds of desert riparian habitat in southern Nevada. Condor, 72:431–436.

AUSTIN, O. L., JR. 1961. *Birds of the World.* New York: Golden Press.

BAGG, A. M. 1955. Airborne from Gulf to Gulf. Bull. Mass. Aud. Soc., 39:106–110, 159–168.

BAILEY, R. E. 1952. The incubation patch of passerine birds. Condor, 54:121–136.

BAIRD, D. 1967. Age of fossil birds from the greensands of New Jersey. Auk, 84:260–262.

BAIRD, J., and A. J. MEYERRIECKS. 1965. Birds feeding on an ant mating swarm. Wilson Bull., 77:89–91.

BAKUS, G. J. 1959. Observations on the life history of the Dipper in Montana. Auk, 76:190–207.

BALDA, R. P., and S. CAROTHERS. 1968. Nest protection by the Brown-headed Cowbird *(Molothrus ater)*. Auk, 85:324–325.

BALDWIN, P. H., and D. E. OEHLERTS. 1964. The status of ornithological literature, 1964. Studies in Biol. Lit. and Communications, No. 4. Philadelphia: Biol. Abstracts, Inc.

BANG, B. G. 1966. The olfactory apparatus of the tubenosed birds (Procellariiformes). Acta Anatomica, 65:391–415.

———. and S. COBB. 1968. The size of the olfactory bulb in 108 species of birds. Auk, 85:55–61.

BANKS, R. C., and R. B. CLAPP. 1973. Birds imported into the United States in 1970. Special Scient. Report—Wildlife, No. 164, Washington, D.C.

BARASH, D. P. 1972. Lek behavior in the Broad-tailed Hummingbird. Wilson Bull., 84:202–203.

BARKER, R. J. 1958. Notes on some ecological effects of DDT sprayed on elms. Jour. Wildl. Mngt., 22:269–274.

BARNES, I. R. 1951. Persecution or freedom? Aud. Mag., 53:282–289.

BARROWS, W. B. 1912. *Michigan Bird Life.* Spec. Bull. Mich. Agr. College, East Lansing.

BARTHOLOMEW, G. A., JR., and T. J. CADE. 1957. The body temperature of the American Kestrel, *Falco sparverius*. Wilson Bull., 69:149–154.

———— and ————. 1963. The water economy of land birds. Auk, 80:504–539.

———— and R. E. MacMillen. 1961. Water economy of the California Quail and its use of sea water. Auk, 78:505–514.

———— and W. R. Dawson. 1952. Body temperatures in nestling Western Gulls. Condor, 54:58–60.

———— and C. H. Trost. 1970. Temperature regulation in the Speckled Mousebird, *Colius striatus*. Condor, 72:141–146.

Bartlett, G. 1952. A wholesale attraction, but not destruction, of migrating birds by the Albany (N.Y.) airport ceilometer. Feathers, 14:61–66.

————. 1956. Albany's ceilometer — killer of migrants. Feathers, 18:57–60.

Barton, A. J. 1958. A releaser mechanism in the feeding of nestling Chimney Swifts. Auk, 75:216–217.

Bartonek, J. C. 1965. Mortality of diving ducks on Lake Winnipegosis through commercial fishing. Can. Field Nat., 7:15–20.

————. and J. J. Hickey. 1969. Food habits of Canvasbacks, Redheads, and Lesser Scaup in Manitoba. Condor, 71:280–290.

Batts, H. L., Jr. 1958. The distribution and population of nesting birds on a farm in southern Michigan. Jack-Pine Warbler, 36:131–149.

Baxter, R. M. and E. K. Urban. 1970. On the nature and origin of the feather colouration in the Great White Pelican *Pelecanus onocrotalus roseus* in Ethiopia. Ibis, 112:336–339.

Beal, F. E. L. 1897. Some common birds in their relation to agriculture. Farmer's Bull. 54, U.S.D.A., Washington, D.C.

Beals, M. V. 1952. New age record for a Blue Jay. Bird-Banding, 23:168.

Beddall, B. G. 1957. Historical notes on avian classification. Syst. Zool., 6:129–136.

————. 1963. Range expansion of the Cardinal and other birds in the northeastern states. Wilson Bull., 75:140–158.

Bédard, J. 1969. The nesting of the Crested, Least, and Parakeet Auklets on St. Lawrence Island, Alaska. Condor, 71:386–398.

Beebe, F. L. 1960. The marine peregrines of the Northwest Pacific Coast. Condor, 62:145–189.

Beecher, W. J. 1969. Possible motion detection in the vertebrate middle ear. Bull. Chicago Acad. Sci., 11:155–210.

Beer, C. G. 1965. Clutch size and incubation behavior in Black-billed Gulls *(Larus bulleri)*. Auk, 82:1–18.

————. 1970. On the responses of Laughing Gull chicks *(Larus atricilla)* to calls of the adults. I. Recognition of the voices of the parents. Anim. Beh., 18:652–660.

Beer, J. R., L. D. Frenzel, and N. Hansen. 1956. Minimum space requirements of some nesting passerine birds. Wilson Bull., 68:200–209.

———— and D. Tibbitts. 1950. Nesting behavior of the Red-winged Blackbird. Flicker, 22:61–77.

———— and W. Tidyman. 1942. The substitution of hard seeds for grit. Jour. Wildl. Mngt., 6:70–82.

Belcher, J. W., and W. L. Thompson. 1969. Territorial defense and individual song recognition in the Indigo Bunting. Jack-Pine Warbler, 47:76–83.

Bell, D. G. 1965. Studies of less familiar birds. Brit. Birds, 58:139–145.

Bellrose, F. C. 1958. Celestial orientation by wild Mallards. Bird-Banding. 29:75–90.

———. 1963. Orientation behavior of four species of waterfowl. Auk, 80:257–289.

———. 1966. Orientation in waterfowl migration. In *Animal Orientation and Navigation*. Proc. 27th Annual Biol. Colloq. Corvallis: Oregon State Univ. Press.

———. 1968. Waterfowl migration corridors east of the Rocky Mountains in the United States. Biol. Notes No. 61, Ill. Nat. Hist. Sur., Urbana.

Bené, F. 1947. The feeding and related behavior of hummingbirds, with special reference to the Black-chin. Memoirs Boston Soc. Nat. Hist., 19:403–478.

Bennett, H. R. 1952. Fall migration of birds at Chicago. Wilson Bull., 64:197–220.

Berger, A. J. 1951a. Nesting density of Virginia and Sora Rails in Michigan. Condor, 53:202.

———. 1951b. Ten consecutive nests of a Song Sparrow. Wilson Bull., 63:186–188.

———. 1952. The comparative functional morphology of the pelvic appendage in three genera of Cuculidae. Amer. Midl. Nat., 47:513–605.

———. 1953a. Protracted incubation behavior of a female American Goldfinch. Condor, 55:151.

———. 1953b. Three cases of twin embryos in passerine birds. Condor, 55:157–158.

———. 1957. Nesting behavior of the House Sparrow. Jack-Pine Warbler, 35:86–92.

———. 1972. *Hawaiian Birdlife*. Honolulu: The Univ. Press of Hawaii.

——— and D. V. Howard. 1968. Anophthalmia in the American Robin. Condor, 70:386–387.

Bergstrom, E. A. 1952. Extreme old age in terns. Bird-Banding, 23:72–73.

———. 1956. Extreme old age in birds. Bird-Banding, 27:128–129.

Berndt, R., and H. Steinberg. 1968. Terms, studies and experiments on the problems of bird dispersion. Ibis, 110:256–269.

Biehn, E. R. 1951. Crop damage by wildlife in California. Game Bull., 5:1–71, Calif. Dept. Fish and Game.

Billings, S. M. 1968. Homing in Leach's Petrels. Auk, 85:36–43.

Bissonnette, T. H. 1937. Photoperiodicity in birds. Wilson Bull., 49:241–270.

———. 1939. Sexual photoperiodicity in the Blue Jay *(Cyanocitta cristata)*. Wilson Bull., 51:227–232.

Black, G. A. 1953. The use of birds among the prehistoric Indians of the Ohio Valley. Kent. Warbler, 29:3–7.

Blackwelder, R. E. 1967. A critique of numerical taxonomy. Syst. Zool., 16:64–72.

Blake, C. H. 1958. Respiration rates. Bird-Banding, 29:38–40.

———. 1959. Terminal migrants and transmigrants. Bird-Banding, 30:233.

———. 1969. An old warbler, Bird-Banding, 40:255.

BLEITZ, D. 1951. Nest of Pygmy Nuthatches attended by four parents. Condor, 53:150–151.

BLUMER, M., H. L. SANDERS, J. F. GRASSLE, and G. R. HAMPSON. 1971. A small oil spill. Environment, 13:2–12.

BOCK, C. E., and L. W. LEPTHIEN. 1972. Winter eruptions of Red-breasted Nuthatches in North America, 1950–1970. Amer. Birds, 26:558–560.

BOCK, W. J. 1961. Salivary glands in the Gray Jays *(Perisoreus)*. Auk, 78:355–365.

——. 1963. The cranial evidence for ratite affinities. Proc. Intern. Ornith. Congress, 13:39–54.

BOND, R. R. 1957. Ecological distribution of breeding birds in the upland forests of southern Wisconsin. Ecol. Monog., 27:351–384.

BONNOT, P. 1928. An outlaw Barn Owl. Condor, 30:320.

BOUDREAU, G. W. 1968. Alarm sounds and responses of birds and their application in controlling problem species. Living Bird, 7:27–46.

BOURNE, W. R. P. 1970. Special review—after the "Torrey Canyon" disaster. Ibis, 112:120–125.

BOWERS, D. E. 1959. A study of variation in feather pigments of the Wrentit. Condor, 61:38–45.

BOYD, E. M. 1958. Birds and some human diseases. Bird-Banding, 29:34–38.

——. 1962. A half-century's changes in the bird-life around Springfield, Massachusetts. Bird-Banding, 33:137–148.

—— and S. A. NUNNELEY. 1964. Banding records substantiating the changed status of ten species of birds since 1960 in the Connecticut Valley. Bird-Banding, 35:1–8.

BRACKBILL, H. 1952. Observations on remating in the American Robin, *Turdus migratorius.* Auk, 69:465–466.

——. 1960. Foot-quivering by foraging Hermit Thrushes. Auk, 77:477–478.

——. 1966. Herons leaving the water to defecate. Wilson Bull., 78:316.

BRAND, A. R. 1938. Vibration frequency of passerine bird song. Auk, 55:263–268.

—— and P. P. KELLOGG. 1939a. Auditory responses of Starlings, English Sparrows, and Domestic Pigeons. Wilson Bull., 51:38–41.

—— and ——. 1939b. The range of hearing of canaries. Science, 90:354.

BREWER, R. 1972. An evaluation of winter bird population studies. Wilson Bull., 84:261–277.

—— and J. A. ELLIS. 1958. An analysis of migrating birds killed at a television tower in east-central Illinois, September 1955–May 1957. Auk, 75:400–414.

BROCKWAY, B. F. 1967. The influence of vocal behavior on the performer's testicular activities in Budgerigars *(Melopsittacus undulatus)*. Wilson Bull., 79:328–334.

BRODKORB, P. 1949 (1950). The number of feathers in some birds. Quart. Jour. Fla. Acad. Sci. 12(4):241–245.

——. 1955. Numbers of feathers and weights of various systems in a Bald Eagle. Wilson Bull., 67:142.

——. 1960. How many species of birds have existed? Bull. Fla. State Mus., 5:41–53.

BROLEY, C. L. 1947. Migration and nesting of Florida Bald Eagles. Wilson Bull., 59:3–20.

———. 1958. The plight of the American Bald Eagle. Aud. Mag., 60:162–163.

BROOKS, W. S. 1968. Comparative adaptations of the Alaskan redpolls to the Arctic environment. Wilson Bull., 80:253–280.

BROUN, M. 1949. *Hawks Aloft: The Story of Hawk Mountain.* New York: Dodd Mead.

——— and B. V. GOODWIN. 1943. Flight-speeds of hawks and crows. Auk, 60:487–492.

BROWER, L. P., W. N. RYERSON, L. L. COPPINGER, and S. C. GLAZIER 1968. Ecological chemistry and the palatibility spectrum. Science, 161:1349–1350.

BROWN, J. L. 1959. Method of head scratching in the Wrentit and other species. Condor, 61:53.

———. 1963. Aggressiveness, dominance and social organization in the Stellar Jay. Condor, 65:460–484.

———. 1970. Cooperative breeding and altruistic behavior in the Mexican Jay, *Aphelocoma ultramarina.* Anim. Beh., 18:366–378.

———. 1972. Communal feeding of nestlings in the Mexican Jay *(Aphelocoma ultramarina)*: interflock comparisons. Anim. Beh., 20:395–403.

BROWN, L. H. 1966. Observations on some Kenya eagles. Ibis, 108:531–572.

——— and A. ROOT. 1971. The breeding behavior of the Lesser Flamingo, *Phoeniconaias minor.* Ibis, 113:147–172.

BROWN, W. H. 1971. Winter population trends in the Red-shouldered Hawk. Amer. Birds, 25:813–817.

BRUSH, A. H. 1969. On the nature of "cotingin." Condor, 71:431–433.

——— and H. SEIFRIED. 1968. Pigmentation and feather structure in genetic variants of the Gouldian Finch, *Poephila gouldiae.* Auk, 85:416–430.

BUCHHEISTER, C. W. 1960. What about problem birds? Aud. Mag., 62:116–118.

BUCKLEY, F. G. 1968. Behavior of the Blue-crowned Hanging Parrot *Loriculus galgulus* with comparative notes on the Vernal Hanging Parrot *L. vernalis.* Ibis, 110:145–164.

——— and P. A. BUCKLEY. 1968. Upside-down resting by young Green-rumped Parrotlets *(Forpus passerinus).* Condor, 70:89.

BUCKLEY, P. A., and F. G. BUCKLEY. 1968. Tongue flicking by a feeding Snowy Egret. Auk, 85:678.

——— and ———. 1972. Individual egg and chick recognition by adult Royal Terns *(Sterna maxima maxima).* Anim. Beh., 20:457–462.

BUMP, G. 1963. Status of the foreign game introduction program. Trans. N. A. Wildl. Conf., 28:240–247.

BURKE, V. E. M., and L. H. BROWN. 1970. Observations on the breeding of the Pink-backed Pelican, *Pelecanus rufescens.* Ibis, 112:499–512.

BURTT, H. E., and M. L. GILTZ. 1966. Recoveries of Starlings banded at Columbus, Ohio. Bird-Banding, 37:267–273.

——— and ———. 1971. Red-winged Blackbirds wintering in a decoy trap. Bird-Banding, 42:287–289.

BUSS, I. O., R. D. CONRAD, and J. R. REILLY. 1958. Ulcerative enteritis in the pheasant, Blue Grouse and California Quail. Jour. Wildl. Mngt., 22:446–449.

CADE, T. J., and J. L. BUCKLEY. 1953. A mass emigration of Sharp-tailed Grouse from the Tanana Valley, Alaska, in 1934. Condor, 55:313.

——— and J. A. DYBAS, JR. 1962. Water economy of the Budgerygah. Auk, 79:345–364.

——— and L. GREENWALD. 1966. Nasal salt secretions in falconiform birds. Condor, 68:338–350.

——— and G. L. MACLEAN. 1967. Transport of water by adult sandgrouse to their young. Condor, 69:323–343.

——— C. M. WHITE, and J. R. HAUGH. 1967. Peregrines and pesticides in Alaska. Raptor Res. News, 1:23–38.

———. 1968. Peregrins and pesticides in Alaska. Condor, 70:170–178.

CAHALANE, V. H. 1944. A nutcracker's search for buried food. Auk, 61:643.

———. 1955. Some effects of exotics on nature. Atlantic Nat., 10:176–185.

———. 1969. World conference on bird hazards to aircraft. Defenders of Wildlife, 44:272–273.

CALDER, W. A. 1968. Respiratory and heart rates of birds at rest. Condor, 70:358–365.

———. 1971. Temperature relationships and nesting of the Calliope Hummingbird. Condor, 73:314–321.

CALDWELL, L. D., E. P. ODUM, and S. G. MARSHALL. 1963. Comparisons of fat levels in migratory birds killed at a central Michigan and a Florida Gulf coast television tower. Wilson Bull., 75:428–434.

CARPENTER, R. E., and M. A. STAFFORD. 1970. The secretory rates and the chemical stimulus for secretion of the nasal salt glands in the Rallidae. Condor, 72:316–324.

CASE, N. A., and O. H. HEWITT. 1963. Nesting and productivity of the Red-winged Blackbird in relation to habitat. Living Bird, 2:7–20.

CASE, R. M. 1973. Bioenergetics of a covey of Bobwhites. Wilson Bull., 85:52–59.

CASEMENT, M. B. 1966. Migration across the Mediterranean observed by radar. Ibis, 108:461–491.

CHAMBERLAIN, B. R. 1954. Safety factor in a hanging nest. The Chat, 18:48–50.

CHAMBERLAIN, D. R., W. B. GROSS, G. W. CORNWELL, and H. S. MOSBY. 1968. Syringeal anatomy in the Common Crow. Auk, 85:244–252.

CHAPIN, J. P., and L. W. WING. 1959. The wideawake calendar, 1953 to 1958. Auk, 76:153–158.

CHETTLEBURGH, M. R. 1952. Observations on the collection and burial of acorns by jays in Hainault Forest. Brit. Birds, 45(10):359–364.

CHOATE, E. A. 1972. Spectacular hawk flight at Cape May, New Jersey, on 16 October 1970. Wilson Bull., 84:340–341.

CLAPP, R. B., and F. C. SIBLEY. 1966. Longevity records of some central Pacific seabirds. Bird-Banding, 37:193–197.

——— and C. D. HACKMAN. 1969. Longevity record for a breeding Great Frigatebird. Bird-Banding, 40:47.

CLARK, G. A., JR. 1969. Oral flanges of juvenile birds. Wilson Bull., 81:270–279.

———. 1970. Avian bill-wiping. Wilson Bull., 82:279–288.

CLENCH, M. H. 1970. Variability in body pterylosis with special reference to the genus *Passer.* Auk, 87:650–691.

COBB, S. 1959. On the angle of the cerebral axis in the American Woodcock. Auk, 76:55–59.

COCHRAN, W. W., G. G. MONTGOMERY, and R. R. GRABER. 1967. Migratory flights of *Hylocichla* thrushes in spring: a radio-telemetry study. Living Bird, 6:213–225.

COLLIAS, E. C., and N. E. COLLIAS. 1964. The development of nest-building behavior in a weaverbird. Auk, 81:42–52.

——— and ———. 1973. Further studies on development of nest-building behavior in a weaverbird *(Ploceus cucullatus).* Anim. Beh., 21:371–382.

COLLIAS, N. E. 1965. Evolution of nest building. Nat. Hist., 74(7):40–47.

——— and R. D. TABER. 1951. A field study of some grouping and dominance relations in Ring-necked Pheasants. Condor, 53:265–275.

COLTHERD, J. B. 1966. The domestic fowl in ancient Egypt. Ibis, 108:217–223.

CONWAY, W. G., and J. BELL. 1968. Observations on the behavior of Kittlitz's Sandplovers at the New York Zoological Park. Living Bird, 7:57–70.

COOCH, F. G. 1964. A preliminary study of the survival value of a functional salt gland in prairie Anatidae. Auk, 81:380–393.

COOCH, G. 1955. Observations on the autumn migration of Blue Geese. Wilson Bull., 67:171–174.

COOKE, W. W. 1915. Bird migration. USDA Bull. 185.

CORNWALL, G. W., and A. B. COWAN. 1963. Helminth populations of the Canvasback *(Aythya valisneria)* and host-parasite environmental relationships. Trans. N. A. Wildl. Conf., 28:173–199.

COTTAM, C. 1962. Is the Attwater Prairie Chicken doomed? Audubon, 64:328–330.

———, C. S. WILLIAMS, and C. A. SOOTER. 1942. Flight and running speeds of birds. Wilson Bull., 34:121–131.

COTTRELL, V. M. 1955. Strange is the Kiwi. Nat. Mag., 48:41–43, 52.

CRACRAFT, J. 1971. A new family of the Hoatzin-like birds (Opisthocomiformes) from the Eocene of South America. Ibis, 113:229-233.

———. 1972. A new Cretaceous charadriiform family. Auk, 89:36–46.

CRAIG, W. 1918. Appetites and aversions as constituents of instincts. Woods Hole Biol. Bull., 34:91–107.

CRAIGHEAD, F., and J. CRAIGHEAD. 1956. *Hawks, Owls and Wildlife.* Harrisburg, Pa.: Stackpole.

CROOK, J. H. 1970. Social organization and the environment: aspects of contemporary social ethology. Anim. Beh., 18:197–209.

CROSBY, G. T. 1972. Spread of the Cattle Egret in the western hemisphere. Bird-Banding, 43:205-212.

CURRY-LINDAHL, K. 1963. New theory on a fabled exodus. Nat. Hist., 72:47–53.

CUTHBERT, N. L. 1954. A nesting study of the Black Tern in Michigan, Auk, 71:36–63.

DAMBACH, C. A., and D. L. LEEDY. 1948. Ohio studies with repellent materials with notes on damage to corn by pheasants and other wildlife. Jour. Wildl. Mngt., 12:392–398.

DANIEL, J. C., JR. 1957. An embryological comparison of the domestic fowl and the Red-winged Blackbird. Auk, 74:340–358.

D'ARMS, E., and D. R. GRIFFIN. 1972. Balloonists' reports of sounds audible to migrating birds. Auk, 89:269–279.

DAVIS, D. E. 1942. Number of eggs laid by Herring Gulls. Auk, 59:549–554.

———. 1955. Determinate laying in Barn Swallows and Black-billed Magpies. Condor, 57:81–87.

———. 1959. Observations on territorial behavior of Least Flycatchers. Wilson Bull., 71:73–85.

DAVIS, J. 1957. Comparative foraging behavior of the Spotted and Brown Towhees. Auk, 74:129–166.

———. 1958. Singing behavior and the gonad cycle of the Rufous-sided Towhee. Condor, 60:308–336.

———. 1971. Breeding and molt schedules of the Rufous-collared Sparrow in coastal Peru. Condor, 73:127–146.

——— and L. WILLIAMS. 1957. Irruptions of Clark Nutcracker in California. Condor, 59:297–307.

——— and ———. 1964. The 1961 irruption of the Clark's Nutcracker in California. Wilson Bull., 76:10–18.

DAVIS, M. 1969. Siberian Crane longevity. Auk, 86:347.

DAVIS, R. 1968. Food requirements of Barn Swallow nestlings. Inland Bird Bird Banding News, 40:63.

DAVISON, V. E. 1962. Taste, not color, draws birds to berries and seeds. Audubon, 64:346–350.

DAWN, W. 1959. Cattle Egrets provoke cattle to move and pick flies off bulls. Auk, 76:97–98.

DAWSON, W. R., and F. C. EVANS 1960. Relation of growth and development to temperature regulation in nestling Vesper Sparrows. Condor, 62:329–340.

DE BENEDICTIS, P. A. 1966. The bill-brace feeding behavior of the Galapagos Finch *(Geospiza conirostris)*. Condor, 68:206–208.

DE KIRILINE, L. 1954. The voluble singer of the treetops. Aud. Mag., 56:109–111.

DELIUS, J. D. 1973. Agnostic behavior of juvenile gulls, a neuroethological study. Anim. Beh., 21:236–246.

DENNIS, J. V. 1951, 1952. A balanced diet for birds. Aud. Mag., 53:398–403; 54:52–57.

———. 1957. Food distributors in tree trunk and tree top. Aud. Mag., 59:36–41.

———. 1964. Woodpecker damage to utility poles: with special reference to the role of territory and resonance. Bird-Banding, 35:225–253.

———. 1967a. Damage by Golden-fronted and Ladder-backed Woodpeckers to fence posts and utility poles in South Texas. Wilson Bull., 79:75–88.

————. 1967b. Fall departure of the Yellow-breasted Chat *(Icteria virens)* in eastern North America. Bird-Banding, 38:130–135.

————. 1971. Species using Red-cockaded Woodpecker holes in northeastern South Carolina. Bird-Banding, 42:79–87.

DEXTER, R. W. 1952. Extra-parental cooperation in the nesting of Chimney Swifts. Wilson Bull., 64:133–139.

————. 1959. Two 13-year-old age records for the House Sparrow. Bird-Banding, 30:182.

————. 1960. Analysis of Chimney Swift returns at Kent, Ohio, 1956–1959. Bird-Banding, 31:87–89.

DICE, L. R. 1945. Minimum intensities of illumination under which owls can find dead prey by sight. Amer. Nat., 79:385–416.

DICKINSON, J. C., JR. 1969. A string-pulling Tufted Titmouse. Auk, 86:559.

DILGER, W. C. 1954. Electrocution of parakeets at Agra, India. Condor, 56:102–103.

————. 1956. Hostile behavior and reproductive isolating mechanisms in the avian genera *Catharus* and *Hylocichla.* Auk, 73:313–353.

————. 1957. The loss of teeth in birds. Auk, 74:103–104.

————. 1960. Agonistic and social behavior of captive redpolls. Wilson Bull., 72:115–132.

DIXON, J. B. 1937. The Golden Eagle in San Diego County, California. Condor, 39:49–56.

DIXON, K. L. 1959. Ecological and distributional relations of desert scrub birds of western Texas. Condor, 61:397–409.

DOUVILLE, C. H., and C. E. FRILEY, JR. 1957. Records of longevity in Canada Geese. Auk, 74:510.

DOW, D. D. 1965. The role of saliva in food storage by the Gray Jay. Auk, 82:139–154.

DOWNER, A. C. 1972. Longevity records of Indigo Buntings wintering in Jamaica. Bird-Banding, 43:287.

DOWNS, J. R. 1963. Can Blue Jays swim? Bird-Banding, 34:221.

DOWNS, W. G. 1959. Little Egret banded in Spain taken in Trinidad. Auk, 76:241–242.

DREIS, R. E., and G. O. HENDRICKSON. 1952. Wood Duck production from nest-boxes and natural cavities in the Lake Odessa area, Iowa, in 1951. Iowa Bird Life, 22:19–22.

DUKE, G. E., G. A. PETRIDES, and R. K. RINGER. 1968. Chromium-51 in food metabolizability and passage rates studies with the Ring-necked Pheasant. Poultry Sci., 47:1356–1364.

DYKSTRA, W. W. 1960. Nuisance bird control. Aud. Mag., 62:118–119.

DZUBIN, A. 1965. A study of migrating Ross Geese in western Saskatchewan. Condor, 67:511–534.

EATON, S. W. 1958. A life history study of the Louisiana Waterthrush. Wilson Bull., 70:211–236.

EDWARDS, E. P. 1943. Hearing ranges of four species of birds. Auk, 60:239–241.

EKLUND, C. R. 1959. Antarctic ornithological studies during the IGY. Bird-Banding, 30:114–118.

ELDER, W. H., and C. M. KIRKPATRICK. 1952. Predator control in the light of recent wildlife management concepts. Wilson Bull., 64:126–128.

ELDER, W. H. and M. W. WELLER. 1954. Duration of fertility in the domestic Mallard hen after isolation from the drake. Jour. Wildl. Mngt., 18:495–502.

EMANUEL, V. L. 1961. Another probable record of an Eskimo Curlew on Galveston Island, Texas. Auk, 78:259–260.

EMLEN, J. T., JR. 1937. Bird damage to almonds in California. Condor, 39:192–197.

———. 1941. An experimental analysis of the breeding cycle of the Tri-colored Red-wing. Condor, 43:209–219.

———. 1952. Social behavior in nesting Cliff Swallows. Condor, 54:177–199.

———. 1957. Display and mate selection in the whydahs and bishop birds. The Ostrich (Dec. 1957):202–213.

———. 1969. The squeak lure and predator mobbing in wild birds. Anim. Beh., 17:515–516.

———. 1973. Territorial aggression in wintering warblers at Bahama agave blossoms. Wilson Bull., 85:71–74.

——— and R. L. PENNY. 1966. The navigation of penguins. Scient. Amer., 215:105–113.

EMLEN, S. T. 1969. The development of migratory orientation in young Indigo Buntings. Living Bird, 8:113–126.

ENDERSON, J. H. 1965. A breeding and migration survey of the Peregrine Falcon. Wilson Bull., 77:327–339.

ERICKSON, J. E. 1969. Banding studies of wintering Baltimore Orioles in North Carolina, 1963–1966. Bird-Banding, 40:181–198.

ERPINO, M. J. 1968. Nest-related activities of Black-billed Magpies. Condor, 70:154–165.

ETKIN, W. (ed.). 1964. *Social Behavior and Organization Among Vertebrates.* Chicago: Univ. of Chicago Press.

EVANS, P. R. 1969. Ecological aspects of migration, and pre-migratory fat deposition in the Lesser Redpoll, *Carduelis flammea cabaret.* Condor, 71:316–330.

EVANS, S. M., and G. R. PATTERSON. 1971. The synchronization of behavior in flocks of estrildine finches. Anim. Beh., 19:429–438.

EVENDEN, F. G. 1957. Observations on nesting behavior of the House Finch. Condor, 59:112–117.

EYSTER, M. B. 1954. Quantitative measurement of the influence of photoperiod, temperature, and season on the activity of captive songbirds. Ecol. Monog., 24:1–28.

FARNER, D. S. 1949. Age groups and longevity in the American Robin: comments, further discussion, and certain revisions. Wilson Bull., 61:68–81.

——— and D. L. SERVENTY. 1959. Body temperature and the ontogeny of thermoregulation in the Slender-billed Shearwater. Condor, 61:426–433.

FEDUCCIA, A. 1972. Variation in the posterior border of the sternum in some tree-trunk foraging birds. Wilson Bull., 84:315–328.

FERRY, P. 1952. The battle of the eggs. Nat. Hist., 61:176–181.

FICKEN, M. S. 1962. Agonistic behavior and territory in the American Redstart. Auk, 79:607–632.

————. 1964. Nest-site selection in the American Redstart. Wilson Bull., 76:189–190.

———— and R. W. FICKEN. 1968. Courtship of Blue-winged Warblers, Golden-winged Warblers, and their hybrids. Wilson Bull., 80:161–172.

FICKEN, R. W. 1963. Courtship and agonistic behavior of the Common Grackle, *Quiscalus quiscala*. Auk, 80:52–72.

FISHER, A. K. 1908. Economic value of predaceous birds and mammals. *Yearbook*, Washington, D.C.: USDA.

FISHER, H. I. 1957. The function of M. depressor caudae and M. caudofemoralis in pigeons. Auk, 74:479–486.

————. 1966a. Hatching and the hatching muscle in some North American ducks. Trans. Ill. State Acad. Sci., 59:305–325.

————. 1966b. Airplane-albatross collisions on Midway Atoll. Condor, 68:229–242.

————. 1966c. Midway's deadly antennas. Audubon, 68:220–223.

————. 1970. The death of Midway's antennas. Audubon, 72:62–63.

————. 1971. Experiments on homing in Laysan Albatrosses, *Diomedea immutabilis*. Condor, 73:389–400.

————. 1972. Sympatry of Laysan and Black-footed Albatrosses. Auk, 89:381–402.

———— and E. E. DATER. 1961. Esophageal diverticula in the Redpoll, *Acanthis flammea*. Auk, 78:528–531.

FISHER, J. 1954. *A History of Birds*. Boston: Houghton.

————. 1964. Endangered species lists, pp. 268–273 in Fisher and Peterson, *The World of Birds*. New York: Doubleday.

———— and S. CHARLTON. 1967. A tragedy of errors. Audubon, 69:72–85.

FORSYTH, B. J., and D. JAMES. 1971. Springtime movements of transient nocturnally migratory landbirds in the Gulf Coastal Bend region of Texas. Condor, 73:193–207.

FOWLER, J. M., and J. B. COPE. 1964. Notes on the Harpy Eagle in British Guiana. Auk, 81:257–273.

FRANKHAUSER, D. P. 1971. Annual adult survival rates of blackbirds and Starlings. Bird-Banding, 42:36–42.

FRENCH, N. R. 1954. Notes on breeding activities and on gular sacs in the Pine Grosbeak. Condor, 56:83–85.

FRIEDMANN, H. 1925. Notes on differential threshold of reaction to vitamin D deficiency in the House Sparrow and the chick. Biol. Bull. 69, 1:71–74.

————. 1934. The instinctive emotional life of birds. Psychoanal. Rev. 21, 384:1–57.

————. 1955. The honey-guides. Bull. 208, U. S. Nat'l Mus., 1–292.

————. 1963. Host relations of the parasitic cowbirds. Bull. 223, U. S. Nat'l Mus., 1–276.

FRINGS, H., and W. A. BOYD. 1952. Evidence for olfactory discrimination by the Bobwhite Quail. Amer. Midl. Nat., 48:181–184.

———— and B. SLOCUM. 1958. Hearing ranges for several species of birds. Auk, 75:99–100.

FRITSCH, L. E., and I. O. BUSS. 1958. Food of the American Merganser in Unakwik Inlet, Alaska. Condor, 60:410–411.

FRY, C. H. 1969. The recognition and treatment of venomous and non-venomous insects by small bee-eaters. Ibis, 111:23–29.

FULLER, R. W., and E. BOLEN. 1963. Dual Wood Duck occupancy of a nesting box. Wilson Bull., 75:94–95.

GANIER, A. F. 1964. The alleged transportation of its eggs or young by the Chuck-will's-widow. Wilson Bull., 76:19–27.

GASTON, S., and M. MENAKER. 1968. Pineal function: the biological clock in the sparrow? Science, 160:1125–1127.

GEIS, M. B. 1956. Productivity of Canada Geese in the Flathead Valley, Montana. Jour. Wildl. Mngt., 20:409–419.

GIER, H. T. 1952. The air sacs of the loon. Auk, 69:40–49.

GILL, F. B. 1967. Birds of Rodriguez Island (Indian Ocean). Ibis, 109:383–390.

———. 1971. Tongue structure of the sunbird *Hypogramma hypogrammica*. Condor, 73:485–486.

GINGERICH, P. D. 1972. A new partial mandible of *Ichthyornis*. Condor, 74:471–473.

GLENNY, F. H. 1954. Antarctica as a center of origin of birds. Ohio Jour. Sci., 54:307–314.

GOCHFELD, M. 1972. The proposed registry for nesting data on neotropical birds. Amer. Birds, 26:18–20.

GOERTZ, J. W. 1962. An opossum-titmouse incident. Wilson Bull., 74:189–190.

GOFORTH, W. R. 1971. The three-bird chase in Mourning Doves. Wilson Bull., 83:419–424.

——— and T. S. BASKETT. 1971. Social organization of penned Mourning Doves. Auk, 88:528–542.

GONZALES, R. B. 1971. Report on the 1969 status of the Monkey-eating Eagle on Mindanao Island, Philippines. Bull. Intern. Council for Bird Pres., XI:154–168.

GOODGE, W. R. 1959. Locomotion and other behavior of the Dipper. Condor, 61:4–17.

GOODMAN, J. M. 1960. *Aves Incendiaria*. Wilson Bull., 72:400–401.

GOWER, C. 1939. The use of the bursa of Fabricius as an indication of age in game birds. Trans. N. A. Wildl. Conf., 4:426–430.

GOODWIN, D. 1965. A comparative study of captive blue waxbills (Estrildidae). Ibis, 107:285–315.

GOSLOW, G. E., JR., 1972. Adaptive mechanisms of the raptor pelvic limb. Auk, 89:47–64.

GOTTLIEB, G. 1965. Components of recognition in ducklings. Nat. Hist., 74:12–19.

———. 1966. Species identification by avian neonates: contributory effect of perinatal auditory stimulation. Anim. Beh., 14:282–290.

GRABER, R. R. 1955. Artificial incubation of some non-galliform eggs. Wilson Bull., 67:100–109.

———. 1965. Night flight with a thrush. Audubon, 67:368–374.

———. 1968. Nocturnal migration in Illinois—different points of view. Wilson Bull., 80:36–71.

————. 1972. Studies of bird hazards to aircraft. Can. Wildl. Report Series, 14:1–105. (Review by R. R. Graber.)

———— and W. C. Cochran. 1960. Evaluation of an aural record of nocturnal migration. Wilson Bull., 72:253–273.

———— and J. W. Graber. 1965. Variation in avian brain weights with special reference to age. Condor, 67:300–318.

Grandy, J. W. 1972. Digestion and passage of the blue mussels eaten by Black Ducks. Auk, 89:189–190.

Greenewalt, C. H. 1960. *Hummingbirds.* New York: Doubleday.

————. 1968. *Bird Song: Acoustics and Physiology.* Washington, D.C.: Smiths. Inst. Press.

Gregory, J. T. 1952. The jaws of the Cretaceous toothed birds, *Ichthyornis* and *Hesperornis.* Condor, 54:73–88.

Griffin, D. R. 1940. Homing experiments with Leach's Petrels. Auk, 57:61–74.

————. 1953. Acoustic orientation in the Oil Bird, *Steatornis.* Proc. Nat. Acad. Sci., 39:884–893.

Grimm, R. J., and W. M. Whitehouse. 1963. Pellet formation in a Great Horned Owl: a roentgenographic study. Auk, 80:301–306.

Grinyer, I., and J. C. George. 1969. Some observations on the ultrastructure of the hummingbird pectoral muscles. Can. J. Zool., 47:771–780.

Griscom, L. 1945. *Modern Bird Study.* Cambridge, Mass.: Harvard Univ. Press.

Gross, A. O. 1952. Nesting of Hicks' Seedeater at Barro Colorado Island, Canal Zone. Auk, 69:433–446.

————. 1958. Life history of the Bananaquit of Tobago Island. Wilson Bull., 70:257–279.

————. 1964. Albinism in the Herring Gull. Auk, 81:551–552.

————. 1965. The incidence of albinism in North American birds. Bird-Banding, 39:67–71.

————. 1968. Albinistic eggs (white eggs) of some North American birds. Bird-Banding, 39:1–6.

Grzimek, B. 1961. The last great herds of Africa. Nat. Hist., 70:8–21.

Guggisberg, C. A. W. 1972. *Crocodiles: Their Natural History, Folklore and Conservation.* Harrisburg, Pa.: Stackpole Books.

Gullion, G. W. 1951. The frontal shield of the American Coot. Wilson Bull., 63:157–166.

Hailman, J. P. 1959. Convergence in passerine alarm calls. Bird-Banding, 30:232.

————. 1960. A field study of the Mockingbird's wing-flashing behavior and its association with foraging. Wilson Bull., 72:346–357.

————. 1968. Visual-cliff responses of newly-hatched chicks of the Laughing Gull *Larus atricilla.* Ibis, 110:197–200.

Hamerstrom, F. 1969. A harrier population study. Chapt. 31 in *Peregrine Falcon Populations,* J. J. Hickey (ed). Madison: Univ. Wisc. Press.

Hamilton, W. J., III. 1962a. Evidence concerning the function of nocturnal call notes of migratory birds. Condor, 64:390–401.

————. 1962b. Does the Bobolink navigate? Wilson Bull., 74:357–366.

------. 1965. Sun-oriented display of the Anna's Hummingbird. Wilson Bull., 77:38–44.

------ and M. C. HAMMOND. 1960. Oriented overland spring migration of pinioned Canada Geese. Wilson Bull., 72:385–391.

HAMRUM, C. L. 1953. Experiments on the senses of taste and smell in the Bob-white Quail *(Colinus virginianus virginianus)*. Amer. Midl. Nat., 49:872–877.

HANN, H. W. 1937. Life history of the Oven-bird in southern Michigan. Wilson Bull., 49:145–237.

------. 1940. Polyandry in the Oven-bird. Wilson Bull., 52:69–72.

------. 1945. *An Introduction to Ornithology*. Ann Arbor, Mich.: Edwards Brothers.

------. 1953. *The Biology of Birds*. Ann Arbor, Mich.: Edwards Brothers.

HANNA, E. C. 1917. Further notes on the White-throated Swifts of Slover Mountain. Condor, 19:1–8.

HANSON, H. C. 1953. Inter-family dominance in Canada Geese. Auk, 70:11–16.

------ and C. W. KOSSACK. 1957. Weight and body-fat relationships of Mourning Doves in Illinois. Jour. Wildl. Mngt., 21:169–181.

HARDY, J. W. 1961. Studies in behavior and phylogeny of certain New World jays (Garrulinae). Univ. Kans. Sci. Bull., 42:13–49.

------. 1963. Epigamic and reproductive behavior of the Orange-fronted Parakeet. Condor, 65:169–199.

------. 1970. Duplex nest construction by Hooded Oriole circumvents cowbird parasitism. Condor, 72:491.

HARRISON, C. J. O. 1965. The chestnut-red melanin in schizochroic plumages. Ibis, 107:106–108.

------. 1966. Alleged xanthochroism in bird plumages. Bird-Banding, 37:121.

HARTMAN, F. A. 1955. Heart weight in birds. Condor, 57:221–238.

------. 1959. Sparrow Hawks attempting to breed in the laboratory. Wilson Bull., 71:384–385.

HATCH, D. E. 1970. Energy conserving and heat dissipating mechanisms of the Turkey Vulture. Auk, 87:111–124.

HAUGEN, A. O. 1952. Trichomoniasis in Alabama Mourning Doves. Jour. Wildl. Mngt., 16:164–169.

HAVERSCHMIDT, F. 1946. Observations on the breeding habits of the Little Owl. Ardea, 34:214–246.

------. 1958. Notes on the breeding habits of *Panyptila cayennensis*. Auk, 75:121–130.

HAYNE, D. W., and H. A. CARDINELL. 1949. Damage to blueberries by birds. Mich. State Coll. Agr. Exp. Sta. Quart. Bull., 32:213–219.

HAYS, H. 1972. Polyandry in the Spotted Sandpiper. Living Bird, 11:43–57.

------ and G. DONALDSON. 1970. Sand-kicking camouflages young Black Skimmers. Wilson Bull., 82:100.

HAYS, H., and R. W. RISEBROUGH. 1971. The early warning of the terns. Nat. Hist., 80:39–47.

------. 1972. Pollutant concentrations in abnormal young terns from Long Island Sound. Auk, 89:19–35.

HAZELHOFF, E. H. 1951. Structure and function of the lung of birds. Poultry Sci., 30:3–10.

HEATH, J. E. 1962. Temperature fluctuation in the Turkey Vulture. Condor, 64:234–285.

HECHT, W. R. 1951. Nesting of the Marsh Hawk at Delta, Manitoba. Wilson Bull., 63:167–176.

HEINTZELMAN, D. S., and R. MacCLAY. 1971. An extraordinary autumn migration of White-breasted Nuthatches. Wilson Bull., 83:129–131.

HELMS, C. W., and R. B. SMYTHE. 1969. Variation in major body components of the Tree Sparrow *(Spizella arborea)* sampled within the winter range. Wilson Bull., 81:280–292.

HENDRICKSON, H. T. 1969. A comparative study of the egg white proteins of some species of the avian order Gruiformes. Ibis, 111:80–91.

HENSLEY, M. M., and J. B. COPE. 1951. Further data on removal and repopulation of the breeding birds in a spruce-fir forest community. Auk, 68:483–493.

HERBERT, R. A., and K. G. S. HERBERT. 1965. Behavior of Peregrine Falcons in the New York City region. Auk, 82:62–94.

HERMAN, C. M. 1962. The role of birds in the epizootiology of eastern encephalitis. Auk, 79:99–103.

———, L. N. LOCKE, and G. M. CLARK. 1962. Foot abnormalities of wild birds. Bird-Banding, 33:191–198.

HESS, E. H. 1959. Imprinting. Science, 130:133–141.

HEYDWEILLER, A. M. 1935. A comparison of winter and summer territories and seasonal variations of the Tree Sparrow. Bird-Banding, 6:1–11.

HICKEY, J. J., and D. W. ANDERSON. 1968. Chlorinated hydrocarbons and eggshell changes in raptorial and fish-eating birds. Science, 162:271–273.

——— and L. B. HUNT. 1960. Initial songbird mortality following a Dutch elm disease control program. Jour. Wildl. Mngt., 24:259–265.

———, J. A. KEITH, and F. B. COON. 1966. An exploration of pesticides in a Lake Michigan ecosystem. J. Appl. Ecol., 3:141–154.

HILL, D. O. 1964. Light sensitivity and the function of the nictitating membrane in a nocturnal owl. Condor, 66:305–306.

HINDE, R. A. 1959. Some recent trends in ethology. In S. Koch, ed., *Phychology, a Study of a Science.* Study I, Vol. 2, pp. 561–610. New York: McGraw-Hill.

———. 1966. Animal behaviour. In H. Mahan, ed., *A Synthesis of Ethology and Comparative Psychology.* New York: McGraw-Hill.

HINDWOOD, K. A. 1959. The nesting of birds in the nests of social insects. Emu, 59:1–36.

HOBSON, R. L. 1952. Wrens and skulls. Kent. Warbler, 28:36.

HOCHBAUM, H. A. 1942. Sex and age determination of waterfowl by cloacal examination. Trans. N. A. Wildl. Conf., 7:299–307.

———. 1944. The Canvasback on a prairie marsh. Washington, D.C.: Amer. Wildl. Inst.

HOFFMEISTER, D. F., and H. W. SETZER. 1947. The postnatal development of two broods of Great Horned Owls. Univ. of Kans. Publ., 1:157–173.

HOFSLUND, P. B. 1954. The hawk pass at Duluth. Flicker, 26:96–99.

———. 1959. Fall migration of Herring Gulls from Knife Island, Minnesota. Bird-Banding, 30:104–114.

———. 1966. Hawk migration over the western tip of Lake Superior. Wilson Bull., 78:79–87.

HÖHN, E. O. 1967a. The relevance of J. Christian's theory of a density-dependent endocrine population regulating mechanism to the problem of population regulation in birds. Ibis, 109:445–446.

———. 1967b. Observations on the breeding biology of Wilson's Phalarope *(Steganopus tricolor)* in Central Alberta. Auk, 84:220–244.

———. 1971. Observations on the breeding behavior of Grey and Red-necked Phalaropes. Ibis, 113:335–348.

HOLCOMB, L. C. 1965. Long nest attentiveness for a Cardinal. Condor, 67:359.

———. 1966. Red-winged Blackbird nestling development. Wilson Bull., 78:283–288.

HOLCOMB, R. W. 1969. Oil in the ecosystem. Science, 166:204–206.

HOLMES, R. T. 1966. Molt cycle of the Red-backed Sandpiper *(Calidris alpina)* in western North America. Auk, 83:517–533.

———. 1971. Latitudinal differences in the breeding and molt schedules of Alaskan Red-backed Sandpipers *(Calidris alpina)*. Condor, 73:93–99.

HOLSTEIN, V. 1942. Duehogen *Astur gentilis dubious* (Sparrman). Biologiske Studier over Danske Rovfugle, I:1–55. English summary.

HOMBERG, L. 1957. Fishing Crows. Fauna och Flora, 5:182–185.

HOPKINS, G. H. E., and T. CLAY. 1952. *A Check List of the Genera and Species of Mallophaga*. London: British Museum.

HOPKINS, M., JR. 1953. The Black Vulture as a predator in southern Georgia. Oriole, 18:15–17.

HORAK, G. J. 1970. A comparative study of the foods of the Sora and Virginia Rail. Wilson Bull., 82:206–213.

HORWICH, R. H. 1965. An ontogeny of wing-flashing in the Mockingbird with reference to other behaviors. Wilson Bull., 77:264–281.

———. 1969. Behavioral ontogeny of the Mockingbird. Wilson Bull., 81:87–93.

HOU, H. C. 1929–1931. Relation of the preen gland *(Glandula uropygialis)* of birds to rickets. Chinese Jour. Physiol., 3:171–182; 4:79–92; 5:11–18.

HOWARD, H. 1952. The prehistoric avifauna of Smith Creek Cave, Nevada, with a description of a new gigantic raptor. Bull. So. Calif. Acad. Sci., 51:50–54.

HOWELL, A. B. 1920. Habits of *Oceanodroma leucorhoa beali* vs *O. socorroensis*. Condor, 22:41–42.

HOWELL, J. C., and J. T. TANNER. 1951. An accident to migrating birds at the Knoxville Airport. Migrant, 22:61–62.

HOWELL, T. R. 1955. A southern hemisphere migrant in Nicaragua. Condor, 57:188–189.

——— and G. A. BARTHOLOMEW. 1961. Temperature regulation in nesting Bonin Island Petrels, Wedge-tailed Shearwaters, and Christmas Island Shearwaters. Auk, 78:343–354.

—— and ——. 1969. Experiments on nesting behavior of the Red-tailed Tropicbird, *Phaëthon rubricauda.* Condor, 71:113–119.

—— and W. R. DAWSON. 1954. Nest temperatures and attentiveness in the Anna Hummingbird. Condor, 56:93–97.

HRUBANT, H. E. 1955. An analysis of the color phases of the Eastern Screech Owl, *Otus asio,* by the gene frequency method. Amer. Nat., 89:223–230.

HUGGINS, R. A. 1941. Egg temperatures of wild birds under natural conditions. Ecology, 22:148–157.

HUGHES, M. R. 1970. Relative kidney size in nonpasserine birds with functional salt glands. Condor, 72:164–168.

HUMPHREY, P. S. 1958. Diving of a captive Common Eider. Condor, 60:408–410.

HUMPHRIES, D. A., and P. M. DRIVER. 1967. Erratic display as a device against predators. Science, 156:1767–1768.

HUNT, J. H. 1971. A field study of the Wrenthrush, *Zeledonia coronata.* Auk, 88:1–20.

HYNDMAN, C. C., and A. S. HYNDMAN. 1972. The shell pigment of Golden Eagle eggs. Condor, 74:200–201.

INGRAM, C. 1959. The importance of juvenile cannibalism in the breeding biology of certain birds of prey. Auk, 76:218–226.

INOUE, Y. 1954. The swallows are remembered. Bull. Mass. Aud. Soc., 38:57–58.

IRVING, L. 1960a. Nutritional condition of Water Pipits on Arctic nesting grounds. Condor, 62:469–472.

——. 1960b. Birds of Anaktuviik Pass, Kobub and Old Crow: a study in Arctic adaptation. U. S. Nat'l. Mus. Bull. 217.

JACKSON, J. A. 1973. Histoplasmosis—an occupational hazard for bird-banders? Inland Bird Banding News, 45:52–57.

JAEGER, E. C. 1948. Does the Poor-will "hibernate"? Condor, 50:45–46.

——. 1949. Further observations on the hibernation of the Poor-will. Condor, 51:105–109.

JEHL, J. R., JR., 1968. The systematic position of the Surfbird, *Aphriza virgata.* Condor, 70:206–210.

——. 1973. Studies of a declining population of Brown Pelicans in northwestern Baja, California. Condor, 75:69–79.

JEMISON, E. S., and R. H. CHABRECK. 1962. Winter Barn Owl foods in a Louisiana marsh. Wilson Bull., 74:95–96.

JENNI, D. A. and G. COLLIER. 1972. Polyandry in the American Jacana (*Jacana spinosa*). Auk, 89:743–765.

JENKINSON, M. A., and R. M. MENGEL. 1970. Ingestion of stones by goat-suckers (Caprimulgidae). Condor, 72:236–237.

JICKLING, L. 1940. Mr. Bob-white sticks it out. Jack-Pine Warbler, 18:114–115.

JOHNS, J. E. 1964. Testosterone-induced nuptial feathers in phalaropes. Condor, 66:449–455.

JOHNSGARD, P. A. 1960. A quantitative study of sexual behavior of Mallards and Black Ducks. Wilson Bull., 72:133–155.

—— and J. KEAR. 1968. A review of parental carrying of young by waterfowl. Living Bird, 7:89–102.

JOHNSON, O. W. 1968. Some morphological features of avian kidneys. Auk, 85:216–228.

—— and J. N. MUGAAS. 1970. Quantitative and organizational features of the avian renal medulla. Condor, 72:288–292.

JOHNSON, R. A. 1941. Nesting behavior of the Atlantic Murre. Auk, 58:153–163.

JOHNSON, R. E. 1965. Reproductive activities of Rosy Finches, with special reference to Montana. Auk, 82:190–205.

JOHNSTON, D. W. 1958. Sex and age characters and salivary glands of the Chimney Swift. Condor, 60:73–84.

——. 1961. Timing of annual molt in the Glaucous Gulls of northern Alaska. Condor, 63:474–478.

——. 1963. Heart weights of some Alaskan birds. Wilson Bull., 75:435–446.

——. 1966. A review of the vernal fat deposition picture in overland migrant birds. Part I: the White-Throated Sparrow at the southern edge of its wintering range. Bird-Banding, 37:172–182.

—— and T. P. HAINES. 1957. Analysis of mass bird mortality in October, 1954. Auk, 74:447–458.

—— and R. McFARLANE. 1967. Migration and bioenergetics of flight in the Pacific Golden Plover. Condor, 69:156–168.

JUBB, R. A. 1966. A longevity record for the Yellowbill Duck *(Anas undulata)*. Bokmakierie, 18:44–45.

KAHL, M. P., JR. 1963. Thermoregulation in the Wood Stork, with special reference to the role of the legs. Physiol. Zool., 36:141–151.

——. 1971. Spread-wing postures and their possible functions in the Ciconiidae. Auk, 88:715–722.

——. 1972. The pink tide. Nat. Hist., 81(5):64–71.

KALE, H. W., II. 1966. Plumages and molts in the Long-billed Marsh Wren. Auk, 83:140–141.

KALMBACH, E. R., and J. F. WELCH. 1946. Colored rodent baits and their value in safeguarding birds. Jour. Wildl. Mngt., 10:353–360.

KEE, D. T. 1964. Natural color preferences of the domestic chicken and European Quail. Unpubl. PhD. dissertation, Mich. State Univ., East Lansing.

KEETON, W. T. 1971. Evidence that birds use magnetic cues in migration. Ind. Aud. Quart., 49:66–67.

KELLY, J. W. 1956. Prolonged incubation by an Anna Hummingbird. Condor, 58:163.

KEMPER, C. A. 1958. Bird destruction at a TV tower. Aud. Mag., 60:270–271. 290–293.

——. 1964. A tower for TV: 30,000 dead birds. Audubon, 66:86–90.

KENDEIGH, S. C. 1932. A study of Merriam's temperature laws. Wilson Bull., 44:129–143.

——. 1940. Factors affecting length of incubation. Auk, 57:499–513.

——. 1947. Bird population studies in the Coniferous Forest Biome during a spruce budworm outbreak. Biol. Bull., Dept. Lands and Forests, Ontario, Can., 1:1–100.

———. 1948. Bird populations and biotic communities in northern lower Michigan. Ecology, 29:101–104.

———. 1969. Tolerance of cold and Bergmann's rule. Auk, 86:13–25.

——— and S. P. BALDWIN. 1928. Development of temperature control in nestling House Wrens. Amer. Nat., 62:249–278.

———, T. C. KRAMER, and F. HAMERSTROM. 1956. Variations in egg characteristics of the House Wren. Auk, 73:42–65.

KENYON, K. W., and D. W. RICE. 1958. Homing of Laysan Albatrosses. Condor, 60:3–6.

KEPLER, C. B., and K. C. PARKES. 1972. A new species of warbler (Parulidae) from Puerto Rico. Auk, 89:1–18.

KERN, J. A. 1968. Quest for the Quetzal. Audubon, 70:28–39.

KESSEL, B. 1953. Distribution and migration of the European Starling in North America. Condor, 55:49–67.

KESSLER, F. W. 1960. Egg temperatures of the Ring-necked Pheasant obtained with a self-recording potentiometer. Auk, 77:330–336.

KILHAM, L. 1958. Sealed-in winter stores of Red-headed Woodpeckers. Wilson Bull., 70:107–113.

———. 1962. Nest sanitation of Yellow-bellied Sapsucker. Wilson Bull., 74:96–97.

———. 1963. Food storing of Red-bellied Woodpeckers. Wilson Bull., 75:227–234.

———. 1964. Interspecific relations of crows and Red-shouldered Hawks in mobbing behavior. Condor, 66:247–248.

———. 1970. Feeding behavior of Downy Woodpeckers. I. Preference for paper birches and sexual differences. Auk, 87:544–556.

———. 1971. Use of blister beetle in bill-sweeping by White-breasted Nuthatch. Auk, 88:175–176.

KLOEK, G. P., and C. L. CASLER. 1972. The lung and air-sac system of the Common Grackle. Auk, 89:817–825.

KLUIJVER, H. N., J. LIGTOVOET, C. VAN DEN OUWELANT, and F. ZEGWAARD. 1940. De Levenwijze van den Winterkoning, *Troglodytes t. troglodytes* (L.). Limosa, 13:1–51.

KNAPPEN, P. 1932. Number of feathers on a duck. Auk, 49:461.

KNORR, O. A. 1957. Communal roosting of the Pygmy Nuthatch. Condor, 59:398.

KOOYMAN, G. L., C. M. DRABEK, R. ELSMER, and W. B. CAMPBELL. 1971. Diving behavior of the Emperor Penguin, *Aptenodytes forsteri*. Auk, 88:775–795.

KREBS, J. R. 1970. The efficiency of courtship feeding in the Blue Tit *(Parus caeruleus)*. Ibis, 112:108–110.

KUSHLAN, J. A. 1972. Aerial feeding in the Snowy Egret. Wilson Bull., 84:199–200.

KUYT, E. 1967. Two banding returns for the Golden Eagle and Peregrine Falcon. Bird-banding, 38:78–79.

LACK, D., and L. LACK. 1933. Territory reviewed. Brit. Birds, 27:179–199.

LANCASTER, D. A. 1964. Life history of the Boucard Tinamou in British Honduras. Part I: Distribution and general behavior. Condor, 66:165–181.

LAND, H. C. 1963. A tropical feeding tree. Wilson Bull., 75:199–200.

LANE, E. N., and S. HARTGEN. 1971. Jungle cocks, trout flies, and smugglers. Audubon, 73:38–43.

LANYON, W. E. 1957. The comparative biology of the meadowlarks *(Sturnella)* in Wisconsin. Publ. Nuttall Ornith. Club, 1:1–67.

LASIEWSKI, R. C. 1963. Oxygen consumption of torpid, resting, active, and flying hummingbirds. Physiol. Zool., 36:122–140.

——— and G. A. BARTHOLOMEW. 1966. Evaporative cooling in the Poor-will and Tawny Frogmouth. Condor, 66:477–490.

——— and W. R. DAWSON. 1964. Physiological responses to temperature in the Common Nighthawk. Condor, 66:477–490.

———, ———, and G. A. BARTHOLOMEW. 1970. Temperature regulation in the Little Papuan Frogmouth, *Podargus ocellatus.* Condor, 72:332–338.

———, S. H. HUBBARD, and W. R. MOBERLY. 1964. Energetic relationships of a very small passerine bird. Condor, 66:212–220.

——— and H. J. THOMPSON. 1966. Field observation of torpidity in the Violet-green Swallow. Condor, 68:102–103.

LASKEY, A. R. 1944. A study of the Cardinal in Tennessee. Wilson Bull., 56:27–44.

———. 1962. Breeding biology of Mockingbirds. Auk, 79:596–606.

LASZLO, R. 1970. Food of nestling Bald Eagles on San Juan Island, Washington. Condor, 72:358–361.

LAW, J. E. 1929. The function of the oil-gland. Condor, 31:148–156.

LAWICK-GOODALL, J. and H. VAN. 1967. Use of tools by the Egyptian Vulture. Nature, 212:1468–1469.

LAWRENCE, L. DE K. 1960. Jays of a northern forest. Aud. Mag., 62:266–267, 286–287.

LAY, D. 1959. Aftermath of waste. Texas Game and Fish, 4 pages.

LEDERER, R. J. 1972. The role of avian rictal bristles. Wilson Bull., 84:193–197.

LEOPOLD, A. S. 1953. Intestinal morphology of gallinaceous birds in relation to food habits. Jour. Wildl. Mngt., 17:197–203.

LEWIS, D. M. 1972. Importance of face mask in sexual recognition and territorial behavior in the Yellowthroat. Jack-Pine Warbler, 50:98–109.

LEWIS, H. 1952. Thistle-nesting Goldfinches. Flicker, 24:105–109.

LEWIS, J. B. 1928. Sight and scent in the Turkey Vulture. Auk, 45:467–470.

LEWIS, R. A. 1967. "Resting" heart and respiratory rates of small birds. Auk, 84:131–132.

LIGON, J. D. 1968. Sexual differences in foraging behavior in two species of *Dendrocopos* woodpeckers. Auk, 85:203–215.

———. 1971. Late summer-autumnal breeding of the Pinon Jay in New Mexico. Condor, 73:147–153.

LINCOLN, F. C. 1939. *The Migration of American Birds.* Garden City: Doubleday.

————. 1950. Migration of birds. Circ. 16:1–102, U. S. Fish and Wildl. Serv., Washington, D. C.

LINDBLAD, J. 1969. Bird of darkness. Nat. Hist., 78(2):80–83.

LORENZ, K. 1950. The comparative method of studying innate behavior patterns. Pages 221–268 in *Physiological Mechanisms in Animal Behavior.* Soc. Exp. Biol. Sympos. 4, Cambridge.

LOVELL, H. B. 1952. Black Vulture depredations at Kentucky woodlands. Wilson Bull., 64:48–49.

————. 1958. Baiting of fish by a Green Heron. Wilson Bull., 70:280–281.

LOW, J. B., L. KAY, and D. I. RASMUSSEN. 1950. Recent observations on the White Pelican on Gunnison Island, Great Salt Lake, Utah. Auk, 67:345–356.

LOWE, V. P. W. 1972. Distraction display by a woodcock with chicks. Ibis, 114:106–107.

LOWE-MCCONNELL, R. H. 1967. Biology of the immigrant Cattle Egret *Ardeola ibis* in Guyana, South America. Ibis, 109:168–179.

LOWERY, G. H., JR., 1945. Trans-Gulf spring migration of birds and the coastal hiatus. Wilson Bull., 57:92–121.

————. 1946. Evidence of trans-Gulf migration. Auk, 63:175–211.

LUDWIG, C. C. 1960. Banding returns of Michigan Mourning Doves. Jack-Pine Warbler, 38:29–33.

MACDONALD, D. 1961. Hunting season weights of New Mexico Wild Turkeys. Jour. Wildl. Mngt., 25:442–444.

MACLEAN, G. L. 1968. Field studies on the sandgrouse of the Kalahari desert. Living Bird, 7:209–235.

————. 1972. Clutch size and evolution in the Charadrii. Auk, 89:299–324.

MACMULLAN, R. A., and L. L. EBERHARDT. 1953. Tolerance of incubating pheasant eggs to exposure. Jour. Wildl. Mngt., 17:322–330.

MACROBERTS, M. H. 1970. Notes on the food habits and food defense of the Acorn Woodpecker. Condor, 72:196–204.

MADSEN, J. 1969. The last dance. Audubon, 71:16–23.

MAHER, W. J. 1966. Predation's impact on penguins. Nat. Hist., 75:42–51.

MADURA, M. L. 1952. Feathered observations. Pass. Pigeon, 14:65–68.

MALCOMSON, R. O. 1960. Mallophaga from birds of North America. Wilson Bull., 72:182–197.

MANVILLE, R. H. 1963. Altitude record for Mallard. Wilson Bull., 75:92.

MARKUS, M. B. 1965. The number of feathers on birds. Ibis, 107:394.

MARLER, P., P. MUNDINGER, M. S. WASER, and A. LUTJEN. 1972. Effects of acoustical stimulation and deprivation on song development in red-winged blackbirds *(Agelaius phoeniceus)*. Anim. Beh. 20:586–606.

MARSHALL, H. 1947. Longevity of the American Herring Gull. Auk, 64:188–198.

MASON, E. A. 1961. Mourning Dove banded in Massachusetts in winter, taken in Florida. Bird-Banding, 32:173.

MAY, J. B. 1963. *Homo sapiens* and the Aves. Mass. Aud., 48:41–51.

MAYER, W. 1952. The matin song of the Eastern Kingbird. Pass. Pigeon, 14:91–94.

MAYFIELD, H. F. 1960. *The Kirtland's Warbler.* Bloomfield Hills, Mich.: Cranbrook Inst. of Sci.

———. 1961. Vestiges of proprietary interest in nests by the Brown-headed Cowbird parasitizing the Kirtland's Warbler. Auk, 78:162–166.

———. 1972a. Third decennial census of Kirtland's Warbler. Auk, 89:263–268.

———. 1972b. Bird bones identified from Indian sites at western end of Lake Erie. Condor, 74:344–347.

MAYHEW, W. W. 1955. Spring rainfall in relation to Mallard production in the Sacramento Valley, California. Jour. Wildl. Mngt., 19:36–47.

———. 1963. Homing of Bank Swallows and Cliff Swallows. Bird-Banding, 34:179–190.

MAYR, E. 1945. Birds of paradise. Sci. Guide 127, Amer. Mus. Nat. Hist., New York.

———. 1946. The number of species of birds. Auk, 63:64–69.

———. 1965. Numerical phenetics and taxonomic theory. Syst. Zool., 14:73–97.

MAZZEO, R. 1953. Homing of the Manx Shearwater. Auk, 70:200–201.

MCATEE, W. L. 1920. The local suppression of agricultural pests by birds. Ann. Report Smith. Inst., 411–438.

———. 1926. The relation of birds to woodlots in New York State. Roosevelt Wild Life Bull., 4:7–152.

———. 1951. Comparative abundance of birds in the District of Columbia region 1861–1942. Atlantic Nat., 7:66–82.

MCCABE, R. A. 1951. The song and song-flight of the Alder Flycatcher. Wilson Bull., 63:89–98.

———. 1961. The selection of colored nest boxes by House Wrens. Condor, 63:322–329.

———. 1965. Nest construction by House Wrens. Condor, 67:229–234.

MCCABE, T. T. 1942. Types of shorebird flight. Auk, 59:110–111.

MCCLURE, H. E. 1943. Ecology and management of the Mourning Dove, *Zenaidura macroura* (Linn.) in Cass County, Iowa. Res. Bull. 310, Iowa State Coll. Agr. Exp. Sta.

MCDOWELL, S. 1948. The bony palate of birds. Part I: The Palaeognathae. Auk, 65:520–549.

MCGAHAN, J. 1968. Ecology of the Golden Eagle. Auk, 85:1–12.

MCILHENNY, E. A. 1937. Life history of the Boat-tailed Grackle in Louisiana. Auk, 54:274–295.

———. 1939. Feeding habits of Black Vulture. Auk, 56:472–474.

MCKINLEY, D. 1964. History of the Carolina Parakeet in its southwestern range. Wilson Bull., 76:68–93.

———. 1965. The Carolina Parakeet in the upper Missouri and Mississippi River Valleys. Auk, 82:215–226.

MCNEIL, R. and J. B. BURTON. 1972. Cranial pneumatization patterns and bursa of Fabricius in North American shorebirds. Wilson Bull., 84:329–339.

MEANLEY, B. 1952. Notes on the ecology of the Short-billed Marsh Wren in the lower Arkansas rice fields. Wilson Bull., 64:22–25.

————. 1955. A nesting study of the Little Blue Heron in eastern Arkansas. Wilson Bull., 67:84–99.

————. 1971. Blackbirds and the southern rice crop. Resource Publ., 100:1–64, Washington, D.C.: U.S.D.I.

MEDWAY, L., and D. R. WELLS. 1969. Dark orientation by the Giant Swiftlet *Collocalia gigas*. Ibis, 111:609–611.

MENDELSSOHN, H. 1971. The impact of pesticides on bird life in Israel. Bull. Intern. Council of Bird Pres., 11:75–104.

MENGEL, R. M. 1952. Certain molts and plumages of Acadian and Yellow-bellied Flycatchers. Auk, 69:273–283.

———— and M. A. JENKINSON. 1971. Vocalizations of the Chuck-will's-widow and some related behavior. Living Bird, 10:171–184.

MEWALDT, L. R. 1952. The incubation patch of the Clark Nutcracker. Condor, 54:361.

————. 1958. Pterylography and natural and experimentally induced molt in Clark's Nutcracker. Condor, 60:165–187.

————. 1964. California sparrows return from displacement to Maryland. Science, 146:941–942.

MEYERRIECKS, A. J. 1959. Foot stirring feeding behavior in herons. Wilson Bull., 71:153–158.

————. 1962. Diversity typifies heron feeding. Nat. Hist., 71(6):48–59.

————. 1972. Tool-using by a Double-crested Cormorant. Wilson Bull., 84:482–483.

———— and D. W. NELLIS. 1967. Egrets serving as "beaters" for Belted King-fishers. Wilson Bull., 79:236–237.

MICHENER, H., and J. R. MICHENER. 1935. Mockingbirds, their territories and individualities. Condor, 37:97–140.

MICHENER, J. R. 1951. Territorial behavior and age composition in a population of Mockingbirds at a feeding station. Condor, 53:276–283.

MIDDLETON, A. L. A. 1972. The structure and possible function of the avian seminal sac. Condor, 74:185–190.

MIDDLETON, R. J. 1960. Banding Robins at Norristown. Bird-Banding, 31:136–139.

MILLER, A. H. 1959. Response to experimental light increments by Andean Sparrows from an equatorial area. Condor, 61:344–347.

————. 1961. Molt cycles in equatorial Andean Sparrows. Condor, 63:143–161.

MILLER, L. 1957. Bird remains from an Oregon Indian midden. Condor, 59:59–63.

MILLER, R. S. and R. E. MILLER. 1971. Feeding activity and color preference of Ruby-throated Hummingbirds. Condor, 73:309–313.

MILLER, W. T. 1952. A bird that walks on water. Nat. Mag., 45:69–71.

MILLIGAN, M. 1966. Vocal responses of White-crowned Sparrows to recorded songs of their own and another species. Anim. Beh., 14:356–361.

MILNE, L. J., and M. J. MILNE. 1950. The eyes of birds. Nat. Mag., 43:121–123.

MISKIMEN, M. 1951. Sound production in passerine birds. Auk, 68:493–504.

————. 1957. Absence of syrinx in the Turkey Vulture *(Cathartes aura)*. Auk, 74:104–105.

MITCHELL, J. H. 1973. From peat to pavement. Mass. Aud. Soc. Newsl., 13(1):3–6, 11 (Aug.–Sept.).

MOLDENHAUER, R. R., and J. A. WIENS. 1970. The water economy of the Sage Sparrow, *Amphispiza belli nevadensis*. Condor, 72:265–275.

MOREAU, R. E. 1966. On estimates of the past numbers and of the average longevity of avian species. Auk, 83:403–415.

MORSE, D. H. 1968. The use of tools by Brown-headed Nuthatches. Wilson Bull., 80:220–224.

MOSSMAN, A. S. 1966. Wildlife and the tsetse fly in Rhodesia. Nat. Parks, 40(228):10–15 (Sept.).

MOUNTFORT, G. 1957. Nest-hole excavation by the bee-eater. Brit. Birds, 50:263–267.

MROSOVSKY, N. 1971. Black Vultures attack live turtle hatchlings. Auk, 88:672–673.

MUELLER, H. C., and D. D. BERGER. 1959. Some long-distance Barn Owl recoveries. Bird-Banding, 30:182.

MUNDINGER, P. C. 1972. Annual testicular cycle and bill color change in the Eastern American Goldfinch. Auk, 89:403–419.

MURCHISON, J. 1968. The Wheeler National Wildlife Refuge. Nat. Parks, 42(253):11–13 (Sept.).

MURPHY, D. A., and T. S. BASKETT. 1952. Bobwhite mobility in central Missouri. Jour. Wildl. Mngt., 16:498–510.

MURPHY, R. C. 1936. *Oceanic Birds of South America*. New York: Macmillan.
——— and L. S. MOWBRAY. 1951. New light on the Cahow, *Pterodroma cahow*. Auk, 68:266–280.

MURRAY, B. G., JR. 1965. On the autumn migration of the Blackpoll Warbler. Wilson Bull., 77:122–133.

MUSSEHL, T. W. 1960. Blue Grouse production, movements, and populations in the Bridger Mountains, Montana. Jour. Wildl. Mngt., 24:60–68.

NAIR, K. K. 1954. A comparison of the muscles in the forearm of a flapping and a soaring bird. Jour. Anim. Morph. Physiol., 1:26–34.

NELSON, A. L., and A. C. MARTIN. 1953. Gamebird weights. Jour. Wildl. Mngt., 17:36–42.

NERO, R. W. 1956. A behavior study of the Red-winged Blackbird. Wilson Bull., 68:5–37, 129–150.
———. 1963. Comparative behavior of the Yellow-headed Blackbird, Red-winged Blackbird, and other icterids. Wilson Bull., 75:376–413.
———. 1964. Detergents—deadly hazards to water birds. Audubon, 66:26–27.

NEWTON, I. 1966. The moult of the Bullfinch *Pyrrhula pyrrhula*. Ibis, 108:41–67.

NICE, L. B., M. M. NICE, and R. M. KRAFT. 1935. Erythrocytes and hemoglobin in the blood of some American birds. Wilson Bull., 47:120–124.

NICE, M. M. 1937. Studies in the life history of the Song Sparrow. I. A population study of the Song Sparrow. Trans. Linn. Soc., 4:1–243.
———. 1943. Studies in the life history of the Song Sparrow. II. The behavior of the Song Sparrow and other passerines. Trans. Linn. Soc., 6:1–328.

———. 1953. The question of ten-day incubation periods. Wilson Bull., 65:81–93.

———. 1954. Problems of incubation periods in North American birds. Condor, 56:173–197.

———. 1957. Nesting success in altricial birds. Auk, 74:305–321.

——— and W. E. SCHANTZ. 1959. Head-scratching movements in birds. Auk, 76:339–342.

NICKELL, W. P. 1951. Studies of habitats, territory, and nests of the Eastern Goldfinch. Auk, 68:447–470.

———. 1958. Variations in engineering features of the nests of several species of birds in relation to nest sites and nesting materials. Butler Univ. Bot. Studies, 13:121–140.

NIERING, W. A. 1968. Wetlands and cities. Mass. Aud., 53:12–18.

NILES, D. M. 1972. Molt cycles of Purple Martins *(Progne subis)*. Condor, 74:61–71.

NISBET, I. C. T. 1963. Measurements with radar of the height of nocturnal migration over Cape Cod, Massachusetts. Bird-Banding, 34:57–67.

———, W. H. DRURY, JR., and J. BAIRD. 1963. Weight-loss during migration. Part I: Deposition and consumption of fat by the Blackpoll Warbler. Bird-Banding, 34:107–138.

NORRIS, R. A. 1960. Density, racial composition, sociability, and selective predation in nonbreeding populations of Savannah Sparrows. Bird-Banding, 31:173–216.

———. 1972. Data on nictitation rates in birds. Bird-Banding, 43:289–290.

——— and F. S. L. WILLIAMSON. 1955. Variation in relative heart size of certain passerines with increase in altitude. Wilson Bull., 67:78–83.

NORRIS-ELYE, L. T. S. 1945. Heat insulation in the tarsi and toes of birds. Auk, 62:455.

ODUM, E. P. 1942. Annual cycle of the Black-capped Chickadee—3. Auk, 59:499–531.

———. 1943. Some physiological variations in the Black-capped Chickadee. Wilson Bull., 55:178–191.

———. 1944. Circulatory congestion as a possible factor in regulating incubation behavior. Wilson Bull., 56:48–49.

———, C. E. CONNELL, and H. L. STODDARD. 1961. Flight energy and estimated flight ranges of some migratory birds. Auk, 78:515–527.

OLSON, H. 1953. Beetle rout in the Rockies. Aud. Mag., 55:30–32.

OLSON, S. L. 1971. Taxonomic comments on the Eurylaimidae. Ibis, 113:507–516.

OPPENHEIM, R. W. 1970. Some aspects of embryonic behavior in the duck *(Anas platyrhynchos)*. Anim. Beh., 18:335–352.

ORENSTEIN, R. J. 1972. Tool-use by the New Caledonian Crow *(Corvus moneduloides)*. Auk, 89:674–676.

ORING, L. W., and M. L. KNUDSON. 1972. Monogamy and polyandry in the Spotted Sandpiper. Living Bird, 11:59–73.

OSBORNE, D. R. 1968. The functional anatomy of the skin muscles in Phasianinae. Unpubl. PhD thesis, Mich. State Univ., East Lansing.

Owen, D. F. 1959. Mortality of the Great Blue Heron as shown by banding recoveries. Auk, 76:464–470.

———. 1963. Variation in North American Screech Owls and the subspecies concept. Syst. Zool., 12:8–14.

Packard, G. C. 1967. House Sparrows: evolution of populations from the Great Plains and Colorado Rockies. Syst. Zool., 16:73–89.

Parkes, K. C. 1953. The incubation patch in males of the suborder Tyranni. Condor, 55:218–219.

Parmalee, P. W. 1954. The vultures: their movements, economic status, and control in Texas. Auk, 71:443–453.

Parmelee, D. F. 1964. Survival in the Painted Bunting. Living Bird, 3: 5–7.

Pavlov, I. P. 1927. *Conditional Reflexes.* New York: Oxford Univ. Press.

Payne, R. S., and W. H. Drury, Jr. 1958. Marksman of the darkness. Nat. Hist., 67:316–323.

Peakall, D. B. 1970a. The Eastern Bluebird: its breeding season, clutch size, and nesting success. Living Bird, 9:239–256.

———. 1970b. *p,p'*-DDT: effects on calcium metabolism and concentration of estradiol in the blood. Science, 168:592–594.

Pearson, A. K., and O. P. Pearson. 1955. Natural history and breeding behavior of the Tinamou, *Nothoprocta ornata.* Auk, 72:113–127.

Pearson, O. P. 1953. The metabolism of hummingbirds. Scient. Amer., 188(1):69–72.

Peller, E. 1963. Operation duck rescue. Audubon, 65:364–367.

Penny, R. L., and J. T. Emlen. 1967. Further experiments on distance navigation in the Adelie Penguin, *Pygoscelis adeliae.* Ibis, 109:99–109.

Percy, L. W. 1963. Further notes on the African Finfoot, *Podica senegalensis* (Vieillot). Bull. Brit. Ornith. Club, 83:127–132.

Peterle, T. J. 1953. An extended incubation period of the Ruffed Grouse. Wilson Bull., 65:119.

Peterson, R. T. 1960. Bird's-eye view: rediscovery on Kauai. Aud. Mag., 62:258–261.

———. 1965. Population trends of Ospreys in the northeastern United States. Chapter 28 in J. J. Hickey, ed., *Peregrine Falcon Populations,* pp. 333–337. Madison: Univ. Wisc. Press.

Petrides, G. A. 1959. Competition for food between five species of East African Vultures. Auk, 76:104–106.

——— and C. R. Bryant. 1951. An analysis of the 1944–1950 fowl cholera epizootic in Texas Panhandle waterfowl. Trans. N. A. Wildl. Conf., 16:193–216.

Pettingill, O. S., Jr. 1942. The birds of a bull's horn acacia. Wilson Bull., 54:89–96.

Phillips, A. R. 1951a. Complexities of migration: a review with original data from Arizona. Wilson Bull., 63:129–136.

———. 1951b. The molts of the Rufous-winged Sparrow. Wilson Bull., 63:323–326.

Phillips, R. E., and H. C. Black. 1956. A winter population study of the Western Winter Wren. Auk, 73:401–410.

PINKOWSKI, B. C. 1971. An analysis of banding-recovery data on Eastern Bluebirds banded in Michigan and three neighboring states. Jack-Pine Warbler, 49:33–50.

PITELKA, F. A. 1942. Territoriality and related problems in North American hummingbirds. Condor, 44:189–204.

PITTMAN, J. A. 1953. Direct observation of the flight speed of the Common Loon. Wilson Bull., 65:213.

PORTER, R., and I. WILLIS. 1968. The autumn migration of soaring birds at the Bosphorus. Ibis, 110:520–536.

POTTER, E. F. 1970. Anting in wild birds, its frequency and probable purpose. Auk, 87:692–713.

PRATT, H. M. 1970. Breeding biology of Great Blue Herons and Common Egrets in Central California. Condor, 72:407–416.

PRESTON, F. W. 1968. The shapes of birds' eggs: mathematical aspects. Auk, 85:454–463.

PROCTOR, V. W. 1968. Long-distance dispersal of seeds by retention in digestive tract of birds. Science, 160:321–322.

QUAY, W. B. 1967. Comparative survey of the anal glands of birds. Auk, 84:379–389.

———. 1972. Infrequency of pineal atrophy among birds and its relation to nocturnality. Condor, 74:33–45.

RAITT, R. J., and R. L. MAZE. 1968. Densities and species composition of breeding birds of a creosotebush community in southern New Mexico. Condor, 70:193–205.

RAMP, W. K. 1965. The auditory range of a Hairy Woodpecker. Condor, 67:183–185.

RAND, A. L. 1953a. Factors affecting feeding rates of anis. Auk, 70:26–30.

———. 1953b. Use of snake skins in birds' nests. Chicago Acad. Sci. Nat. Hist. Misc., 125:1–5.

———. 1954. On the spurs on birds' wings. Wilson Bull., 66:127–134.

——— and R. M. RAND. 1943. Breeding notes on the Phainopepla. Auk, 60:333–341.

RANER, L. 1972. Forekommer polyandri hos smalnabbad simsnappe *(Phalaropus lobatus)* och suartsnappa *(Tringa erythropus)*? Fauna och Flora, 67:135–138.

RATCLIFFE, D. A. 1970. Changes attributable to pesticides in egg breakage frequency and eggshell thickness in some British birds. J. Appl. Ecol., 7:67–115.

RECHER, H. F., and J. A. RECHER. 1969. Comparative foraging efficiency of adult and immature Little Blue Herons *(Florida caerulea)*. Anim. Beh., 17:320–322.

———. 1972. Herons leaving the water to defecate. Auk, 89:896–897.

REEDER, W. G. 1951. Stomach analysis of a group of shorebirds. Condor, 53:43–45.

REICHSTEIN, T., J. VON EUW, J. A. PARSONS, and M. ROTHSCHILD. 1968. Heart poisons in the monarch butterfly. Science, 161:861–866.

RICE, D. W. 1959. Birds and aircraft on Midway Islands. Spec. Scient. Report—Wildlife, No. 44, Washington, D.C.

RICHARDSON, F. 1972. Accessory pygostyle bones of Falconidae. Condor, 74:350–351.

RICHDALE, L. E. 1951. *Sexual Behavior in Penguins.* Lawrence, Kans.: Univ. of Kans.

———. 1954. The starvation theory in albatrosses. Auk, 71:239–252.

———. 1963. Biology of the Sooty Shearwater *Puffinus griseus.* Proc. Zool. Soc. London, 141, Part I:1–117.

RICKLEFS, R. E. 1965. Brood reduction in the Curve-billed Thrasher. Condor, 67:505–510.

———. 1971. Foraging behavior of Mangrove Swallows at Barro Colorado Island. Auk, 88:635–651.

RIPLEY, D. 1950. Strange courtship of birds of paradise. Nat'l Geog. Mag., 97(2):247–278.

RIVOLIER, J. 1959. Polar realm of the Emperors. Nat. Hist., 68:66–81.

ROBINS, J. D., and G. D. SCHNELL. 1971. Skeletal analysis of the *Ammodramus*–Ammospiza grassland sparrow complex: a numerical taxonomic study. Auk, 88:567–590.

ROBINSON, G. G., and D. W. WARNER. 1964. Some effects of prolactin on reproductive behavior in the Brown-headed Cowbird *(Molothrus ater).* Auk, 81:315–325.

ROBSON, F. D. 1948. Kiwis in captivity. Bull. Hawkes Bay Art Gallery and Mus., Napier, New Zealand, 1–8.

RODGER, G., and J. RODGER. 1960. The last great animal kingdom I: a portfolio of Africa's vanishing wildlife. Nat'l Geog. Mag., 118:390–409.

ROGERS, D. T. 1965. Fat levels and estimated flight-ranges of some autumn migratory birds killed in Panama during a nocturnal rainstorm. Bird-Banding, 36:115–116.

ROGERS, D. T., JR., and E. P. ODUM. 1966. A study of autumnal post-migrant weights and vernal fattening of North American migrants in the tropics. Wilson Bull., 78:415–433.

ROHWER, S. A. 1971. Molt and the annual cycle of the Chuck-will's-widow. *Caprimulgus carolinensis.* Auk, 88:485–519.

ROMANOFF, A., and A. ROMANOFF. 1949. *The Avian Egg.* New York: Wiley.

ROSEBERRY, J. L., and W. D. KLIMSTRA. 1970. The nesting ecology and reproductive performance of the Eastern Meadowlark. Wilson Bull., 82:243–267.

RUMSEY, R. L. 1970. Woodpecker nest failures in creosoted utility poles. Auk, 87:367–369.

RYDER, J. P. 1970. A possible factor in the evolution of clutch size in Ross' Goose. Wilson Bull., 82:5–13.

RYDER, R. A. 1959. Interspecific intolerance of the American Coot in Utah. Auk, 76:424–442.

SABINE, W. S. 1955. The winter society of the Oregon Junco: the flock. Condor, 57:88–111.

———. 1959. The winter society of the Oregon Junco: intolerance, dominance and the pecking order. Condor, 61:110–135.

SADLER, K. C. 1961. Grit selectivity by the female pheasant during egg production. Jour. Wildl. Mngt., 25:339–341.

SAFRIEL, U. 1968. Bird Migration at Elat, Israel. Ibis, 110:283–320.

SALT, G. W. 1957. Observations on Fox, Lincoln, and Song Sparrows at Jackson Hole, Wyoming. Auk, 74:258–259.

SALYER, A., and A. WOLFSON. 1968. Avian pineal gland: progonadotropic response in Japanese Quail. Science, 158:178–179.

SALYER, J. C., and K. LAGLER. 1940. The food and habits of American Merganser during winter in Michigan, considered in relation to fish management, Jour. Wildl. Mngt., 4:186–219.

SARGENT, T. D. 1959. Winter studies on the Tree Sparrow, *Spizella arborea*. Bird-Banding, 30:27–37.

———. 1962. A study of homing in the Bank Swallow *(Riparia riparia)*. Auk, 79:234–246.

SAUER, E. G. F., and E. M. SAUER. 1967. Yawning and other maintenance activities in the South African Ostrich. Auk, 84:571–587.

SAUER, FRANZ. 1957. Die Steinenorientierung nächtlich ziehender Grasmücken *(Sylvia atricapilla, borin und curruca)*. Zeitschrift für Tierpsychologie, 14:29–70. (In German with English summary.)

SAUNDERS, A. A. 1951. The song of the song sparrow. Wilson Bull., 63:99–109.

———. 1959. Forty years of spring migration in southern Connecticut. Wilson Bull., 71:208–219.

SAVILE, D. B. O. 1950. The flight mechanism of swifts and hummingbirds. Auk, 67:499–504.

———. 1957a. The primaries of *Archaeopteryx*. Auk, 74:99–101.

———. 1957b. Adaptive evolution in the avian wing. Evolution, 11:212–224.

SAWYER, E. J. 1959. Unusual adaptations of wildlife. Aud. Mag., 61:212–213.

SCHAEFER, E. 1953. Contribution to the life history of the Swallow-Tanager. Auk, 70:403–460.

———. 1954. Zur Biologie des Steisshuhnes, *Nothocercus bonapartei*. Jour. für Ornith., 95:219–232.

SCHLEIDT, W. M., and M. D. SHALTER. 1972. Cloacal foam gland in the quail *Coturnix coturnix*. Ibis, 114:558.

SCHMID, F. C. 1963. Record longevity of a wild Red-shouldered Hawk. Bird-Banding, 34:160.

SCHMIDT-NIELSON, K., and R. FANGE. 1958. The function of the salt gland in the Brown Pelican. Auk, 75:282–289.

———, J. KANWISHER, R. C. LASIEWSKI, J. E. COHN, and W. L. BRETZ. 1969. Temperature regulation and respiration in the Ostrich. Condor, 71:341–352.

SCHNELL, J. H. 1958. Nesting behavior and food habits of Goshawks in the Sierra Nevada of California. Condor, 60:377–403.

SCHORGER, A. W. 1947. The deep diving of the Loon and Old-squaw and its mechanism. Wilson Bull., 59:151–159.

———. 1960. The crushing of *Carya* nuts in the gizzard of the turkey. Auk, 77:337–340.

SCHREIBER, R. W. 1970. Breeding biology of Western Gulls *(Larus occidentalis)* on San Nicolas Island, California, 1968. Condor, 72:133–140.

———— and R. L. DE LONG. 1969. Brown Pelican status in California. Aud. Field Notes, 23:57–59.

SCHREIBER, R. W., and R. W. RISEBROUGH. 1972. Studies of the Brown Pelican. Wilson Bull., 84:119–135.

SCHUMACHER, D. M. 1964. Ages of some captive wild birds. Condor, 66:309.

SCHWARTZ, P. 1964. The Northern Waterthrush in Venezuela. Living Bird, 3:169–184.

SCHWARTZKOPFF, J. 1955. On the hearing of birds. Auk, 72:340–347.

SCOTT, D. M. 1967. Postjuvenal molt and determination of age of the Cardinal. Bird-Banding, 38:37–51.

SCOTT, J. W. 1942. Mating behavior of the Sage Grouse. Auk, 59:477–498.

SCOTT, T. G., Y. L. WILLIS, and J. A. ELLIS. 1959. Some effects of a field application of dieldrin on wildlife. Jour. Wildl. Mngt., 23:409–427.

SCOTT, W. E. (ed). 1947. Silent wings: a memorial to the Passenger Pigeon. Madison: Wisc. Soc. for Ornith., 42 pages.

SELANDER, R. K. 1960. Failure of estrogen and prolactin treatment to induce brood patch formation in Brown-headed Cowbirds. Condor, 62:65.

———— and D. R. GILLER. 1959. Interspecific relations of woodpeckers in Texas, Wilson Bull., 71:107–124.

———— and ————. 1960. First-year plumages of the Brown-headed Cowbird and Red-winged Blackbird. Condor, 62:202–214.

———— and ————. 1963. Species limits in the woodpecker genus *Centurus* (Aves). Bull. Amer. Mus. Nat. Hist., 124:217–273.

———— and D. K. HUNTER. 1960. On the functions of wing-flashing in Mockingbirds. Wilson Bull., 72:341–345.

———— and L. L. KUICH. 1963. Hormonal control and development of the incubation patch in icterids, with notes on behavior of cowbirds. Condor, 65:73–90.

———— and S. Y. YANG. 1966. Behavorial responses of Brown-headed Cowbirds to nests and eggs. Auk, 83:207–232.

SERVENTY, D. L. 1963. Egg-laying timetable of the Slender-billed Shearwater *Puffinus tenuirostris*. Proc. Intern. Ornith. Congress, 13:338–343.

SETON, E. T. 1940. *Trail of an Artist-Naturalist.* New York: Scribner.

SHANK, M. C. 1959. The natural termination of the refractory period in the Slate-colored Junco and in the White-throated Sparrow. Auk, 76:44–54.

SHARP, B. 1971. A transcontinental Mourning Dove recovery. Auk, 88:924.

SHARP, W. M. 1957. Social and range dominance in gallinaceous birds — pheasants and prairie grouse. Jour. Wildl. Mngt., 21:242–244.

SHELFORD, V. E. 1932. Life zones, modern ecology and the failure of temperature summing. Wilson Bull., 44:144–157.

SIBLEY, C. G. 1951. Notes on the birds of New Georgia, Central Solomon Islands. Condor, 53:81–92.

———— and C. FRELIN. 1972. The egg white protein evidence for ratite affinities. Ibis, 114:377–387.

SIEGFRIED, W. R., and B. D. J. BATT. 1972. Wilson's Phalaropes forming feeding associations with Shovelers. Auk, 89:667–668.

SKUTCH, A. F. 1940. Social and sleeping habits of Central American wrens. Auk, 57:293–312.

————. 1952. Kingfishers—sovereigns of the watercourses. Nat. Mag., 45:461–464, 500.

————. 1956. Life history of the Ruddy Ground Dove. Condor, 58:188–205.

————. 1957. Life history of the Amazon Kingfisher. Condor, 59:217–229.

————. 1961. Helpers among birds. Condor, 63:198–226.

————. 1962. The constancy of incubation. Wilson Bull., 74:115–152.

SLESSERS, M. 1970. Bathing behavior of land birds. Auk, 87:91–99.

SLUD, P. 1957. The song and dance of the Long-tailed Manakin, *Chiroxiphia linearis*. Auk, 74:333–339.

SMITH, C. R., and M. E. RICHMOND. 1972. Factors influencing pellet egestion and gastric pH in the Barn Owl. Wilson Bull., 84:179–186.

SMITH, L. H. 1965. Changes in the tail feathers of the adolescent Lyrebird. Science, 147:510–513.

SMITH, N. G. 1968. The advantage of being parasitized. Nature, 219:690–694.

SMITH, R. I. 1968. The social aspects of reproductive behavior in the Pintail. Auk, 85:381–396.

SMITH, R. L. 1959. The songs of the Grasshopper Sparrow. Wilson Bull., 71:141–152.

SMITH, R. W., I. L. BROWN, and L. R. MEWALDT. 1969. Annual activity patterns of caged non-migratory White-crowned Sparrows. Wilson Bull., 81:419–440.

SMITH, S. M. 1972. Roosting aggregations of bushtits in response to cold temperatures. Condor, 74:478–479.

————. 1973. An aggressive display and related behavior in the Loggerhead Shrike. Auk, 90:287–298.

SMITH, W. J. 1959. Movements of Michigan Herring Gulls. Bird-Banding, 30:69–104.

SMYTHE, M., and G. A. BARTHOLOMEW. 1966. The water economy of the Black-throated Sparrow and the Rock Wren. Condor, 68:447–458.

SNOW, B. K. 1970. A field study of the Bearded Bellbird in Trinidad. Ibis, 112:299–329.

————. 1972. A field study of the Calfbird *Perissocephalus tricolor*. Ibis, 114:138–162.

SNOW, D. W. 1968. The singing assemblies of little hermits. Living Bird, 7:47–55.

————. 1971a. Social organization of the Blue-backed Manakin. Wilson Bull., 83:35–38.

————. 1971b. Evolutionary aspects of fruit-eating by birds. Ibis, 113:194–202.

———— and B. K. SNOW. 1960. Northern Waterthrush returning to same winter quarters in successive winters. Auk, 77:351–352.

SNYDER, D. E. 1960. Dovekie flights and wrecks. Mass. Aud., 44:117–121.

SNYDER, D. P., and J. F. CASSEL. 1951. A late summer nest of the Red Crossbill in Colorado. Wilson Bull., 63:177–180.

SNYDER, L. L. 1948. Additional instances of paired ovaries in raptorial birds. Auk, 65:602.

SOIKKELI, M. 1967. Breeding cycle and population dynamics in the Dunlin (*Calidris alpina*). Ann. Zool. Fenn., 42:158–198.

SOKOLOWSKI, J. 1960. *The Mute Swan in Poland.* Warsaw: State Council for Conserv. of Nat., 28 pages.

SOUCIE, G. 1972. Oil shale—Pandora's new box. Audubon, 74:106–112.

SOUTHERN, W. E. 1959. Homing of Purple Martins. Wilson Bull., 71:254–261.

———. 1968. Experiments on the homing ability of Purple Martins. Living Bird, 7:71–84.

———. 1972. Influence of disturbances in the earth's magnetic field on Ring-billed Gull orientation. Condor, 74:102–105.

SPARKS, J. H. 1965. Clumping and allo-preening in the Red-thighed Falconet *Microhierax caerulescens burmanicus.* Ibis, 107:247–248.

SPEIRS, J. M. 1945. Flight speed of the Old-squaw. Auk, 62:135–136.

———. 1963. Survival and population dynamics with particular reference to Black-capped Chickadees. Bird-Banding, 34:87–93.

SPENCER, O. R. 1943. Nesting habits of the Black-billed Cuckoo. Wilson Bull., 55:11–22.

SPRING, L. W. 1965. Climbing and pecking adaptations in some North American woodpeckers. Condor, 67:457–488.

SPRUNT, A., JR. 1960. Bird boarders in the southeast. Aud. Mag., 62:236–239.

STAEBLER, A. E. 1941. Number of contour feathers in the English Sparrow. Wilson Bull., 53:126–127.

STAMM, A. L. 1951. Four species choose same nesting tree. Kent. Warbler, 27:23–24.

STANTON, P. B. 1970. Birds and oil. Mass. Aud. Soc. Newsl., 10:3–6.

STEIN, R. C. 1963. Isolating mechanisms between populations of Traill's Flycatchers. Proc. Amer. Phil. Soc., 107:21–50.

STEFANSKI, R. A. 1967. Utilization of the breeding territory in the Black-capped Chickadee. Condor, 69:259–267.

STETTENHEIM, P., A. M. LUCAS, E. M. DENNINGTON, and C. JAMROZ. 1963. The arrangement and action of the feather muscles in chickens. Proc. Intern. Ornith. Congress, 13:918–924.

STEVENSON, H. M. 1957. The relative magnitude of the trans-Gulf and circum-Gulf spring migrations. Wilson Bull., 69:39–77.

———. 1972. The recent history of Bachman's Warbler. Wilson Bull., 84:344–347.

STEWART, P. A. 1952. Dispersal, breeding behavior, and longevity of banded Barn Owls in North America. Auk, 69:227–245.

———. 1955. An audibility curve for two Ring-necked Pheasants. Ohio J. Sci., 55:122–125.

———. 1958. Locomotion of Wood Ducks. Wilson Bull., 70:184–187.

STEWART, R. E. 1949. Ecology of a nesting Red-shouldered Hawk population. Wilson Bull., 61:26–35.

———. 1951. Clapper Rail populations of the middle Atlantic states. Trans. N. A. Wildl. Conf., 16:421–430.

———. 1952. Breeding populations of Clapper Rail at Chincoteague, Virginia. Special Scient. Report—Wildlife, 18:55.

————. 1953. A life history study of the Yellow-throat. Wilson Bull., 65:99–115.

———— and J. W. ALDRICH. 1951. Removal and repopulation of breeding birds in a spruce-fir forest community. Auk, 68:471–482.

STEWART, R. E., and H. A. KANTRUD. 1972. Population estimates of breeding birds in North Dakota. Auk, 89:766–788.

STINE, P. M. 1959. Changes in the breeding birds of Bird Haven Sanctuary over a period of forty-five years. Wilson Bull., 71:372–380.

STODDARD, H. L., SR., and R. A. NORRIS. 1967. Bird casualties at a Leon County, Florida, TV tower. Bull. No. 8, Tall Timbers Res. Sta., 1–104.

STOKES, A. W. 1950. Breeding behavior of the Goldfinch. Wilson Bull., 62:107–127.

————. 1971. Parental and courtship feeding in Red Jungle Fowl. Auk, 88:21–29.

———— and H. W. WILLIAMS. 1971. Courtship feeding in gallinaceous birds. Auk, 88:543–559.

STONE, C. P., D. F. MATT, J. F. BESSER, and J. W. DE GRAZIO. 1972. Bird damage to corn in the United States in 1970. Wilson Bull., 84:101–105.

STORER, R. W. 1956. The fossil loon, *Colymboides minutus.* Condor, 58:413–426.

————. 1958. Loons and their wings. Evolution, 12:262–263.

————. 1966. Sexual dimorphism and food habits in three North American accipiters. Auk, 83:423–436.

————. 1969. The behavior of the Horned Grebe in spring. Condor, 71:180–205.

————. 1970. Independent evolution of the Dodo and the Solitaire. Auk, 87:369–370.

STRESEMANN, E. 1959. The status of avian systematics and its unsolved problems. Auk, 76:269–280.

————. 1963a. The nomenclature of plumages and molts. Auk, 80:1–8.

————. 1963b. Variations in the number of primaries. Condor, 65:449–459.

SULLIVAN, J. O. 1965. "Flightlessness" in the Dipper. Condor, 67:535–536.

SUOMALAINEN, H., and E. ARHIMO. 1945. On the microbial decomposition of cellulose by wild gallinaceous birds (Family Tetraonidae). Ornis Fennica, 22:21–23.

SUTHERS, R. A. 1960. Measurement of some lake-shore territories of the Song Sparrow. Wilson Bull., 72:232–237.

SUTTON, G. M. 1940. Roadrunner. Pages 36–51 in A. C. Bent's Life Histories of North American Cuckoos, Goatsuckers, Hummingbirds, and Their Allies. Bull. 176, U. S. Nat'l Mus.

———— and D. F. PARMELEE. 1954a. Nesting of the Greenland Wheatear on Baffin Island. Condor, 56:295–306.

————. 1954b. Nesting of the Snow Bunting on Baffin Island. Wilson Bull., 66:159–179.

SVÄRDSON, G. 1957. The "invasion" type of bird migration. Brit. Birds, 50:314–343.

SWAN, L. W. 1970. Goose of the Himalayas. Nat. Hist., 79:68–75.

TABLER, F. B. 1956. Notes on the Barn Swallow. Kent. Warb., 32:43–46.

TANNER, J. T. 1952. Black-capped and Carolina Chickadees in the southern Appalachian Mountains. Auk, 69:407–424.

THAPLIYAL, J. P., and R. N. SAXENA. 1964. Absence of a refractory period in the Common Weaver Bird. Condor, 66:199–208.

THOMAS, W. 1969. The buzzards of Hinckley. Nat'l Wildl., 7:34–35.

THOMPSON, W. L. 1960. Agonistic behavior in the House Finch. Part I: Annual cycle and display patterns. Condor, 62:245–271. Part II: Factors in aggressiveness and sociability. Condor, 62:378–402.

———. 1965. A comparative study of bird behavior. Jack-Pine Warbler, 43:110–117.

THOMSON, A. L., and E. P. LEACH. 1952. Report on bird-ringing for 1951. Brit. Birds, 45:265–277.

THORPE, W. H. 1951. The definition of terms used in animal behavior studies. Bull. Anim. Beh., 9:34–40.

———. 1963. *Learning and Instinct in Animals.* London: Methuen.

TINBERGEN, N. 1936. The function of sexual fighting in birds; and the problem of the origin of "territory." Bird-Banding, 7:1–8.

———. 1960. The natural control of insects in pine woods. I. Factors influencing the intensity of predation by songbirds. Arch. Nederl. Zool., 13:265–343.

——— and A. C. PERDECK. 1951. On the stimulus situation releasing the begging response in the newly hatched Herring Gull chick (*Larus argentatus argentatus* Pont.). Behaviour, 3:1–39.

TINKER, F. A. 1958. Avian botulism—the battle at Bear River. Aud. Mag., 60:116–119, 140, 174–177, 224–227.

TOOKE, A. I. 1961. The birds that helped Columbus. Aud. Mag., 63:252–253, 287, 296.

TORDOFF, H. B. 1954. Social organization and behavior in a flock of captive, nonbreeding Red Crossbills. Condor, 56:346–358.

——— and R. M. MENGEL. 1956. Studies of birds killed in nocturnal migration. Univ. Kans. Publ., 10:1–44.

TRAINER, D. O., S. D. FOLZ, and W. M. SAMUEL. 1968. Capillariasis in the Gyrfalcon. Condor, 70:276–277.

TRAUGER, D. L., A. DZUBIN, and J. P. RYDER. 1971. White geese intermediate between Ross' Geese and Lesser Snow Geese. Auk, 88:856–875.

TRAYLOR, M. A. 1968. Winter molt in the Acadian Flycatcher, *Empidonax virescens.* Auk, 85:691.

TUCKER, R. 1958. Taxonomy of the salivary glands of vertebrates. Syst. Zool., 7:74–83.

TURČEK, F. J. 1958. On bird banding in the USSR. Bird-Banding, 29:111–112.

UDVARDY, M. D. F. 1958. Ecological and distributional analysis of North American birds. Condor, 60:50–66.

ULLMANN, H. S. 1963. The most prosperous bird colony. Audubon, 65:150–151.

VAN TYNE, J. 1932. Winter returns of the Indigo Bunting in Guatemala. Bird-Banding, 3:110.

———. 1951. A Cardinal's, *Richmondena cardinalis,* choice of food for adult and for young. Auk, 68:110.

VERBEEK, N. A. M. 1971. Hummingbirds feeding on sand. Condor, 73:112–113.

VERNER, J. 1965. Breeding biology of the Long-billed Marsh Wren. Condor, 67:6–30.

WAGNER, H. O. 1957. Variation in clutch size at different latitudes. Auk, 74:243–250.

WALKER, L. W. 1967. Standing room only. Nat'l. Wildl., 5:42–46.

WALKINSHAW, L. H. 1945. Aortic rupture in Field Sparrow due to fright. Auk, 62:141.

———. 1952. Chipping Sparrow notes. Bird-Banding, 23:101–108.

———. 1953. Life-history of the Prothonotary Warbler. Wilson Bull., 65:152–168.

———. 1972. Kirtland's Warbler—endangered. Amer. Birds, 26:3–9.

WALLACE, G. J. 1939. Bicknell's Thrush, its taxonomy, distribution and life history. Proc. Boston Soc. Nat. Hist., 41:211–402.

———. 1941. Winter studies of color-banded chickadees. Bird-Banding, 12:49–67.

———. 1942. Returns and survival rate of wintering Tree Sparrows. Bird-Banding, 13:81–83.

———. 1948. The Barn Owl in Michigan. Mich. State Coll. Agr. Exp. Sta. Tech. Bull., 208:1–61.

———. 1956. A case of micropthalmia in the American Robin. Wilson Bull., 68:151–152.

———. 1969. Endangered and declining species of Michigan birds. Jack-Pine Warbler, 47:70–75.

WARBURTON, F. E. 1952. Sparrow Hawk, *Falco sparverius,* eats bread. Auk, 69:85.

WARD, P. 1965. The breeding biology of the Black-faced Dioch, *Quelea quelea,* in Nigeria. Ibis, 107:326–349.

———. 1971. The migration patterns of *Quelea quelea* in Africa. Ibis, 113:275–297.

——— and D. D. CRUZ. 1968. Seasonal changes in the thymus gland of a tropical bird. Ibis, 110:203–205.

WARHAM, J. 1963. The Rockhopper Penguin, *Eudyptes chrysocome,* at Macquarie Island. Auk, 80:229–256.

WATSON, G. E. 1963. The mechanism of feather replacement during natural molt. Auk, 80:486–495.

——— and G. J. DIVOKY. 1971. Identification of *Diomedea leptorhyncha* Coues 1866, an albatross with remarkably small salt glands. Condor, 73:487–489.

WATSON, J. B., and K. S. LASHLAY. 1915. An historical and experimental study of homing. Marine Biol. Papers, Carnegie Inst., Washington, D.C., 7:1–60.

WAYBURN, P. 1972. *Edge of Life: The World of the Estuary.* New York: Sierra Club Books.

WEAVER, R. L. 1943. Reproduction in English Sparrows. Auk, 60:62–74.

WEEDEN, R. B. 1961. Outer primaries as indicators of age among Rock Ptarmigan. Jour. Wildl. Mngt., 25:337–339.

———. 1966. Molt of primaries of adult Rock Ptarmigan in Central Alaska. Auk, 83:587–596.

———. 1967. Seasonal and geographic variation in the foods of adult White-tailed Ptarmigan. Condor, 69:303–309.

———. 1969. Food of Rock and Willow Ptarmigan in central Alaska with comments on interspecific competition. Auk, 79:161–172.

WEISE, C. M. 1962. Migratory and gonadal responses of birds on long-continued short day-lengths. Auk, 79:161–172.

WELLER, M. W. 1957. Growth, weights, and plumages of the Redhead, *Aythya americana.* Wilson Bull., 69:5–38.

———. 1965. Chronology of pair formation in some Nearctic *Aythya* (Anatidae). Auk, 82:227–235.

———. 1968. The breeding of the parasitic Black-headed Duck. Living Bird, 7:169–207.

———. 1971. Experimental parasitism of American Coot nests. Auk, 88:108–115.

WEST, G. C., L. J. PEYTON, and L. IRVING. 1968. Analysis of spring migration of Lapland Longspurs to Alaska. Auk, 85:639–653.

WETHERBEE, D. K. 1958. Unilateral microphthalmia in *Quiscalus quiscula* and synophthalmia in *Mimus polyglottos.* Auk, 75:101–103.

WETMORE, A. 1936. The number of contour feathers in passeriform and related birds. Auk, 53:159–169.

WHARTON, W. P. 1959. Homing by a female cowbird. Bird-Banding, 30:228.

WHITAKER, L. M. 1957. Lark Sparrow oiling its tarsi. Wilson Bull., 69:179–180.

WHITE, C. M. 1963. Botulism and myiasis as mortality factors in falcons. Condor, 65:442–443.

———. 1969. Functional gonads in Peregrines. Wilson Bull., 81:339–340.

WIBLE, M. 1960. Notes on feeding and fecal-sac disposal of sapsuckers. Wilson Bull., 72:399.

WILBUR, S. R., W. D. CARRIER, J. C. BORNEMAN, and R. W. MALLETTE. 1972. Distribution and numbers of the California Condor. Amer. Birds, 26:819–823.

WILCOX, H. H. 1952. The pelvic musculature of the loon, *Gavia immer.* Amer. Midl. Nat., 48:513–573.

WILEY, R. H. 1971. Cooperative roles in mixed flocks of antwrens (Formicariidae). Auk, 88:881–892.

WILLIAMS, G. G. 1950. The nature and causes of the "coastal hiatus." Wilson Bull., 62:175–182.

———. 1952. Birds on the Gulf of Mexico. Auk, 69:428–432.

WILLIAMS. G. R. 1952. The California Quail in New Zealand. Jour. Wildl. Mngt., 16:460–483.

————. 1953. The dispersal from New Zealand and Australia of some introduced European passerines. Ibis, 95:676–692.

WILLIS, E. 1960. A study of the foraging behavior of two species of Ant-Tanagers. Auk, 77:150–170.

————. 1966. Competitive exclusion and birds at fruiting trees in western Colombia. Auk, 83:479–480.

WILLOUGHBY, E. J., and T. J. CADE. 1964. Breeding behavior of the American Kestrel (Sparrow Hawk). Living Bird, 3:75–96.

WILLSON, M. F. 1971. Seed selection in some North American finches. Condor, 73:415–429.

————. 1972. Seed size preference in finches. Wilson Bull., 84:449–455.

WILLUGHBY, F. 1676. . . . *Ornithologiae . . . London.* Translated into English, and enlarged . . . by John Ray in 1678.

WILSON, E. S. 1969. Personal recollections of the Passenger Pigeon. Ind. Aud. Quart., 47:112–121.

WING, L. W. 1952. Number of contour feathers on a cowbird, *Molothrus ater.* Auk, 69:90.

WOLF, L. L. 1969. Breeding and molting periods in a Costa Rican population of the Andean Sparrow. Condor, 71:212–219.

———— and J. S. WOLF. 1971. Nesting of the Purple-throated Carib Hummingbird. Ibis, 113:306–315.

WOLFE, L. R. 1950. Notes of the birds of Korea. Auk, 67:433–455.

WOLFSON, A. 1948. Bird migration and the concept of continental drift. Science, 108:23–30.

WOODBURY, A. M. 1941. Animal migration—periodic-response theory. Auk, 58:463–505.

WOOLFENDON, G. E. 1956. Comparative breeding of *Ammospiza caudacuta* and *A. maritima.* Univ. Kans. Publ. Mus. Nat. Hist., 10:45–75.

————. 1967. Selection for a delayed simultaneous wing molt in loons (Gaviidae). Wilson Bull. 79:416–420.

WORTH, C. B. 1949. Birds and human disease. Wilson Bull., 61:183–186.

————, V. HAMPARIAN, and G. RAKE. 1957. A serological survey of ornithosis in bird banders. Bird-Banding, 28:92–97.

WRIGHT, B. S. 1953. The relation of Bald Eagles to breeding ducks in New Brunswick. Jour. Wildl. Mngt., 17:55–62.

WURSTER, C. F., and D. B. WINGATE. 1968. DDT residues and declining reproduction in the Bermuda Petrel. Science, 159:979–981.

WYNNE-EDWARDS, V. C. 1952. Zoology of the Baird expedition (1950) I: The birds observed in central and south-east Baffin Island. Auk, 69:353–391.

YAPP, W. B. 1962. Some physiological limitations on migration. Ibis, 104:86–89.

YARROW, R. M. 1970. Changes in redstart breeding territory. Auk, 87:359–361.

YEAGLEY, H. L. 1947, 1951. A preliminary study of a physical basis of bird navigation. Jour. Appl. Physics, 18:1035–1036; 22:746–760.

YOCOM, C. F. 1952. Techniques used to increase nesting of Canada Geese. Jour. Wildl. Mngt., 16:425–428.

528]   *Literature Cited*

YOUNG, S. P. 1952. The bounty system. Atlantic Nat., 8:10–17.
ZIMMERMAN, J. L., and J. V. MORRISON. 1972. Vernal testes development in tropical-wintering Dickcissels. Wilson Bull., 84:475–481.
ZUSI, R. 1967. The role of the depressor mandible muscle in kinesis of the avian skull. Proc. U. S. Nat'l Mus., 123:1–28.

Index

A

Acadia National Park, 444
"Accidental," 331
Accipiter, 139
Accipitridae, 39, 143, 359
Adaptations
  anatomical, 395–96
  behavioral, 397–98
  of bills, 126–27
  for climbing, 97–99
  of feet, 129–30
  for flight, 91, 106–108
  physiological, 396–97
  for running, 99–100
  skeletal, 88–92, 234
  for swimming, 101–105
  for wading, 100–101
Adherent coloration, 80
Adherent nest, 254
Adrenal glands, 184–85
Africa, 370–71
Aftershaft, 59, 62
Age records, 314–16
Agonistic behavior, 198–202
Aigrette, 63, 403
Airplanes, collisions with birds, 414–15
Air sac, 150–51, 157
Albatross, 2, 35, 42, 88, 91, 112, 114, 116,
    157, 162, 291, 302, 310, 338
  Black-footed, 33, 414
  Laysan, 33, 344, 414, 456
  Royal, 281
  Short-tailed, 403
  Wandering, 65, 89, 302
Albinism, 78, 79, 269
Albumen, 165, 166, 267
Alcid, 103, 229, 270, 365
  see also Auk; Murre
Alcoholic, 474
Allelomimetic behavior, 187
Allopatric species, 33

Altricial bird, 159, 267, 278, 280, 287,
    288–91, 313
Amber list, 429
American Birding Association, 12
American Ornithologists' Union, 37, 466
Anabrus simplex, 10
Anarhynchus frontalis, 127
Anas crecca crecca, 32, 36
Ani, 48
  Groove-billed, 208
Anophthalmia, 177
Aptenodytes patagonica, 105
Anser indicus, 341
Anting, 213–14
Ant-tanager, 208
Appetitive behavior, 188
Apteria, 67, 68, 105
Apteryx, 41, 170
Apus, 301
Aquila heliaca, 435
Aransas Wildlife Refuge, 432, 445
Arbib's Blue List, 423, 429, 438, 484–85
Archaeopteryx, 26, 27, 28, 35, 43, 97, 359
Archaeornis, 26
Arctic Research Station, 21
Arrival, spring, 220
Aspect ratio, 109
Atavism, 338
Athene noctua, 239
Atticora cyanoleuca, 329
Attracting birds, 15–18
Audubon, John James, 8, 9, 10, 12, 15,
    203, 431, 467
Auk, 47, 99
  Great, 7, 35, 423–24
Auklet, Cassins's, 140, 394
Auricular feather, 180
Australian region, 373
Automatic apportionment, 208, 303
Aves, 30, 39
Avocet, 46, 266

B

Babbler, 2, 36, 52, 173, 215, 304
Bananaquit, 263–64
Banding, 15, 314, 315, 316–17, 318–19,
    328, 331–34
    organizations, 15, 466
    techniques, 15, 16
Barb (barbule, barbicel), 59, 64
Barbet, 36, 50, 137, 368
    Crimson-breasted, 369
Bathing, 83, 215, 306, 311, 312
Bee-eater, 49, 266, 268, 372
Behavior
    agonistic, 198–202
    allelomimetic, 187
    appetitive, 188
    breeding, 198–207, 338
    feeding, 207–13
    flocking, 215–16, 317, 320
    foraging, 214, 319, 390
    innate, 188–92, 310–11
    mobbing, 200, 201
    nesting, 202–207
    use of, in taxonomy, 188
Bergmann's rule, 158
Bill, 100, 126–30, 148, 170, 171, 185, 212,
    264, 266, 395–96
    adaptations, 126–27
    color change, 185–86
    falconiform, 38
Bill sweeping, 215
Binocular vision, 174–75, 176
Binomial nomenclature, 7, 37
Biochrome, 76
Biological stations, 20, 21
Biological Survey, 130
Biome, 376
Biotic communities, 373–85
Bird banding (*see* Banding)
Bird City, 447
Bird Haven Sanctuary, 392, 394
Bird listing, 11, 12
    life list, 12
Bird of paradise, 36, 51, 63, 65, 82, 232,
    237, 247, 373, 403
    Lesser, 236
Birds as food, 7, 399–402
Birds as sport, 18–20
Bittern, 43, 80
    American, 391
    Least, 176, 260, 387, 391
Blackbird, 53, 99, 134, 137, 178, 216, 410
    Brewer's, 231, 245

Blackbird [*cont.*]
    Red-winged, 79, 177, 208, 215, 223,
        230, 245, 260, 278–80, 319, 387, 391
        394, 407
    Tri-colored, 272
    Yellow-headed, 245
Blood, 154, 157
Bluebird, 77, 215, 258
    Eastern, 243, 335
    Fairy, 371
Blue List, Arbib's, 423, 429, 438,
    484–85
Bobolink, 82, 224, 234, 318, 335, 353,
    356, 400, 407
Bobwhite, 139, 268, 286, 326, 331, 397,
    400
    Masked, 438
Bones, 86–92
    reduction and fusion, 88–90
    special feature, 90–91
    structure, 88
Booby, 42, 157, 162, 270, 404
Botulism, 447–48
Bounties, 417
Bowerbird, 51, 178, 232, 237, 240,
    247
    Fawn-breasted, 236
Brailing, 74
Brain, 168–69
Brant, 208
Breathing rate, 151–52, 155
Breeding behavior, 198–207, 338
Broadbill, 50, 258, 371
    Black and Yellow, 372
Bronchus, 147
Brooding, 296–97
Brood parasitism, 204–206, 410
Brood patch, 184, 186, 270, 272–74,
    276
Budgerygah (Budgerigar), 163, 193
Bulbul, 19, 36, 52
Bulla, 149
Bullfinch, 73
Bunting, 54, 195
    Indigo, 69, 77, 190, 221, 247, 319, 320,
        348
    Lazuli, 204
    Painted, 299, 475
    Snow, 158, 181, 239, 263, 365, 379
Buoyancy, 88
Bureau of Biological Survey, 376, 406
Bursa fabrieii, 161
Bush-tit, 2

Bush-tit [*cont.*]
  Common, 320
Bustard, 35, 45, 77, 122
  Great, 3
Buteo, 139
Button-quail, 245, 276
  Barred, 246
Buzzard, 143

C

Caecum, 124
Calcium, 123, 165, 184, 458
Calfbird, 280
Camber, 110, 114
Campephilus principalis, 431
Camptorhynchus labradium, 424
Canary, 76, 181
Cannibalism, 207, 284
Canvasback, 67–68, 103, 116, 136, 181,
  267, 450, 455
Capillary, 154
Caprimulgid, 74, 159, 160
  see also Goatsucker
Cardinal, 62, 82, 135, 193, 220, 252, 286,
  303, 306, 311, 360, 475
Carotenoid, 76
Cassowary, 41, 64, 99, 269, 276, 373, 375
Catbird, Gray, 148, 208, 252, 269, 391
Catharus, 212
Cathartidae, 39
Ceilometer, 454
Censuses, 393–94
Center of origin, 358
Centropelma micropterum, 368
Cere, 147, 148
Cerebellum, 168
Cerebrum, 168
Cervical, 90, 91
Charadrius pecuriaris, 203
Chat, Yellow-breasted, 328
Chickadee, 157, 194, 211, 231, 278, 329
  Black-capped, 154, 177, 193, 225,
    226–27, 228, 230, 247, 266, 297,
    307, 316, 320
  Carolina, 225
Chicken
  Attwater's Prairie, 438
  Greater Prairie, 235
  prairie, 62, 83, 218, 230, 247, 383, 426,
    428, 438
  see also Fowl, domestic
Chlamydera cerviniventris, 237

Christmas count, 12
Chuck-will's-widow, 75, 203
Chukar, 100, 411, 418
Ciconia ciconia, 349
Circadian rhythm, 169, 396
Circulatory system, 152–60
Circus approximans, 139
Classification, 34–36, 37–54
  and behavior, 188
  and skeleton, 86
Cliff-dwellers, 228–29
Climax community, 376, 390–91
Climbing birds, 97–99
Clipped wing, 74
Cockfighting, 19
Coloration, plumage, 76–82
Color vision, 177–79
Columella, 180
Comb, 82
Competitive exclusion, 390
Condor, 2, 116
  Andean, 3, 65, 403
  California, 270, 310, 338, 359, 393, 432,
    433, 434
Cone (of eye), 174, 177
Coniferous forest, 376, 381
Continental drift, 336, 359
Contour feathers, 61, 62, 64
Conurpsis carolinensis (*ludovicianus*), 428
Convergent evolution, 39
Copulation, 164, 239–40
Coot, 45, 75, 139, 224, 262, 387
  American, 66, 83, 204, 260
  Horned, 260
Cormorant, 28, 42, 60, 64, 82, 105, 141,
    157, 162, 168
  Double-crested, 215, 268
  Guanay, 9, 393, 394, 404
Cornea, 173
Corvid, 135, 169, 209, 215, 278
  see also Crow; Jay; Magpie
Corvus, 31, 215
Cosmetic coloration, 80
Coturnix, 5, 161, 178, 281, 418
Cotylosaur, 24
Countershading, 80
Courtship and mating, 230, 233–48
  copulation, 164, 239–40
  pair formation, 238
  sex recognition, 82, 238–39
  types of pair bonds, 240–47
Cowbird, 208, 313, 345
  Brown-headed, 204, 233, 274, 439

Crane, 29, 35, 45, 82–83, 100, 147, 149,
      341, 368, 394, 414, 441, 445
   Sandhill, 80, 82, 275
   Whooping, 3, 147, 393, 432–33, 441, 445
Creeper, 52, 98, 270, 368
   Brown, 211, 365, 381
Cricket plague, 10, 406
"Crocodile bird," 144
Crop, 120–21
Crossbill, 135, 218, 278, 326
   Red, 181, 275, 329
Crow, 51, 82, 144, 194, 200, 280, 319,
      337, 407, 410
   Carrion, 31, 241, 242
   Common, 31, 131, 148, 149, 181, 210,
      216, 318, 390
   Hooded, 209
   New Caledonia, 215
   Northwestern, 31
Cryptic coloration, 80
Cuckoo, 48, 100, 133, 204, 267, 269
   Black-billed, 281
   North American, 252, 284
Cuculus canorus, 204
Curlew, Eskimo, 429–30
Cursorial birds, 99–100

D

Damage, bird-caused, 135, 137, 142, 208,
      214, 410–15
DDT, 8, 184, 437, 438, 457–60
Deciduous forest, 376, 381
Deflective coloration, 81
Dendroica angelae, 36
Desert, 383–85
Determinate layer, 272
Diatryma, 29
Dicaeum pygmaeum, 2
Dickcissel, 338, 383
Digestive system, 118–26, 144
Dioch, Black-faced, 280, 328
Dipper, 52, 104, 228, 261, 262, 307, 360
   European, 75
"Disaster" species, 394
Disease, 419–20, 447–51
Display, 192
   agonistic, 198–99
   arena, 230
   courtship, 233–37
   distraction, 206–207
   use of tail for, 65
Dispersal, 358–64
Distribution, 358–85

Distribution [*cont.*]
   by family, 40–54
   list, 366–67
Diurnal migrant, 340
Divergent evolution, 39
Diving birds, 105–107, 173, 414
Dodo, 7, 399
Dominance, 217–18, 234
Dormancy, 156, 159–60
   see also Torpidity
Dove, 5, 47, 163, 172, 213, 265, 281
   Inca, 160
   Mourning, 151, 226, 230, 238, 250, 252,
      253, 276, 333, 339, 400, 442, 449
Dovekie, 394, 399
Down (feather), 61, 62, 64, 402
   natal, 68–71, 290, 291, 62–63
Drainage, 453
Duck, 15, 19, 29, 31, 32, 34, 35, 43, 74,
      75, 105, 106, 109, 115, 139, 141, 144,
      149, 162, 170, 216, 230, 232, 238,
      239, 243, 254, 265, 272, 284, 291,
      309, 380, 387, 394, 400, 401, 410,
      414, 450, 453, 454
   Black, 33, 124, 296
   Black-headed, 204
   botulism in, 447–48
   Labrador, 424–25
   lead poisoning in, 448–49
   Lesser Scaup, 137, 411
   Mexican, 33
   Pekin, 190
   Ring-necked, 103
   Ruddy, 67, 68, 267, 268
   Sea, 103
   Wood, 104, 123, 179, 190, 270
Dunlin, 243

E

Eagle, 8, 44, 139, 270, 291, 310, 410
   Bald, 3, 56, 69, 141–42, 144, 145, 252,
      264, 301, 317, 387, 435–36, 442
   Crowned, 301, 302
   Golden, 3, 56, 226, 268–69, 435, 436
   Harpy, 301, 435
   Monkey-eating, 434
   Spanish Imperial, 435
Ear, 180
Economic value of birds, 63 402–407
   see also Birds as food; Birds as sport;
      Endangered species
Ectoparasite, 214, 450–51
"Edge effect," 389

Egg, 165–66, 202–206, 251, 253, 261, 266–72, 273, 280, 313
  clutch size, 165, 250, 267, 270–72, 281, 294–95
  color, 268–69
  hatching, 166, 190, 207, 282–86
  incubation, 165–66, 272–82, 294–95
  laying sequence, 271–72
  pipping, 282
  retrieval, 203
  shape, 268
  shell thinning, DDT and, 184–85, 438, 458
  size, 266–68, 280
  structure, 166
Egg tooth, 282
Egret, 43, 210–11, 252, 403, 440
  Cattle, 208, 209, 360
  Great, 63, 101
  Little, 214, 360
  Reddish, 210
  Snowy, 214, 447
Eider, 254, 402
  Common, 103
Elephantbird, 268, 399
Empidonax, 31
Emu, 41, 99, 269, 276, 281, 373
Encephalitis, 419
Endangered species, 429–41
  lists, 483–89
  protection for, 441–47
Endocrine gland, 183–86
Endoparasite, 450–51
Epigamic coloration, 81
Epimeletic behavior, 187
Erythrism, 79
Ethiopian region, 369–71
Ethology, 188
Et-epimeletic behavior, 187
Eurylaimus ochromalus, 372
Evolution, 33, 39
Excretory system, 160–64
Extinct birds, 421–29
Eye, 173–79, 290

F

Falcon, 44, 91, 114, 414
  Peregrine, 116, 140, 165, 189, 333–34, 436, 448, 456
  Prairie, 19, 38, 140, 148
Falconry, 19–20
Falco peregrinus, 436
False crop, 121

Families, 34, 37–39
  food habits of North American, 131–33
  orders and, of living birds, 40–54
Fat storage (reserves), 95, 221, 339
Feather, 1, 164, 169, 170, 213
  auricular, 180
  development, 56–58
  distribution (pterylosis), 66–68
  function, 60–61, 156–57, 158
  kinds, 57, 58, 61–65, 67, 68
  maintenance, 83–84, 215
  as nest material, 254–57, 263
  number, 57, 66, 67
  origin, 26, 56
  structure, 57–60
  as tool, 215
  see also Plumage
Feather tracts, 67, 68
Fecal sac, 305–306
Feces, 161
  removal from nest, 178, 305–306
Feeders, types, 18
Feeding
  courtship, 237–38
  habits, 130–46, 163, 172–73, 207–13, 397, 406–10
  of mate, 278–79
  by regurgitation, 300, 301, 302, 303
  schedules, 121, 299–302
  techniques, 126–28, 210–14, 302–304, 310
  of young, 63, 207–208, 227, 297–304
  see also Food habits
"Feeding tree," 137
Field studies (research), 13–15, 20–22
Filoplume, 61, 62, 64
Finch, 54, 407
  Galapagos, 209, 215
  House, 218, 299
  Purple, 135, 193
  Rosy, 329
  Winter, 319
Finfoot, 98
Fish-eating birds, 141–43
Fish and Wildlife Service, 13–14, 15, 16, 21, 131, 178, 349, 399, 406, 407, 415, 529
Fixed action patterns, 189
Flamingo, 43, 76, 84, 91, 100, 126, 254, 303, 394, 401
Flapping flight, 110, 111–12
Fledgling period, 290, 291, 294–95, 307

Flicker, 81, 119, 120, 272, 291
  Common, 239
Flight, 106–17
  adaptations for, 106–108
  definitions, 108–109
  flapping, 110, 111–12
  gliding, 109, 111
  hovering, 114
  muscles, 93, 111
  soaring, 112–13
  speed, 114–17
  use of tail in, 114–15
  wind factor, 113, 114–16
Flipper, 105
Flocking behavior, 215–16, 317, 320
Flowerpecker, 2, 53, 144
Flycatcher, 51, 52, 53, 74, 127, 215, 270,
    280, 289, 318, 336, 366, 381
  Acadian, 75, 258
  Alder, 31
  Derby, 265
  Great Crested, 263, 274
  Least, 193, 196
  Willow (Traill's), 31, 195
Flyways, 350–51
Food habits, 130–46, 163, 172–73,
    207–13, 406–10
  fish-eating, 141–43
  fruit-eating, 137–38, 278, 407
  granivorous (seed-eating), 123, 124,
    134–37, 218, 319, 407
  insectivorous, 10, 133–34, 172, 278,
    406–407
  of nestlings, 303
  of North American bird families,
    131–33
  predatory, 138–41, 390
  in relation to man, 406–10
  scavengers, 122, 143–45, 390
  special diets, 144, 208, 361
  storage, 211
  see also Feeding
Foot, 98, 101, 157, 171, 264
  adaptations, 129–30
  falconiform, 38
Foot paddling, 214
Foot quivering, 214
Foot stirring, 214
Foraging behavior, 214, 226, 227, 319,
    390
Forpus passerinus, 206
Fossil, 2, 24–30, 97, 359

Fovea, 174
Fowl, 44
  domestic, 56, 82, 87, 141, 156, 177, 178,
    218, 240, 267, 269, 272, 278–80
  Mallee, 281
  perching mechanism, 96
Fowl cholera, 447
Francolin, 19
Frigatebird, 20, 42, 60, 83, 112, 113, 114,
    157
Frogmouth, 48, 159
"Frounce," 450
Fruit-eaters, 124, 137–38, 278, 407
Fulica cornuta, 260
Fulmar, 45, 404

G

Gallinule, 45, 260
Gallus, 30, 372
Gannet, 42, 105, 229, 394, 404
  North Atlantic, 106
Generic names, 37
Genus (genera), 34
Geologic time table, 25
Gizzard, 122–23, 124
Glaciation, 335–36, 359
Gland
  endocrine, 183–86
  oil (preen), 60, 83–84, 101, 215, 453
  salivary, 118
  salt, 161–62, 395
Gliding flight, 109, 111
Glottis, 147
Gnatcatcher, 52
  Blue-gray, 254
Goatsucker, 48, 133, 156
Goldfinch, 133, 135, 203, 230, 278, 300
  American, 66, 76, 165, 186, 193, 198,
    238, 254, 256, 257, 263, 264, 303, 306
Gonads, 184, 185, 220, 317, 337
"Gooney bird," 414
Goose, 43, 103, 116, 136, 271, 276, 348,
    380, 414
  Bar-headed, 341
  Blue, 318, 341, 342
  Canada, 3, 190, 242, 260, 309–10, 311,
    313
  Lesser Snow, 33, 318
  Ross', 33, 441
  Snow, 13, 365
  White-fronted, 13

Goshawk, 139, 207, 297, 417, 426
European, 281
Grackle, 80, 93, 137, 208, 410
Boat-tailed, 247, 284
Common, 150, 177, 215, 230, 233, 238
Granivorous birds, 123, 124, 134–37, 318, 319, 407
Grassland, 376–83
Grassquilt, 234
Grebe, 41, 75, 105, 141, 262, 270, 368, 391, 414, 447
Pied-billed, 66, 129, 203, 260, 261
Western, 102
Gregariousness (*see* Flocking behavior)
Grit, 123
Grosbeak, 54, 135, 407
Evening, 124, 127, 320, 329, 330
Pine, 121, 135
Rose-breasted, 69, 82, 276
Grouse, 4, 8, 29, 31, 44, 121, 122, 124, 135, 141, 190, 232, 240, 245, 271, 272, 319, 400
Blue, 247, 329
Ruffed, 62, 72, 123, 130, 234–35, 244, 247, 286, 310, 326, 391
Sage, 230, 235, 240, 247, 385
Sharp-tailed, 22, 70, 136, 218, 230, 247, 325–26, 438, 440
Spruce, 136, 247, 381–82
Growth, and development, 288–97
Grus americana, 432
Guan, Horned, 438
Guano, 9, 160, 306, 404–405
Gillemot, 453
Guineafowl, 44, 83, 369
Gular flutter, 157
Gular sac, 82, 83, 121
Gull, 28, 47, 79, 101, 114, 115, 141, 144, 159, 162, 192, 203, 210, 229, 231, 242, 252, 270, 282, 284, 301, 328, 352, 401, 410, 447
Black-backed, 198
California, 10, 329, 406
Franklin's, 208
Glaucous, 75
Heerman's, 338
Herring, 189, 198, 199, 225, 272, 316, 331, 347, 387–88
Laughing, 190
Ring-billed, 225, 347, 387–88
Western, 274
Gygis alba, 76, 250

Gymnogyps californianus, 432
Gyrfalcon, 9, 140, 380

H

Habitats, 386–90
Habituation, 191–92
Hammerhead, 43, 264, 369
Harpia harpyja, 31, 301
Harrier, 44, 139
Hatching, 166, 190, 207, 282–86
Hatching muscle, 97, 282
Hawk, 29, 35, 44, 88, 112–14, 126, 127, 138–40, 147, 165, 175, 179, 252, 265, 272, 284, 291, 319, 340, 365, 450
Broad-winged, 139, 334
Cooper's, 139, 310, 437
feeding habits, 210–12
Marsh, 131, 139, 140, 207, 226, 245, 285, 297, 361–62, 390, 437
Red-shouldered, 38, 139, 200, 226, 437
Red-tailed, 110, 139, 200, 334, 437
Rough-legged, 79, 139, 271, 326
Sharp-shinned, 139, 310, 437
Swainson's, 139
Hawk Mountain Sanctuary, 113, 340, 447
Head scratching, 214–15
Hearing range, 180–82
Heart
beating rate, 154–56
embryonic, 165
weight, 152–54
Heat (*see* Temperature)
Hen, Heath, 383, 426–28, 438
Heron, 6, 29, 43, 63, 84, 100, 141, 143, 214, 252, 265, 303, 307, 317, 328, 403, 440
Black-crowned Night, 309
Great Blue, 3, 302, 387, 395, 445
Green, 210
Little Blue, 212, 240
Night, 438
sanctuary, 447, 452
Hesperornis, 26, 28, 102
Heteronetta atricapilla, 204
Hibernation, 151, 159–60
Histoplasmosis, 419
History, 2–11
fossil, 24–30
Hoatzin, 44, 45, 98, 121, 171, 367

Holarctic region, 365–67
Homing experiments, 344–48
Homoiothermous, 159
Honeycreeper, 53, 120, 121, 144
Honey-eater, 53, 144
Honeyguide, 9, 10, 50, 133, 144, 205, 206,
    247
Hormones, 184–86, 220, 274
Hornbill, 49, 137, 278, 279, 372
Hovering flight, 114
Huia, 128
Hummingbird, 11, 34–35, 49, 65, 80, 91,
    111–12, 114, 120, 121, 127, 144, 154,
    158, 172, 230, 237, 243, 265, 266,
    270, 302, 336, 366, 403
    Anna, 286
    Calliope, 160
    Cuban Bee, 2,3
    Ruby-Throated, 66, 124, 159, 178, 179,
        254, 256, 267, 291, 395
Hybridization, 33–36
Hyoid, 119, 120

I

Ibis, 4, 43, 370, 420
    Sacred, 4, 420
Ibis ibis, 370
Icthyornis, 28
Identification, 11, 12
Image-fighting, 200
Imperforate nostril, 147
Imprinting, 190–91
Incubation, 165–66, 272–82
    attentive and inattentive periods, 274
    periods, 278–82, 294–95
    role of the sexes, 266, 276–78, 294–95
Incubation patch, 272–73
Indian region, 371–72
Injury-feigning, 206
Insect-eaters, 10, 133–34, 172, 278,
    406–407
Instinctive (innate) behavior, 188–92, 311
Integumentary structures, 82–83
Intentional movement, 189, 204
International Commission on Zoological
    Nomenclature, 37
International Council for Bird
    Preservation, 401, 429, 459
International Ornithological Conference,
    465
Intestines, 123–24

Introduced species, 417–19
Inventories, 393–94
Iridoprocne albilinea, 227
Iridescence, 80
Irruption, 325–28, 331
"Island-hopping," 360

J

Jabiru, 43
Jaçana, 46, 82, 130, 245, 250, 276
    American, 246
Jacana spinosa, 246
Jackdaw, 241
Jaeger, 47, 410
    Long-tailed, 271
Jaw, 86
    classification by, 86, 88
    muscles, 93
Jay, 51, 83, 194, 211, 218, 278, 329, 410
    Blue, 77, 82, 101, 137, 173, 208, 320,
        332, 334, 337
    Canada (Gray), 118, 157, 248, 286, 381
    Piñon, 338–39
Junco, 218, 337
    Dark-eyed (Slate-Colored), 154, 320,
        339
    Oregon, 320

K

Kalbfleisch Field Research Station, 21
Kea, 144, 210
Kestrel, 114
    American (Sparrow Hawk), 140, 181,
        208, 239, 340
Kidney, 160, 161, 163, 184
Killdeer, 81, 135, 278, 280
Kingbird, 127, 133
    Eastern, 194–95, 226, 234
Kingfisher, 49, 106, 114, 141, 154, 174,
    179, 258, 263, 264, 266, 301, 306,
    373, 390
    Belted, 82, 211
    Ringed, 276
Kinglet, 52, 181, 234, 270, 329
    Golden-crowned, 62, 365, 381
    Ruby-crowned, 381
Kite, 44, 139
    Everglade, 208, 361–62, 390, 434–35,
        440
    White-tailed, 440
Kittiwake, 192

Kiwi, 2, 35, 41, 147, 156, 160, 170, 171, 258, 265, 267–68, 276, 281, 375, 421
Kiwi pie, 402

L

Land use, 452
Lapwing, 341, 401
Lark, 51, 80, 99, 234
 Horned, 32, 134, 135, 180, 366, 383
Larus atricilla, 190
Larynx, 147
Lead poisoning, 447, 448–49
Leafbird, 52, 371
Learning, 189–92, 310
Leipoa ocellata, 281
Lens, 173
Leopold Committee, 416
Life zones, 373–85
 map, 377
 theory, 376
Limpkin, 45, 359, 445, 446
Linnaeus, 7, 37
Linnet, 135
Lipochrome, 76
"Loafing bar," 230
Locomotion, 86, 97–117
 adaptations for, 97–117
 cursorial, 99–100
 diving, 105–107
 flight, 106–117
 scansorial, 97–99
 swimming, 101–105
 wading, 100–101
Locust plague, 406
Longevity, 316–17
Longspur, 158, 234, 247
 Lapland, 340
Loon, 35, 41, 102, 105, 141, 270, 365, 414, 447
 Common, 88, 106, 116, 150, 151, 228
Loriculus, 206, 213
Lung, 149–51
Lyre bird, 63, 65, 280, 373, 375

M

Magpie, 51, 149, 181, 241, 278, 280, 410
 Black-billed, 261, 264, 272
Maintenance activities, 83–84, 213–15
 use of tools, 215
Maleo, 281

Mallard, 33, 124, 126, 129, 190, 240, 274, 341, 347, 449
Mamo, 403
Manakin, 51, 237
Manomet Bird Observatory, 21
Man-o'-war Bird (*see* Frigatebird)
Martin, Purple, 75, 345, 392
Massachusetts Audubon Society, 446
Mating (*see* Courtship and mating)
Meadowlark, 93, 252
 Eastern, 190, 223, 224, 226, 252, 400
 Western, 190, 226, 383
Megalaema haemacephalia, 369
Megapode, 44, 264, 281, 310
Melanin, 76, 77, 79
Melanism, 79
Mellisuga helenae, 2
Melopsittacus undulatus, 163, 193
Melospiza, 37
Merganser, 103, 141
 American (Common), 142
 Red-breasted, 126
Merlin, 140
Metabolism, 147
Michigan State University, 11, 140
 W. W. Kellogg Station, 20, 21
Microhabitat, 388
Micromacronus leytensis, 2
Migration
 altitude, 340–411
 altitudinal, 329
 cyclic migrations, 325–26
 daily, 325
 definitions, 324–25
 diurnal, 340
 emigrations, 325–28
 fall, 317–18, 329
 latitudinal, 329
 longitudinal, 329
 lunar, 325
 mechanics of, 339–43
 moisture rhythm, 328
 nocturnal, 340
 orientation, 343–49
 origin and causes of, 334–39, 343
 patterns, 324–34
 postbreeding wandering, 317, 328
 reverse, 334
 routes, 349–56
 seasonal (annual), 74–75, 220–23, 328–34
 speed, 221–22, 318, 341–42

Migration theories
  continental drift, 336–37
  food and weather, 334–35
  northern ancestral home, 336
  photoperiodism, 337–39
  southern ancestral home, 336
Migratory Bird Conservation Act, 443
Migratory Bird Hunting Stamp Act, 443
Mites, feather, 213, 257
Mobbing behavior, 200, 201
Mockingbird, 52, 177, 192, 193, 208, 214,
    228, 242, 332, 360
Moisture rhythm, 328
Molt(s)
  postjuvenal, 68, 71–75, 317
  postnatal, 68–69
  postnuptial, 68, 72–75, 317
  prealternate, 69, 73
  prebasic, 69–73
  prenuptial, 68
  in relation to breeding and migration,
    74–76
  sequence, 70–74
Monocular vision, 174–75
Monogamy, 240, 241–44
Monotypic species, 32, 36
Montana State Biological Station, 20
Mortality
  causes, 310, 311–15, 451–60
  longevity, 316–17
Mound-builders, 71, 281, 373
Mounts, 474–75
Mousebird, 49, 160
Murre, 8, 47, 268
  Common, 203, 229, 268, 394
Muscles, 86, 92–97, 100, 157
  adductor mandibulae, 93
  breast, 95
  Crampton's, 173
  depressor mandibulae, 93
  eye, 173
  feather, 92
  flight, 93, 111
  gastrocnemius, 96
  hatching, 97, 282
  jaw, 93
  pectoralis major (and minor), 93–95
  perching mechanism, 95–96
  protractor, 93
  supracoracoideus, 93, 111, 114
  syringeal, 148, 149
Museums, 20, 474

Museums [*cont.*]
  American Museum of Natural History,
    20, 21, 375
  fossils in, 26, 28
Myiopsitta monachus, 418

N

Nares, 147–48, 170
  see also Nostrils
Natal down, 68–74
National Audubon Society, 13, 446, 467
  Christmas count, 12–13
National Wildlife Federation, 18, 468
Navigation theories, 347–49
Nearctic region, 365–67
Necrosyrtes monachus, 143
Neossoptile, 57
Neotropical region, 367–69
Nest, 202, 205, 240–66
  aquatic, 102, 260, 261, 313
  boxes, 15–17, 320
  building, 118, 263–66
  "dump," 270
  habitats, 387
  materials, 202, 203, 230, 252–57, 261,
    262, 263–64
  sites, 202, 228–29, 250–52, 255,
    258–62, 273, 279
  structures, 249–52, 257–58, 259, 264,
    273
Nesting behavior, 202–206, 230
  density, 228–29, 392–94
  table of comparative data, 294–95
Nest leaving, 290, 291, 294–95, 306–309
Nestor notabilis, 144
Nest records, 14
Nest sanitation, 282, 304–306
New Zealand habitat group, 375
Nictitating membrane, 176–77
Nidiculous birds, 287, 291
Nidifugous birds, 287, 291
Nighthawk, 80, 159, 270, 274
  Common, 208
Nocturnal migrants, 340
Nomenclature, 36–37
Nostrils, 147, 148, 396
Notornis, 421–22
Numenius borealis, 429
Nutcracker, Clark's, 75, 211, 274, 326, 329
Nuthatch, 51, 98, 178, 208, 211, 258, 270,
    329, 368
  Brown-headed, 215

Nuthatch [*cont.*]
  Coral-billed, 52
  Pygmy, 248, 320
  Red-breasted, 263, 326, 331, 365
  White-breasted, 320, 331

O

Oilbird, 48, 183, 260, 291, 404
Oil (preen) gland, 60, 64, 83–84, 101, 215, 453
Oil pollution, 453–54
Oldsquaw, 106, 116, 151, 414
Oo, Hawaiian, 403
Operculum, 147, 148
Orders, 34, 39–40
Orders and families
  distinguishing characters, 40–54
  distribution of, 40–54
  number of species in, 40–54
Oriental region, 371–72
Origin (of birds), 24–30
  and dispersal, 358–64
Oriole, 51, 202, 234
  Hooded, 204
  Northern, 137, 258, 265, 360
  Orchard, 213, 226
Ornithischia, 24
Ornithological collections, 474–82
Ornithological organizations and journals, 464–72
  foreign, 472–73
  national, 465–69
  state, 469–72
Ornithosis, 419
Oropendola, 245, 258, 259
Oscine, 2, 64
Osprey, 38, 44, 106, 113, 130, 141, 253, 301, 361, 387, 436, 458
Ostrich, 2, 29, 30, 33, 35, 41, 64, 67, 91, 99, 124, 151, 154, 158, 160, 245, 265, 266, 267, 270, 276, 281, 359, 369, 403–404
Otus, 32
Ouzel, water (*see* Dipper)
Ovenbird, 50, 193, 195, 234, 247, 252, 264, 307
Owl, 7, 29, 48, 66, 77, 91, 114, 127, 138, 156, 165, 169, 175, 179, 180, 272, 284, 291, 319, 365, 395
  Barn, 124, 125, 140, 141, 177, 182, 207, 212, 263, 271, 299, 328, 331–32, 338

Owl [*cont.*]
  Burrowing, 116, 141, 260, 383
  Elf, 260, 384
  feeding habits, 212
  Grass, 48
  Great Horned, 124, 131, 141, 181, 201, 252, 268, 270, 271, 286
  hearing, 181–82
  Little, 239
  Screech, 32, 79, 141, 332
  Short-eared, 207, 331, 366
  Snowy, 141, 158, 271, 326, 327, 331, 379
  vision, 174
Oxpecker, 144

P

Pairing bonds, types of, 240–47, 294–95
Palaearctic region, 365–67
Palaeospiza bella, 29
Pandionidae, 39
Pantropical species, 368
Panyptila cayennensis, 264
Papilla, 56, 58
Paradisaea minor, 236
Parakeet, 98, 163, 419
  Carolina (Louisiana), 428–29
  Monk, 418
Parasites, 214, 447–51
Parathyroid, 183, 184–85
Parks, 443–44
Parrot, 35, 47, 84, 98, 120, 137, 169, 207, 241, 317, 368, 373, 390
  Blue-crowned Hanging, 213
  hanging, 206
  Pygmy, 144
Parrot fever, 419
Parrotlet, Green-rumped, 206
Partridge, Hungarian (Gray), 396, 400, 418, 451
Passerine, 35, 40, 57, 65, 156, 160, 179, 181, 265, 271, 280, 290–91, 307, 313, 439
  see also individual birds (refer to Passeriformes, p. 50)
Patuxent Wildlife Research Center, 15, 21
Pavo, 372
Peacock, 65, 82
"Pebble stealing," 203
Peck order, 217–18
Pecten, 176
Pelecanus, 80, 302

Pelican, 42, 82, 141, 157, 303
  Brown, 106, 107, 142, 162, 438, 458
  Great White, 80
  Pink-backed, 302
  White, 3, 283, 301
Pellets, 124–25, 140
Penguin, 2, 33, 35, 40, 67, 91, 99, 103,
    104, 157, 162, 281, 291, 303, 346,
    348, 404
  Adelie, 203, 270, 276, 394, 395, 396
  Emperor, 3, 250
  King, 105
  Yellow-eyed, 238, 242, 248, 299
Periodicity, 387–88
Perforate nostril, 147, 148
Permanent resident, 11, 330
Pesticide, 456–60
  see also DDT
Petrel, 8, 141, 158, 281, 291, 302, 404
  Bermuda, 7, 399
  Cahow, 399
  Diving, 42
  Leach's, 344
  Storm, 42
  Wilson's, 89, 101
Pewee, Eastern Wood, 254
Phaëthon rubricauda, 203
Phainopepla, 265
Phalacrocorax, 404, 424
Phalarope, 46, 186
  Northern, 246
  Wilson's, 277
Phasianid, 31
Pheasant, 44, 63–64, 82, 99, 121, 124,
    139, 141, 161, 173, 178, 218, 244,
    272, 274–75, 283, 400
  Ring-necked, 100, 130, 181, 396,
    418
  Silver, 269
Phoebe, 202, 257
  Eastern, 15, 66, 254, 292–93, 299–300,
    334, 335, 396
Phoenicopteridae, 91
Phororhacos, 29, 30
Photoperiodism, 337–39
Phyla, 34
Phylogenetic relationships, 34, 86
Picture windows, 456
Piculet, 50, 98
Pigeon, 8, 15, 35, 47, 80, 84, 91, 97, 121,
    122, 133, 151, 170, 174, 177, 181,
    184, 213, 270, 373, 410, 436, 449
  brain, 169

Pigeon [cont.]
  Common, 95, 119, 148, 150, 152, 153,
    161, 180, 281
  fruit, 137
  homing, 20, 347
  muscles, 95
  Passenger, 10, 399, 400, 425–27, 430
  racing, 19
  urogenital system, 161
"Pigeon's milk," 121, 184, 303
Pigment, 76–78
Pineal body, 169
Pinioned bird, 74
Pinna, 62, 180
Pintail, 66, 334
Pipit, 52, 99, 158, 234
  Meadow-, 178
  Water, 340
Pipping of egg, 282
Pithecophaga jefferyi, 434
Pituitary, 169, 183, 184
Plantain-eater, 48, 77, 369
  feather, 59
Pleistocene era, 30, 335, 359
Ploceid, 53, 245
Ploceus, 203, 217, 338
Plover, 46, 144, 284, 365, 401
  Black-bellied, 380
  Egyptian, 203
  Golden, 333, 341, 353, 355
  Piping, 207, 208, 288
  White-fronted, 203
  Wilson's, 273
  Wrybill, 127
Plumage, 158, 171, 177, 239, 402–404
  adaptation, 60, 106, 395–96
  alternate, 69
  basic, 69
  breeding, 63, 73, 236
  coloration, 76–82
  eclipse, 74
  juvenal, 68, 69, 71–73, 290, 291
  natal down, 62–63, 68–71, 290, 291
  nuptial, 68, 73
  ornamental, 62–63
  swimming, 103
  value to man, 402–403
  winter, 68, 74, 191
  see also Feather
Pluvianus aegyptius, 203
Pochard, 103
Podilymbus gigas, 368
Pods, 303

Poikilothermous, 159, 289
Polyandry, 245–46
Polygamy, 231, 241, 244–47
Polygyny, 244–45
Polytypic species, 32, 36
Poor-will, 151, 159
Population, 228–29, 232, 325–28, 331, 392–95
Porphyrin, 77
Postbreeding wanderings, 317, 328
Powder-down feathers, 61, 64
Powdermill Nature Reserve, 21
Precocial birds, 159, 267, 280, 282, 287, 291, 309–10, 313
Predator control, 415–17
Predatory birds, 138–41, 390
Preening (*see* Oil gland)
Proave, 24
Procellariiformes, 39, 42, 147, 170
Promiscuity, 230, 240, 241, 245, 247
Protective legislation, 441–43
Psaltria exilis, 2
Psarocolius angustifrons, 258
Pseudogyps africanus, 143
Pseudosuchia, 24
Psilopaedic birds, 287
Psittacosis, 419
Ptarmigan, 44, 75, 81, 135, 140, 141, 318, 319, 326, 365, 380, 395–96
  Rock, 72, 129, 379
  Willow, 129, 158, 379
Pterocles namaqua, 269
Pterosaur, 24, 56
Pteryla, 67
Pterylosis, 66–68
Ptilopaedic birds, 288
Puffinus tenuirostris, 272, 402
Pulled primary, 74
Pyrrhula pyrrhula, 73
Pyrruloxia, 475

Q

Quail, 44, 99, 101, 141, 154, 169, 178, 245, 281, 400
  California, 124, 139, 163
  Gambel's, 62, 111
  migratory, 5, 418
  see also Bobwhite
Quail disease, 450
Quelea quelea, 280, 328, 338
Quetzal, 4–5
Quill (calamus), 58–59

R

Rail, 29, 35, 45, 75, 100, 231, 260, 400
  Clapper, 313
  Virginia, 129, 230
Raven, 209, 241, 280
Rectrice, 61, 63, 64–65, 72
Redhead, 116, 231, 329, 414
Redpoll, 121, 135, 218, 326, 380
  Common, 395
Redstart, 227, 240, 316
  American, 69, 193, 234, 249
Refractory period, 337, 338
Refuges, 443
Regurgitation, 300, 301, 302, 303
Releasers, 189
Remige, 61, 64, 65, 72, 105
Reproductive failure, DDT and, 438, 458
Reproductive system, 160, 161, 163–66
  egg structure, 166
Reptile, 24–26
Respiratory system, 147–52
Retina, 173–74
Rhachis, 58, 59
Rhea, 41, 99, 246, 269, 276, 359, 367, 399, 402
Rictal bristle, 61, 64, 170
Roadrunner, 48, 100, 384, 395
Robin, 130, 194, 216, 243, 264, 278, 280, 290
  American, 66, 71, 78, 79, 123, 137, 177, 226, 248, 254, 316, 335, 391, 400, 459
  English, 192, 193, 198, 200, 228, 242, 316
Rockhopper, 276
Rocky Mountain Biological Station, 20
Rod (of eye), 174, 177
Roller, 35, 49
Rook, 214
Roost, night, 262, 319–20
Rostrhamus sociabilis plumbeus, 434
Ruff, 230, 240
  European, 234
Running speed, 99, 100

S

Salivary gland, 118
Salmonella, 419
Salt gland, 161–62, 395
Salyer, J. Clark, National Wildlife Refuge, 445
Sanctuaries, 443, 446–47
Sanderling, 277, 318

Sandgrouse, 47, 164, 213
   Namaqua, 269
   Pallas, 325
Sandpiper, 29, 46, 75, 100, 116, 144, 234,
      243, 246, 284, 365
   Baird's, 75
   Least, 135
   Red-backed, 75
   Semipalmated, 75
   Spotted, 246, 277
Sandplover, Kittlitz's, 203
Sapsucker, 98, 119
   Yellow-bellied, 306, 381, 412
Saurischia, 24
Scansorial birds, 97–99
Scaup (*see* Duck)
Scavenger, 122, 143–45, 390
Schemochrome, 76
Sclerotic plate, 173
Scoter, 103
Screamer, 43, 82
Secretarybird, 29, 33, 44, 139, 369
Seed-eater, 265
   Hicks', 247
Seed-eating birds, 123, 124, 134–37, 318,
      319, 407
Semiplume, 61
Seney National Wildlife Refuge, 446
Sense organs, 169
   hearing, 180–83
   smell, 168, 169, 170–72
   touch, 169–70
   vision, 173–79
Seral field community, 376, 391
Seral stage, 391
Sere, 391
Serengeti, 371
Shaft, 58
Shearwater, 42, 101, 158, 162, 281, 302
   Cory's, 148
   Manx, 344, 345
   Slender-billed, 272, 296, 394, 402
   Sooty, 275
Shorebirds, 35, 81, 109, 133, 154, 168,
      214, 250, 265, 266, 270, 317, 318,
      335, 340
   see also individual birds (refer to
         Charadriiformes, p. 46)
Shrike, 53, 127, 173, 200, 211, 368
   Loggerhead, 193, 212, 228, 242, 384
   Northern, 326, 380
Sialia, 77
Siskin, 135

Singing (*see* Songs and calls)
Size
   birds, 2–3, 291
   eggs, 266–68, 280
Skeleton, 26, 86–92
   pneumaticity, 88, 113
Skimmer, 47
   Black, 84, 126, 368
Skins, 474
   directions for preparing, 475–82
Skua, 47, 270, 410
Slotting, 59, 110, 114
Smell, 168, 169, 170–72
Snake-bird, 42, 60, 368
Snipe, 80, 168
   Common, 234, 440
   Painted, 46
Soaring flight, 112–13
Sociability, 387–88
Social organization, 215–18
Solitaire, 422
Songs and calls, 151, 181, 189–91, 192–98,
      224, 311
   for claiming territory, 224
   daily cycle, 194–96
   learning, 189–91
   seasonal cycle, 196–98
   as sex attraction, 232, 233–34
   song analysis, 181, 194, 195
Sora, 129, 230
Sound production, 148–49
Southwestern Research Station, 21
Sparrow, 54, 79, 80, 99, 116, 181, 215,
      252, 270, 282
   Andean, 76, 198, 338
   Barn, 255
   Black-throated, 163
   Cape Sable, 36, 439
   Chipping, 239, 263, 280
   Dusky Seaside, 247, 439
   European Tree, 362
   Field, 224, 391
   Fox, 390, 396
   Golden-crowned, 345–46
   Grasshopper, 193, 391
   House, 15, 33, 84, 134, 141, 178, 180,
         185, 214, 239, 240, 280, 289, 319,
         338, 362, 407, 410, 417
   Ipswich, 362, 439
   Lark, 84
   Lincoln's, 390
   Rufous-collared, 198
   Rufous-winged, 338

Sparrow [*cont.*]
  Sage, 164
  Savannah, 162, 164, 218, 362, 439
  Sharp-tailed, 162, 231, 247
  Song, 11, 32, 37, 162, 164, 165, 190,
      193, 224, 225, 226, 228, 243, 264,
      275, 307, 311, 312, 316, 333, 362,
      381, 390
  Tree, 318–19, 320, 339, 380
  White-crowned, 338, 345–46, 407
  White-throated, 196–97, 320, 337, 339,
      343, 363–64, 407
  Vesper, 159, 290, 391
Speciation, 30–34
Spectrogram, 194–95
Speed
  flight, 114–17
  migration, 221, 318, 341–42
  running, 99, 100
  swimming, 103–104
  walking, 348
Spoonbill, 43
Sporophilia aurita, 247
Spreading, 358
Spur, 82
Starling, 14, 15, 53, 116, 117, 137, 141,
    148, 149, 180, 181, 191, 208, 214,
    215, 216, 316, 332–33, 337, 347, 352,
    390, 407, 410, 411, 417, 436
Starvation hypothesis, 302
Statant nest, 254
Status group, 330
Steatornis, 183, 291, 404
Stephanoaetus coronatus, 301
Stereotype behavior, 189
Stint, 277
Stomach, 122–23
Stork, 2, 43, 157, 372
  Marabou, 144
  Whale-headed, 43, 369
  White, 349
  Wood, 446
  Yellow-billed, 370
"Straggler," 331
Stratification, 387–88
Streptopelia senegalensis, 66
Struthio camelus, 369
Struthiomimus, 24
Subclimax communities, 376
Subspecies, 36
Succession, 390–92
Suet, 18
Summer resident, 11, 12, 330

Sun-bathing, 84
Sunbird, 53, 120, 144, 372
Sunbittern, 35, 45
Supersedence, 217
Supracoracoideus, 93, 111, 114
Swallow, 10, 51, 74, 91, 111, 114, 116,
    127, 154, 159, 179, 208, 215, 257,
    260, 268, 270, 317, 318, 340
  Bank, 229, 258, 299, 345
  Barn, 63, 228, 254, 255, 263, 272, 299,
    306, 364
  Blue and White, 329
  Cliff, 160, 202, 221, 229, 254, 255, 334
  Mangrove, 227
  Tree, 195, 227, 228, 248, 258, 263, 334,
    335
  Violet-green, 160
Swallowtanager, 54, 239
Swan, 43, 91, 103, 136, 149, 174, 276, 317
  Black-necked, 403
  Trumpeter, 3, 400, 439–40, 441
  Whistling, 3, 318, 380, 441
Swift, 35, 48, 74, 99, 111–12, 114, 116
    127, 133, 171, 183, 260–61, 268, 340
  Cayenne, 264
  Chimney, 63, 115, 118, 160, 202, 248,
    260, 288, 289, 291, 302
  crested, 274
  European, 151, 160, 301
  White-throated, 160
Swimming birds, 101–105, 395
Swimming speed, 103–104
Sympatric species, 33
Synchronous hatching, 284
Synophthalmia, 177
Syringeal muscle, 148, 149
Syrinx, 147–49
Syrrhaptes parodoxus, 325

T

Tactile organs, 169–70
Taiga, 381
Tail
  feather, 72
  growth, 290, 292–93
  muscle, 97
  role in flight, 114–15
  use in display, 65
Tailor-bird, 264
Tanager, 36, 54, 137, 234, 336, 366
  Hepatic, 76
  Scarlet, 69, 76

"Tasmanian squab," 402
Taste bud, 118–19, 172
Taxonomy (classification), 34–36, 37–54,
    86, 188
Teal
  Blue-winged, 334, 348
  Common, 32
  Green-winged, 32
Teleoptile, 57
Television tower, 169, 317, 341, 454–56
Telmatornis, 29
Temperature
  body, 156, 160, 396
  egg, 274–75, 281
Temperature control, 156–59, 290, 291,
    294, 396
Teratornis incredibilis, 90
"Terminal" migrant, 347
Tern, 47, 102, 141, 159, 162, 229, 231,
    242, 252, 301, 328, 401–404
  Arctic, 111, 353, 354
  Black, 260, 262, 299, 391
  Common, 276, 284, 313
  Fairy (Elegant), 250, 251
  Gull-billed, 338
  Noddy, 344
  Royal, 203, 269, 310
  Sooty, 325, 338, 344, 394, 456
  White, 76
Territory, 193, 194, 197, 223–33, 238
  function, 232
  types, 231–32
"Thermometer bird," 281
Thermoregulation, 156–59, 296
  see also Temperature Control
Thrasher, 52, 284
  Brown, 71, 165, 289, 385, 391
Threskiornis aethiopica, 4
Thrush, 52, 99, 137, 169, 212, 252, 264,
    269, 270, 300
  Gray-cheeked, 172, 173, 190, 193, 195,
      214, 284, 298, 308, 310, 312, 337,
      342, 363, 380
  Hermit, 138, 214, 337
  Wood, 381, 391–92
Thyroid, 183, 184
Tinamou, 35, 41, 99, 265, 269, 276
  Boucard's, 245, 246
Tit, Great, 316
Titmouse, 51, 82, 160, 193, 209–10, 211,
    238, 250, 258, 270, 331, 332, 368
  Tufted, 177, 210, 263, 307, 360, 381
Todirostrum cinereum, 257

Tody-flycatcher, Common, 258
Tongue, 119–20, 121, 170
Tools, use of, 215
Topography of bird, 56, 57
Torpidity, 151, 156, 159–60, 301, 396
Toucan, 50, 82, 84, 127, 128, 137, 269,
    367
Towhee, 278, 391
  Brown, 226, 390
  Spotted (Rufous-sided), 196, 226, 300,
    301
Tract, feather, 67, 68
Transient, 11, 12, 331
Transplantation, 440
Trial-and-error learning, 191
Trichomoniasis, 449
Trinomial, 34, 37
Trogon, 49, 368
Tropicbird, 42
  Red-tailed, 203
Tundra, 376, 379–81
Turkey, 8, 45, 63, 82, 123, 154, 400, 440
  Ocellated, 4
  Wild, 3, 399
Turnix suscitator, 246
Tympanum, 148, 180

U

Ulcerative enteritis, 450
Umbilicus, 58, 59
University of Michigan Biological Station,
    20
Urogenital system, 161
Uropygium, 84

V

Vagrants, 328
Vane, 58, 59
Veerie, 391
Veins, 152, 153, 165
Vireo, 53, 80, 127, 137, 154, 194
  Red-eyed, 196, 299, 387, 392
  Warbling, 226, 257, 299
  Yellow-throated, 258, 381
Vision, 170, 173–79
Vitality, 387–88
Vanellus vanellus, 401
Volatina, 234
Vulture, 29, 43, 44, 83, 88, 112, 116, 121,
    154, 157, 170, 359, 390
  Black, 143, 171, 172, 397
  Egyptian, 4, 210

Vulture [*cont.*]
 Hooded, 143
 King, 4
 Turkey, 38, 122, 143, 148, 158, 171,
  221, 250
 White-backed, 143

W

Wading birds, 100–101
Wagtail, 52
 Yellow, 178
Warbler, 52, 53, 80, 127, 154, 169, 181,
  215, 252, 269, 270, 317, 318
 Bachman's, 363, 439
 Bay-breasted, 226
 Blackpoll, 222, 339–40, 353
 Black-throated Green, 381
 Black-and-White, 227
 Blue-winged, 34, 233
 Brewster's, 34
 Cape May, 226
 Colima, 36
 Connecticut, 352
 Elfin Woods, 36
 Golden-cheeked, 263, 363
 Golden-winged, 23, 233
 Hooded, 298
 Kirtland's, 362–63, 389, 390, 393, 439,
  447, 470
 Lawrence's, 34
 Nashville, 134
 Palm, 352
 Parula, 254, 257
 Prothonotary, 177, 249
 "Sutton's," 36
 Tennessee, 137, 226
 Yellow, 204, 205, 254, 257, 391
Warm-bloodedness, 56
Water economy and balance, 163–64
Waterfowl, 8, 18, 43, 120, 126, 136, 174,
  215, 221, 270, 340, 349
 see also Duck; Goose; Swan
Waterthrush
 Louisiana, 228, 335
 Northern, 228, 319
Wattle, 82
Waxwing, 52, 76, 172
 Cedar, 121, 124, 137, 208, 228, 230,
  231, 243, 249, 275, 284, 289,
  297–98, 300, 303, 386
Weather, effect on migration, 221,
 343

Weaverbird, 53, 217, 258, 261, 264
 Common, 338
 Village, 203
Web, 58, 59
Weight of birds, 2, 3, 290, 291
Wheatear, 265, 309, 365
Whip-poor-will, 80, 196, 251, 270
White-eyes, 144
Wind, effect on flight, 113, 114–17
 effect on migration, 343
Wing (*see* Flight)
Wing flashing, 214
Wingspread, 2, 3
Winter habits, 318–20
Winter residents, 331
Winter visitor (visitant), 11, 12, 331
Woodcock, 126, 206, 243, 244, 400
 American, 123, 127, 168, 175, 196, 197,
  234, 318, 440
Woodhewer, 50, 98
Wood Ibis, 446
Woodpecker, 34, 35, 50, 63, 91, 98, 119,
  129, 130, 133, 135, 169, 228, 234,
  250, 258, 265, 270, 291, 306, 319,
  390, 406–407, 411–12, 416
 Black-backed Three-toed, 365–66,
  381
 California (Acorn), 137, 211
 Downy, 66, 127, 151, 242, 320
 Gila, 137, 260, 384
 Golden-fronted, 412
 Hairy, 66, 181, 242, 315, 320
 Ivory-billed, 402, 431–32
 Ladder-backed, 412
 Northern, 381
 Pileated, 137, 412–13
 Red-bellied, 137, 211, 381
 Red-cockaded, 137, 263
 Red-headed, 82, 137, 211
Wren, 52, 156, 194, 250, 257, 260, 262,
  270, 319
 Cactus, 228, 239, 262, 275, 319
  384
 House, 151, 156, 159, 178, 193, 243,
  258, 262, 263, 273, 275, 282, 299,
  410
 Long-billed Marsh, 72, 260, 262–63,
  268, 305, 319, 391
 marsh, 80
 Short-billed Marsh, 81, 338
 Winter, 151, 245, 320, 381, 439
Wrentit, 52, 79, 215, 228, 241, 331
Wryneck, 50, 98

X

Xanthochroism, 80
Xerophyte, 383

Y

Yellowstone National Park, 208
Yellowthroat, Common, 68, 195, 197, 239,
 391
Yolk sac, 289
Young, 206–207
 altricial, 287
 brooding of, 296–97
 carrying of, by adults, 206
 feeding of, 63, 207–208, 297–304
 growth and development, 287–96
 mortality and nesting success, 311–17

Young [*cont.*]
 nest leaving, 291, 306–309
 postnest life, 307–11
 precocial, 287
 table of nesting data, 294–95

Z

Zonotrichia, 76, 198, 346
Zoological (faunal) regions, 364
 Australian, 373
 Ethiopian, 369–71
 Holarctic (palaearctic and nearctic),
  365–67
 Neotropical or South American, 367–69
 Oriental or Indian, 371–72
Zugdisposition, 339
Zugunruhe, 337–38